D0850062

CONTEMPORARY PROBLEMS IN PLANT ANATOMY

Professor Katherine Esau

CONTEMPORARY PROBLEMS IN PLANT ANATOMY

Edited by

Richard A. White
Department of Botany
Duke University
Durham, North Carolina

William C. Dickison
Department of Biology
University of North Carolina at Chapel Hill
Chapel Hill, North Carolina

1984

AP

ACADEMIC PRESS, INC.
(Harcourt Brace Jovanovich, Publishers)

Orlando San Diego San Francisco New York London
Toronto Montreal Sydney Tokyo São Paulo

Academic Press Rapid Manuscript Reproduction

ACADEMIC PRESS, INC.
Orlando, Florida 32887

United Kingdom Edition published by
ACADEMIC PRESS, INC. (LONDON) LTD.
24/28 Oval Road, London NW1 7DX

Library of Congress Cataloging in Publication Data
Main entry under title:

Contemporary problems in plant anatomy.

Papers from a plant anatomy symposium held at Duke University and the University of North Carolina at Chapel Hill in early 1983.
Includes index.
Contents: Development of the stem conducting tissues in monocotyledons / P. Barry Tomlinson -- Comparative development of vascular tissue patterns in the shoot apex of ferns / Richard A. White -- The role of subsidiary trace bundles in stem and leaf development of the dicotyledons / Philip R. Larson -- [etc.]
1. Botany--Anatomy--Congresses. I. White, R. A. (Richard Alan), Date. II. Dickison, William C.
QK640.3.C66 1983 581.4 83-21400
ISBN 0-12-746620-7 (alk. paper)

PRINTED IN THE UNITED STATES OF AMERICA

84 85 86 87 9 8 7 6 5 4 3 2 1

This volume is dedicated with admiration and appreciation to Professor Katherine Esau. Few individuals have so influenced anatomical research in the United States during this century. Her standards of excellence, productivity, and generosity to her colleagues provide us with a model for which we are all grateful.

CONTENTS

CONTRIBUTORS

Numbers in parenthesis indicate the pages on which the authors' contributions begin.

William C. Dickison, *Department of Biology, The University of North Carolina, Chapel Hill, North Carolina 17514 (495)*

John Dransfield, *Royal Botanic Gardens, Kew, England (397)*

Ray F. Evert, *Department of Botany, The University of Wisconsin, Madison, Wisconsin 53706 (145)*

Jack B. Fisher, *Fairchild Tropical Garden, Miami, Florida 33156 (541)*

Wolfgang Hagëmann, *Institut für Systematische Botanik der Universität, 6900 Heidelberg, Germany (301)*

Donald R. Kaplan, *Department of Botany, University of California, Berkeley, Berkeley, California 94720 (261)*

Philip R. Larson, *Forest Sciences Laboratory, P. O. Box 898, Rhinelander, Wisconsin 54501 (109)*

Elizabeth M. Lord, *Department of Botany and Plant Sciences, University of California at Riverside, Riverside, California 92521 (451)*

R. Scott Poethig, *Department of Agronomy, College of Agriculture, University of Missouri, Columbia, Missouri 65211 (235)*

Phillip M. Rury, *Botanical Museum, Harvard University, Cambridge, Massachusetts 02138 (495)*

P. Barry Tomlinson, *Harvard Forest, Petersham, Massachusetts 01366 (1)*

Shirley C. Tucker, *Department of Botany, Louisiana State University, Baton Rouge, Louisiana 70703 (351)*

Natalie W. Uhl, *L. H. Bailey Hortorium, 467 Mann Library, Cornell University, Ithaca, New York 14853 (397)*

Richard A. White, *Department of Botany, Duke University, Durham, North Carolina 27706 (53)*

PREFACE

The chapters of this book represent the presentations at a plant anatomy symposium held at Duke University and The University of North Carolina at Chapel Hill in early 1983 and which was supported by the National Science Foundation. The basic objective of the symposium was to bring together plant anatomists to discuss and outline problems in four basic research areas in contemporary anatomy. Differentiation of cells and tissues, leaf development, floral development, and systematic and ecological anatomy represent areas of currently active research in which new data are changing or modifying long-held views. New problems in each of these areas have required new techniques and approaches.

There has been a revival of interest in the interrelationships between the leaf and stem. Detailed studies have been made of leaf initiation relative to the establishment and development of the vascular pattern in the shoot. Careful analyses of vascular strands and examinations of early vascular cell differentiation have demonstrated the correlation between leaf development with xylem pattern differentiation by documenting the close relationship of leaf phyllotaxy to protoxylem maturation at the shoot apex. This fundamental association between leaf development at the shoot apex and the establishment of primary vascular pattern, provascular tissue , and primary xylem in particular has been elaborated upon for gymnosperms, dicotyledons, monocotyledons, and the ferns. The connection between the shoot meristem and the proto- and metaxylem in monocotyledons and the ferns is discussed by Tomlinson and White, and the putative functional role(s) of subsidiary bundles in dicotyledons are described in detail by Larson. Refined techniques of electron microscopy have produced substantial progress in our understanding of phloem cell diffentiation and maturation. The survey of phloem structure provided in this book by Evert is the most extensive and complete review of the subject available.

One of the long-standing topics in plant anatomy relates to the development of the leaf at the shoot meristem. Numerous descriptive studies have indicated considerable variation in patterns among major plant groups. In addition, many distantly or unrelated groups have similar forms of mature leaves, but developmental patterns have been shown to be different. Important new methods of data analysis, exemplified by Kaplan's contribution, have demonstrated that different developmental mechanisms or patterns can be described that clarify the processes whereby similar mature forms are arrived at through different means. In addition, approaches from the developmental and biological perspective have been initiated through clonal analysis of leaf morphogenesis. Studies of this kind by Poethig provide another way of examining leaf ontogeny and the development of mature leaf form. In practice, leaf development is frequently isolated from shoot development. In reality, of course, leaf development is inextricably related to shoot development as a whole. As a result of this artificial separation, however necessary it may be for simplification of the study, a comprehensive view of shoot ontogeny remains elusive in the current anatomical literature. An attempt has been made by Hagëmann to develop a comprehensive theory of the morphological, developmental, and phylogenetic aspects of the leaves of vascular plants.

The study of floral morphology and anatomy continues to provide essential information for interpreting the phylogenetic relationships of flowering plants, and clarifying the evolutionary origin of, and homologies between floral parts of diverse form. In addition to employing the necessary standard techniques of serially sectioned and cleared flowers, Tucker and Uhl and Dransfield have combined the older methods with new approaches including the techniques of microdissection of floral buds and subsequent photography and the scanning electron microscopy of buds at various developmental stages. The obvious immediate advantage of these techniques is that they enable an individual to obtain directly a three-dimensional picture of floral bud development without the intermediate process of reconstruction. These studies pay careful attention to differences in the order of initiation of floral parts and variation in the size and shape of developing organs. Studies of this type will eventually clarify floral terminology and homologies, and provide a basis for understanding how differences in floral structure have evolved. Following careful developmental analyses, Lord proposes a synthetic approach to the study of floral development and diversity in order to explain form in an evolutionary context.

In recent years, a totally new awareness has developed among plant anatomists with regard to the importance of correlating plant structure with function, habit, and habitat. Rury and Dickison point out that solutions to the difficult problems posed will require the attention of very broadly trained individuals and collaborative efforts among plant anatomists, ecologists, and physiologists. Tree architectural models as they relate to structure and function in trees are discussed by Fisher.

A central theme present throughout all of the contributions in this volume is to view the plant as a developmental, structural, and functional whole. The mature anatomy and function of organs, parts, and cells are interrelated from development through maturity and the influences of the environment cannot be ignored. As new information is being uncovered and correlated with existing data, new trends of structural evolution in vascular plants are emerging. Trends and concepts that were once believed to be well-founded are now being seriously questioned. The future of plant anatomy promises to be both challenging and exciting as new questions are raised and answers to the difficult questions outlined in this volume are uncovered.

Richard A. White
William C. Dickison

CONTEMPORARY PROBLEMS IN PLANT ANATOMY

DEVELOPMENT OF THE STEM CONDUCTING TISSUES IN MONOCOTYLEDONS

P.B. Tomlinson

Harvard University
Harvard Forest
Petersham, MA

Frame-by-frame ciné analysis of serial sections allows complete objectivity in the unravelling of complex 3-dimensional structures. The principle approaches include either microscopic methods, e.g. the shuttle microscope, or direct surface photography. Commercially available films have been prepared. The method was developed to analyze the vascular system in the stem of larger monocotyledons, which are impossible to analyze quantitatively by any other method. Other applications, where the analytical method suits the complexity of the system, include: tree ferns with well-developed medullary systems; the resin canals of conifers; anomalous vascular systems in dicotyledons; rhizomatous axes in Nymphaeales, floral vasculature (e.g. Uhl, this volume). Shuttle microscopy has been used to supplement more orthodox techniques in recent work on the dicotyledonous stem vascular system (Larson, this volume). Integrating structure, development and hydraulic function, a detailed description of the mature axis and crown of the palm *Rhapis excelsa* is presented. Vascular differentiation within the procambial template is described leading to the distinction between the *morphological* and the *vascular* insertion of the leaf on the stem. Linkage between leaf trace protoxylem and stem metaxylem is timed to coincide with expansion of the leaf blade. The interconnection protects the hydraulic integrity of the main axis at times of physical stress, important in the palm

Contemporary Problems in
Plant Anatomy

ISBN 0-12-746620-7

which has no system of vascular renewal. The extended distances over which vascular developmental processes occur in the palm crown (several centimeters) clearly demonstrate requirements which must occur in all vascular plants. Discontinuous protoxylem differentiation in the leaf-stem union, which seems universally present in vascular plants, is explicable in these same functional terms.

I. INTRODUCTION

In the early 1960's Martin Zimmermann and I developed the method of analysis of complex vascular systems in plants which involves frame-by-frame ciné photography of serial or sequential sections (Zimmermann and Tomlinson, 1966). The method was originally devised to elucidate the overall course of vascular bundles in large monocotyledons, notably palms, and particularly to recognize the extent of interconnection between bundles, an observation of direct importance to understanding transport pathways in such stems. In this account I wish to review, very briefly, some of the significant observations using the techniques, of both our own and of other workers, but more specifically to present an integrated view of the crown structure of large monocotyledons, uniting the approaches of descriptive and developmental anatomy with aspects of physiology, all of which has come from application of the ciné technique.

The technique is a powerful analytical tool for unravelling complex three-dimensional structures because it is fully objective, provides quantifiable information and produces completely reproducible documentation of intricate internal anatomy. The approach, in one sense, is not new since it was proposed and apparently even attempted in 1907, but it was never applied to a given biological problem and the method of precisely superimposing successive images was never clearly indicated. If there is any disadvantage to ciné analysis, in our development of the method, it is, paradoxically, that the results presented conventionally in the pages of scientific journals cannot do justice to the information made available (cf. Bell, 1980a). Furthermore, the results provide a highly integrated demonstration of complex micromorphological structures at a holistic level which represents something of a departure from reductionist approaches in classical plant anatomy. It is perhaps for this reason that the methods have not been more widely applied, coupled with the somewhat necessary tedium of section preparing and filming. However, the method is in fact very economical of effort because of the complete objectivity and certainty of positive results. Since developmental as well as structural information is obtained (i.e. the disposition of structures in time as well as space) we have the opportunity for a highly integrated view of organ ontogeny in plants. The later discussion of the palm crown is intended to exemplify this; it includes previously unpublished results.

II. METHODS OF CINEMATOGRAPHIC ANALYSIS

A brief survey of methods is appropriate especially as the range of techniques has been developed as new materials and problems demanded new approaches. Three main techniques have been developed, respectively referred to as the "drawing" method, the "shuttle" method and the "surface" method. The first two are microscopic and use compound or stereo microscopes, the last is macroscopic and uses the camera lens directly. The principles of the method have been outlined by Zimmermann, 1976.

The original technique photographed serial or sequential sections, cut on a sliding microtome, through the compound (less often stereo) microscope, usually at fairly low magnification. Two methods of superimposing serial images are used. In the *drawing method* a fixed outline drawing of distinctive and more constant features, executed with a drawing apparatus (camera lucida), is used to superimpose successive sections. As orientation is gradually lost, new drawings need to be made. In the *optical shuttle* an optical bridge links two microscopes in such a way that the images produced separately by the two microscopes can be superimposed in one eyepiece outlet and photographed in sequence. By this method, alignment is precise and any subjectivity of the drawing method is eliminated. The shuttle method has the slight disadvantage of requiring single or a few sections mounted on each slide. The drawing method is particularly suited to filming sequences of embedded material, since there is a primary alignment which results from ribboning, but it can be adapted to deal with large numbers of sections stained and mounted temporarily, as in the work of French and Tomlinson (1980; 1981a-d; 1983).

In the *surface method* the ciné camera is used to photograph directly (through telephoto lenses mounted on extension tubes to increase working distance) the surface of a specimen planed at regular intervals on a sliding microtome. Increased depth of field and precise light control was later improved by use of a specially designed short-cycle electronic flash synchronous with the camera shutter, although this is not absolutely essential to the method. The specimen clamp of the microtome has to be replaced by a modified attachment which allows the specimen to be advanced along the optical axis of the camera. We have a diversity of attachments for doing this, the most precise being a jack which forces the specimens through a series of clamping rollers. In very crude analyses, such as branch attachments of large axes, the microtome can be replaced by a radial arm saw which cuts thin slices from the specimen as it advances along simple guides. The surface method may depend on the existence of natural internal contrast in the specimen, or some optical contrast, as between open vessels and ground tissue in wood, or between lignified support tissue and ground parenchyma in fern rhizomes. Surface staining can be used, but is often limited by the corrosive nature of preferred reagents, such as phloroglucinol and concentrated HCl. A useful technique involves analysis of specimens after dye injection; this is very useful in analyzing the course of vessels. The main limitation of the method is then the resolving power of the camera lens.

Details of the methods and the several variations they permit are published in many individual papers, and the Harvard Forest Annual Reports. The total range of structural possibilities which the plant anatomist may need to investigate are amenable to these techniques, with the special advantage, already mentioned, of holistic approaches. As examples of the range of materials which have been investigated we can mention detailed morphological and anatomical investigation of floral structures, analysis of large and small monocotyledonous stems, analysis of fern rhizomes, of the course of vessels in wood and the extent of vessel embolism in diseased elm trees. Some of our films are used routinely in our own teaching and three specially-prepared films have been made available commercially (Zimmermann, 1971; Zimmermann and Mattmüller, 1982).

A. *Vascular Systems in Monocotyledonous Stems*

The following is a brief review of some of the results of our application of the method both by ourselves and other workers. Where possible these are made comparative.

1. *Palms* (Fig. 1). A fundamental result of our early attempt to unravel the vasculature of the palm stem (Zimmermann and Tomlinson, 1965) was the recognition of the basic constructional principle which underlies this complexity, and which increasingly we can recognize as characterizing monocotyledons generally. The principle (referred to as the "*Rhapis*-principle", since it was first observed in the small palm *Rhapis excelsa*) demonstrates that all leaf traces branch in an upward direction as they diverge towards their associated leaf. Since all bundles intrinsically behave alike, we can treat the single leaf trace as a unit of construction, not forgetting, however, that in palms each leaf trace may be supplied by a hundred or more such traces. One branch of each leaf trace becomes a continuing axial bundle, other branches become short bridges connecting the leaf trace with adjacent axial bundles, while there can be traces to axillary structures like buds or inflorescences. The process is repeated at regular intervals within each trace (leaf contact distance). This branching of the leaf trace establishes the axial and lateral vascular continuity in the palm stem. Within a single stem there is considerable range of quantitative variation, but never enough to obscure the underlying pattern. By virtue of the cinématographic method of analysis, which makes

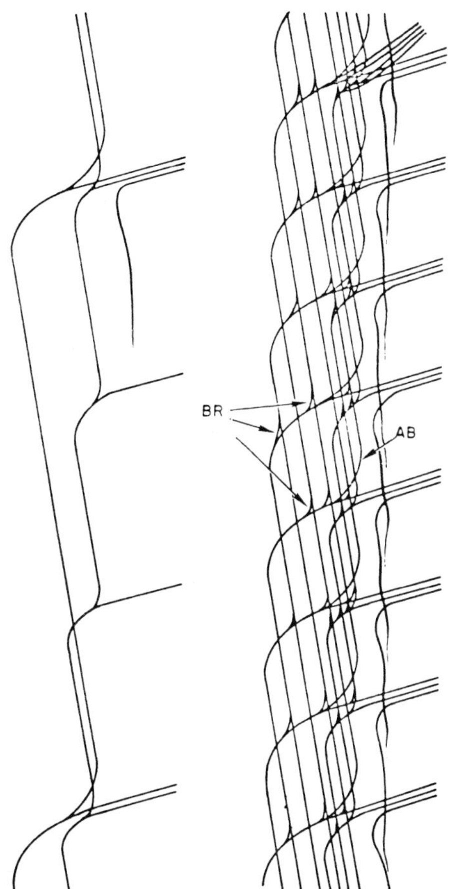

Fig. 1. Elements of vascular construction in the stem of the palm Rhapis excelsa. *Longitudinal course of vascular bundles represented diagrammatically in one plane.*
LEFT: Course of a major, minor and cortical bundle diverging to four widely separated leaves. Vascular continuity is maintained by an axial bundle which branches from an outgoing leaf trace. Major bundles differ from minor bundles in the greater radial displacement and longer distance (leaf contact distance) between successive bundle branchings. Cortical bundles end blindly below.
RIGHT: The system more nearly representing the complexity of an actual stem with 3 traces to each successive leaf. Vascular continuity is further provided by short bridge bundles (BR) *connecting from leaf trace to axial bundles* (AB) *(after Zimmermann, 1973).*

it possible to look at hundreds or even thousands of departing leaf traces, our generalized statements are made unequivocable.

Within the palms themselves detailed analysis of a diversity of genera shows the existence of this structural principle throughout the family and without major exceptions, but with interesting quantitative variation (Zimmermann and Tomlinson, 1974). Some of the variation is related to differences in stem diameter, other seems to a limited extent to be diagnostic for a genus. Variation can occur in a single species, as in *Rhapis* itself where axes of different orientation (erect-leafy versus horizontal-scaly) are compared (Tomlinson and Zimmermann, 1966).

An important systematic principle which comes from these studies, supplemented by more extensive examination of other histological features by orthodox techniques, is that in their stem structure palms are highly characteristic and any palm can be diagnosed as a member of the family by its stem structure. This supports the recognition of the naturalness and relative isolation of the family (Tomlinson, 1979). This is perhaps not a surprising conclusion, but suggests that other groups of monocotyledons need to be compared in the same way.

We have subsequently established that the *Rhapis*-principle, in its simplest form, occurs in such families as Juncaceae, Bromeliaceae, Cyperaceae and some Araceae, but is more usually expressed in some modified form, as in Cyclanthaceae, Pandanaceae and some other Araceae. The *Rhapis*-principle thus seems a useful descriptive "type" for comparison among monocotyledons, and can be used to compare monocotyledonous vasculature with that of dicotyledons (Zimmermann and Tomlinson, 1972). Of particular interest are those families considered to be related to palms.

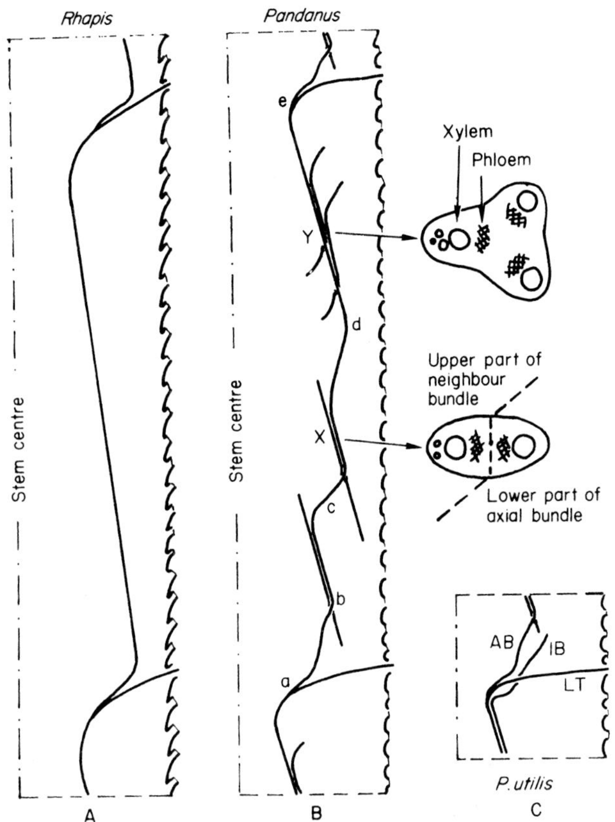

Fig. 2. Course of stem vascular bundles in Rhapis *and* Pandanus *compared diagrammatically in one plane. A.* Rhapis excelsa *(cf. Fig. 1). B.* Pandanus *represented through one leaf contact distance. Traced acropetally an axial bundle diverges from a leaf trace at* a, *becomes temporarily associated with other axial bundles at* b, c *and* d *and branches distally into a leaf trace and a further axial bundle at* e, *to repeat the cycle. At* X *the compound bundle is bipolar and represented diagrammatically to the right in transverse section; at* Y *the bundle is tripolar since it consists of 3 fused bundles. In each compound bundle the innermost strand is always that which will diverge first as a leaf trace and includes protoxylem. C. Detail of axial bundle* (AB) *and inverted bundle* (IB) *at level of divergence of leaf trace* (LT) *(after Zimmermann et al., 1974).*

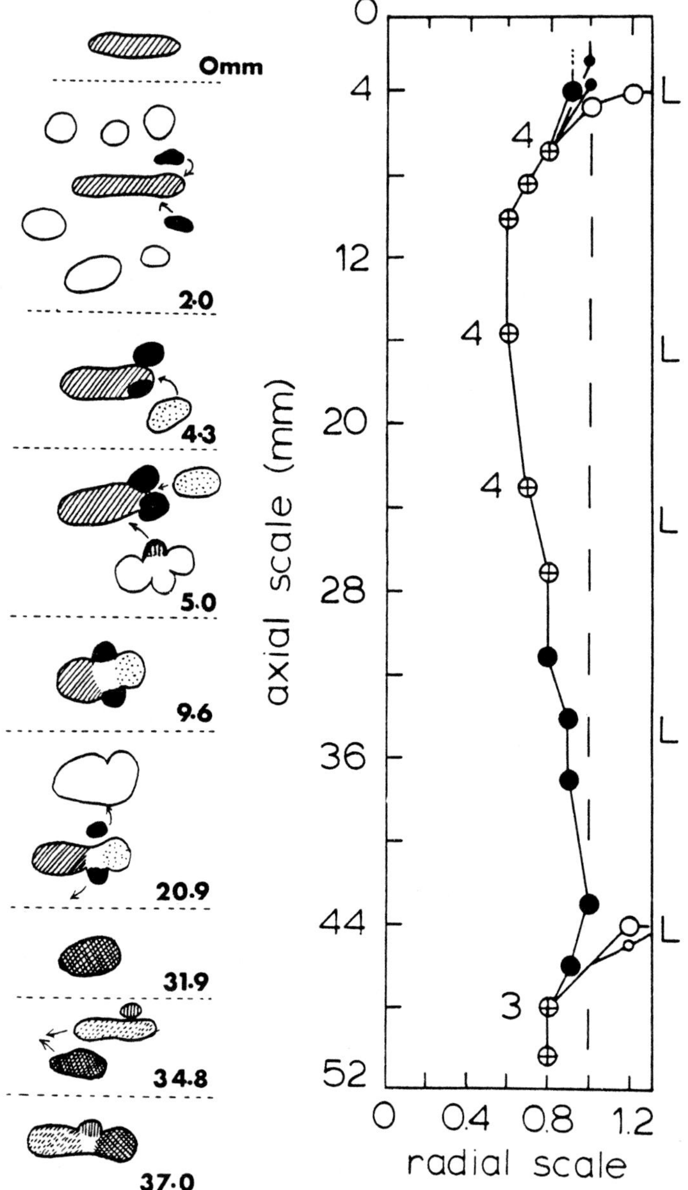

Fig. 3. Dicranopygium crinitum *(Cyclanthaceae). Diagrammatic course of a single stem vascular bundle represented in one plane. Left: transverse sectional view of a leaf trace (hatched) which forms a multipolar bundle. Right: plot of the same bundle; ⊕ - multipolar bundle; ● - simple bundle (after French et al., 1983).*

2. *Pandanaceae* (Fig. 2). In this family, although the *Rhapis*-principle is demonstrable, there is superimposed upon it a characteristic association of stem bundles leading to the formation of distinctive "compound" bundles which are conspicuous in single transverse sections (Zimmermann *et al.*, 1974). The three-dimensional analysis which ciné methods facilitates shows that bundle association follows a precise pattern. In following any leaf trace in a downward direction (i.e. the reverse of the general direction of development, but convenient for descriptive purposes) it becomes associated with one, two or more axial bundles but without establishment of continuity between xylem and phloem. The relative orientation of bundles in each individual complex is always the same and has a fairly complex developmental basis. Since the pattern has been demonstrated in all members of the family which have been examined it seems to be diagnostic for the group, further emphasizing the naturalness of the group and its discreteness from the palms. Three-dimensional vascular construction thus becomes a "taxonomic character". It is not therefore unreasonable to make further comparative surveys using this character, as is shown below.

3. *Cyclanthaceae* (Fig. 3). In view of the putative affinities between cyclanths and pandans and the known occurrence of compound bundles in the two families, it is appropriate to make three-dimensional analyses for comparative purposes. A recently-completed survey (French *et al.*, 1983) which has included representatives of all genera of cyclanths shows again a characteristic pattern of stem vasculature, but one which differs significantly from that of the Pandanaceae. Compound bundles with varying degrees of complexity occur in all taxa examined, but always based on the same constructional principle. A leaf trace, followed in a downward (basipetal)

direction becomes associated with one or more axial bundles (which itself may be compound). However, unlike the Pandanaceae, the aggregation of originally discrete bundles is but a precursor to the partial or complete fusion of all bundles in the complex i.e. vascular tissues eventually become continuous in a basal direction. In the simplest condition the bundle complex at its base in each cycle is represented by a simple collateral (sometimes amphivasal) vascular bundle.

One can easily equate (perhaps homologize is an appropriate word) this pattern with that found in the *Rhapis* principle (Fig. 4). In *Rhapis*, again using a descriptive convention, the leaf trace usually fuses with one or two bridges and an axial bundle when it is traced basipetally. Vascular elements are incorporated into the simple bundle almost immediately and can no longer be recognized as independent vascular strands. In Cyclanthaceae, however, the strands remain together but as independent strands for a considerable distance below the level of insertion of the leaf trace. This aggregation of separate vascular strands is seen in section as a distinctive compound bundle. However, in the basal part of such a bundle independent strands fuse and the compound bundle typically becomes a simple (though sometimes amphivasal) bundle before it is included as an axial bundle in a new constructional cycle. This basic principle governs vascular construction in the Cyclanthaceae; there is, of course, considerable quantitative variation at the generic level which may be diagnostic. In *Ludovia*, for example, bundles are usually bipolar, in *Thoracocarpus* they are multipolar. An important conclusion, therefore, is that the superficial appearance of similar compound vascular bundles in Pandanaceae and Cyclanthaceae is misleading and dependent on rather dis-

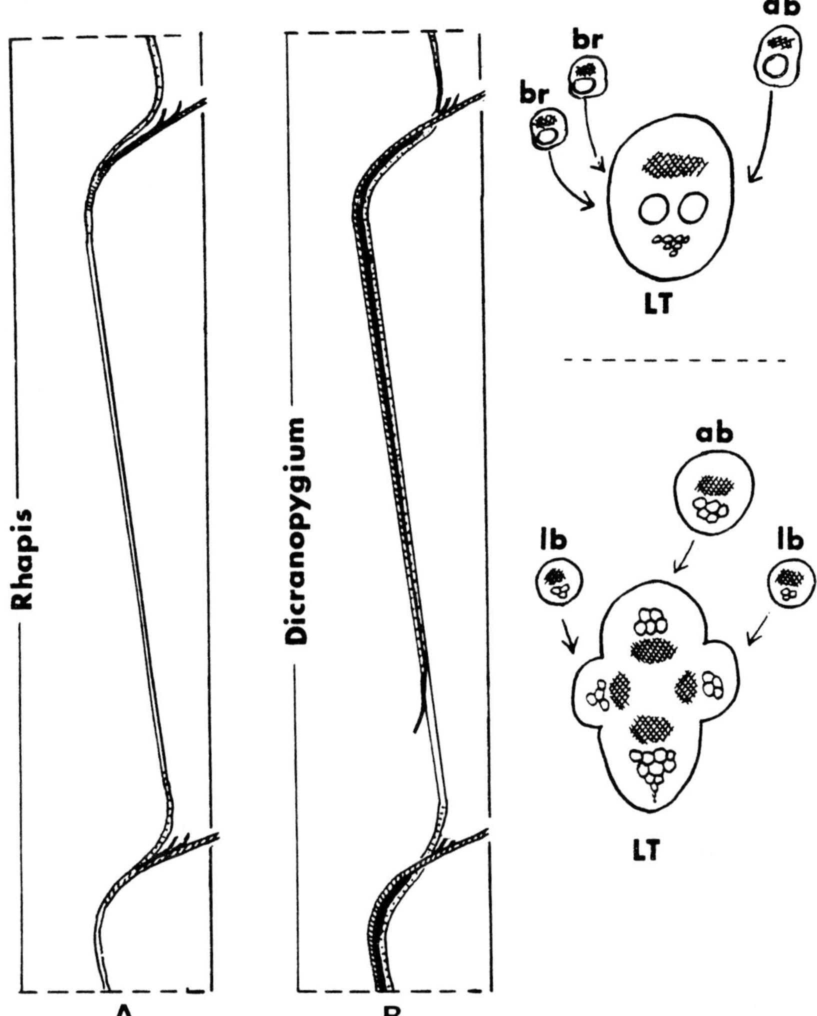

Fig. 4. Course of stem vascular bundles in Rhapis *and* Dicranopygium *(Cyclanthaceae) compared. A.* Rhapis excelsa *(cf. Fig. 1). B.* Dicranopygium crinitum. *In* Rhapis, *when traced basipetally bridges* (br) *and axial bundles* (ab) *unite with the leaf trace* (LT) *and do not remain discrete, as shown in the diagrammatic transverse section (upper right); in* Dicranopygium *the axial bundles* (ab), *lateral bundles* (lb), *which seem homologous with bridges, and leaf trace* (LT) *remain discrete over many internodes before fusing basally towards the end of the cycle, as shown in the diagrammatic transverse section (lower right) (after French and Tomlinson, 1983).*

similar constructional principles. This conclusion is only possible when detailed three-dimensional analysis is made.

4. *Araceae* (Fig. 5). It has been known since the time of van Tieghem (1867) that some aroids possess "compound" vascular bundles. It therefore becomes of interest to establish how this feature, evident in single transverse sections, can be used to establish the affinities of aroids with Palmae, Pandanaceae and Cyclanthaceae.

A recently completed survey of the Araceae, involving 3-dimensional analysis of stem vasculature in representatives of over 60 genera has lead to a number of conclusions, some systematic, some functional (French and Tomlinson, 1980; 1981a-d; 1983). Since the objective was to make a broad survey, some streamlining of procedures was necessary; most analyses were made on section series mounted temporarily in Karo syrup after bleaching in hydrochloric acid and staining in basic fuchsin. This emphasizes lignified structures at the expense of other features. In the absence of two microscopes, the drawing method was used throughout.

In contrast to the three previous families, the aroids proved to have a highly diversified vascular pattern. In part this is related to the extremely wide range of growth habits and architecture in the family, with geophytic stems including those of rhizomatous, cormous, or tuberous species as well as species with lianescent or sub-arborescent aerial stems with extended internodes. However, none of the axes are markedly woody, as in palms and pandans. In axes which have congested internodes the stem vascular system is without recognizable pattern and little useful comparative information was made available. This conclusion applies to isolated genera in some subfamilies (e.g. *Gonatopus*, *Zamioculcas* in Pothoideae, French and Tomlinson, 1981a) or

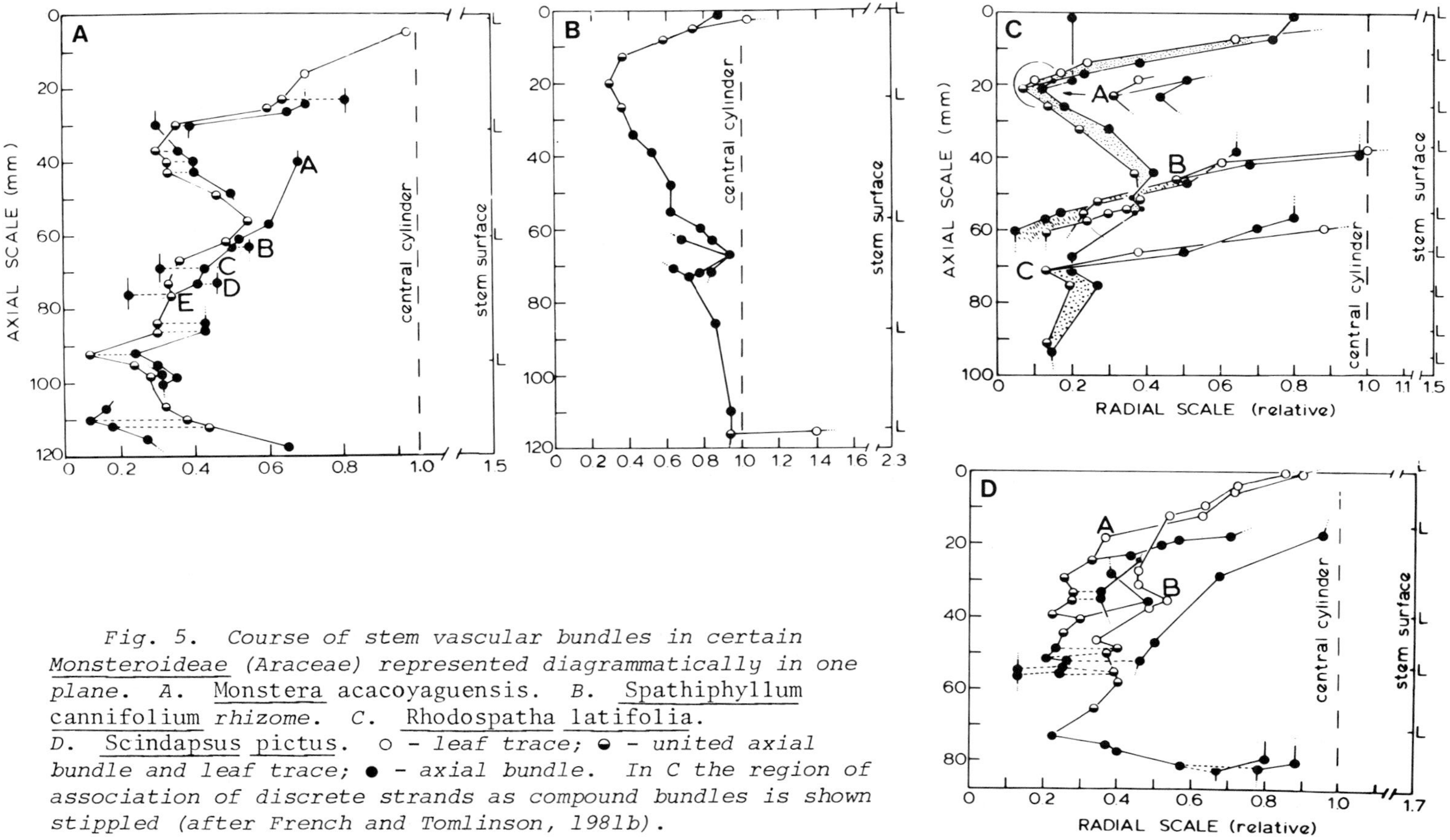

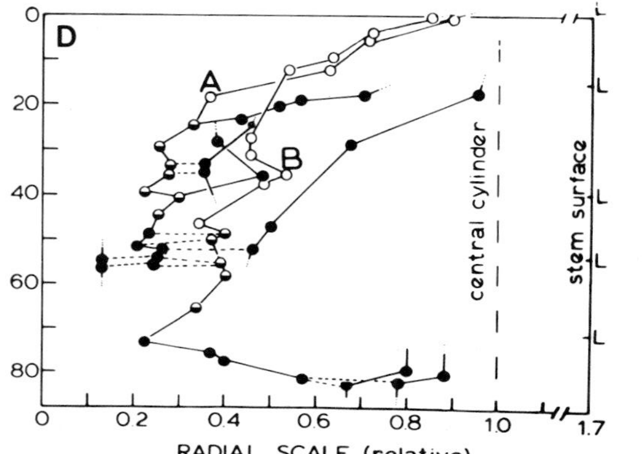

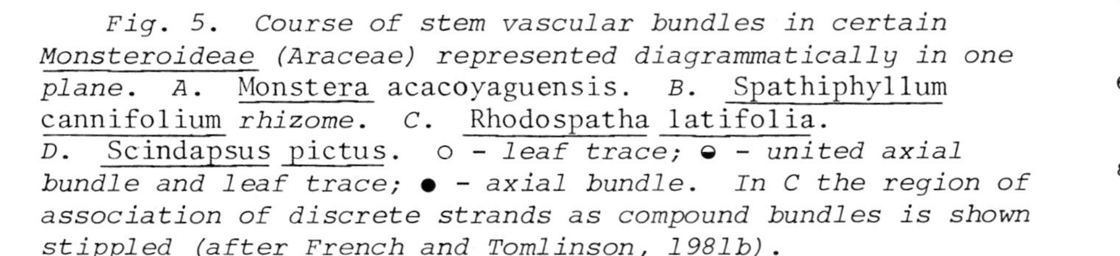

Fig. 5. Course of stem vascular bundles in certain Monsteroideae *(Araceae) represented diagrammatically in one plane. A.* Monstera acacoyaguensis. *B.* Spathiphyllum cannifolium *rhizome. C.* Rhodospatha latifolia. *D.* Scindapsus pictus. ○ *- leaf trace;* ◒ *- united axial bundle and leaf trace;* ● *- axial bundle. In C the region of association of discrete strands as compound bundles is shown stippled (after French and Tomlinson, 1981b).*

even to larger assemblages of genera (e.g. most Aroideae, French and Tomlinson, 1983). This is not to say that systematically diagnostic information was not observed, but there was none with reference to vascular pattern. In more significant directions, stem vasculature was shown to be very heterogenous in certain subfamilies, notably Pothoideae, confirming the generally heterogenous nature of this assemblage (French and Tomlinson, 1981a). Some groups of genera, in contrast, in other subfamilies were relatively homogenous (e.g. *Rhaphidophora* and related genera in Monsteroideae, French and Tomlinson, 1981b). In part this evidence has been used as a basis for some generic realignment and the general impression has been given that Engler's scheme of subdivision of the family is capable of considerable improvement. However, these systematic matters take us beyond our present restricted approach.

An important observation which arose out of this general survey was that compound vascular bundles could be constructed in different ways, and it has been suggested that they have originated independently in at least 3 ways. No example quite comparable to either that found in the Pandanaceae or the Cyclanthaceae was observed. In *Lasia, Cercestis* and *Rhektophyllum*, however, "compound" bundles can be recognized to a limited extent, with the basipetal fusing of initially separate but aggregated strands (French and Tomlinson, 1981c). This diversity within the Araceae and contrast with cyclanths and pandans demonstrates that vascular pattern can be informative of systematic relationships, even if the conclusions are somewhat negative. Compound vascular bundles can be either highly diagnostic of a family (e.g. Pandanaceae) and result from a distinctive developmental process, or can represent developmental variation of the basic *Rhapis*-principle, but still diagnostic for the taxon in which they occur. Only three-dimensional analyses could have lead to these conclusions.

An observation of functional significance, but not without systematic application, was that the method of attachment of branch traces in aroids varied considerably and could be regarded as diagnostically useful at the generic level. Variation includes the extent to which branch traces associate closely with leaf traces, form complexes which attach more directly to axial bundles, and are distributed circumferentially in the outer layers of the central cylinder, rather than the inner layers (e.g. French and Tomlinson, 1981a, Figs. 24-32; 1981c, Figs. 25-28). This variation, of course, has a developmental basis but a common feature found in extended aerial axes is the attachment of traces to buds which normally remain undeveloped. The functional significance of this association becomes evident when one appreciates that extended axes of aroids, notably the lianescent forms, have a marked capability for self repair after damage. When a stem is broken, the distal-most axillary bud below the break is activated and replaces the lost shoot. Sometimes two shoots are produced and the axis proliferates. Since the stem vascular system is determinate in the absence of a secondary vascular cambium, there is no opportunity to establish vascular continuity between an old axis and its new branch via development of new vascular tissue - the existing branch vasculature must "anticipate" the possibility of an accident if the whole shoot system is to be maintained. Undoubtedly the root-climbing habit of many aroids supplements this process of self repair, since adventitious roots above the level of damage can eventually supply the new shoot, but it is the inherent deterministic nature of the process which guarantees its functional success. Developmental processes during the establishment of the new branch connection still need to be analyzed and, in particular, the changes in vascular tissues at the time these opportunistic unions are established need to be investigated. These are complex structures and can only be appreciated fully by 3-dimensional analysis.

5. *Monocotyledons with Secondary Growth* (Fig. 6). We have published the results of only one detailed study of the course of vascular bundles in monocotyledons with secondary development of a vascular system, using *Dracaena fragrans* (Agavaceae), a commonly cultivated African species, as an example (Zimmermann and Tomlinson, 1969, 1970). The course of bundles in primary development corresponds in principle to that of *Rhapis* except that the axial bundle diverges in a peripheral position and may initially anastomose directly with other axial strands. The secondary system itself consists of an anastomosing series of vascular bundles. In the phase of establishment growth the two systems are confluent via bundles at the inner limit of the secondary vascular cylinder which distally are continuous directly as axial bundles of the primary system. Traced in a basipetal direction, some bundles of the primary system become confluent with the anastomosing secondary system. Here there seems to be a developmental homology between cortical bundles (in monocotyledons without secondary growth) and secondary bundles (in monocotyledons with secondary growth). This developmental continuum between primary and secondary tissues is, in part, a consequence of the continuity between primary and secondary meristematic tissues which has been noted by many authors in other representatives of the group.

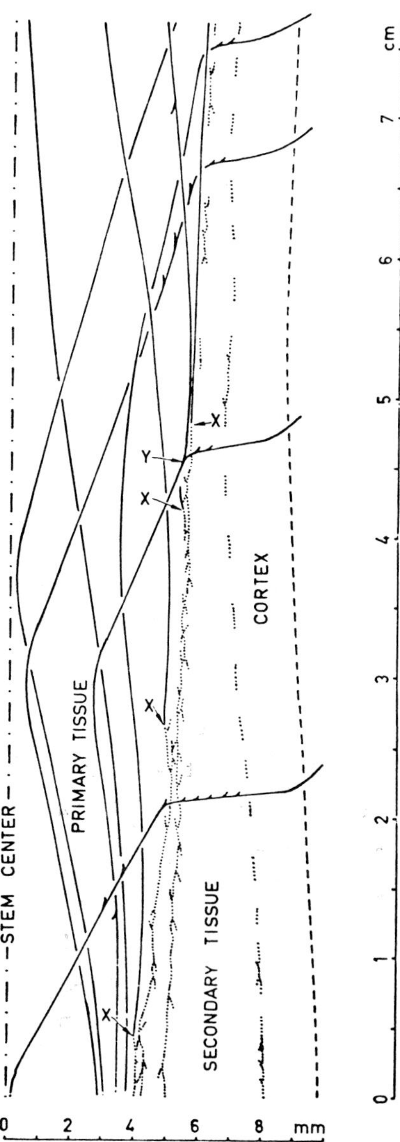

Fig. 6. Dracaena fragrans (Agavaceae), a monocotyledon with secondary thickening. Diagram of course of stem vascular bundles represented in one plane. At Y are examples of axial bundles diverging from leaf traces within the primary tissue. At X are examples of vascular bundles from the inner limit of the secondary tissue becoming incorporated into the primary tissue. Numerous bridges to both primary and secondary bundles are represented by short upwardly projecting bundles (after Zimmermann and Tomlinson, 1970).

Unpublished observations on other monocotyledons with secondary thickening growth, such as *Beaucarnea, Cordyline, Lomandra, Klattia* and related genera, show the same constructional principles. Since these genera belong to several unrelated systematic groups (Agavaceae, Lomandraceae, Iridaceae) it seems reasonable to conclude that this developmental step is a relatively easy one and is polyphyletic in its origin.

Development of secondary vascular tissues in monocotyledons is readily perceived as an extension of that part of the initiation of the primary vasculature which gives rise to the axial system i.e. the meristematic cap, as recognized by Zimmermann and Tomlinson (1972). The region may be continuously meristematic so that cap and the secondary vascular cambium are topographically continuous (the usual condition); otherwise the two meristems may be topographically (but probably only temporarily) discontinuous in the early ontogeny of the stem, even though continuous subsequently as shown by Stevenson (1980) for *Beaucarnea* (cf. also Stevenson and Fisher, 1980). This topic is often rendered unnecessarily contentious because investigators do not take into consideration holistic aspects of shoot construction in these plants, in particular differences observed during ontogenetic development, as well as architectural differences between different taxa.

The phenomenon of secondary development of vascular tissues in monocotyledons seems therefore to be of morphogenetic, architectural and ecological significance but is not informative of phylogenetic relationships.

One structural feature which seems common to this group of monocotyledons with secondary growth, but which has not been explained functionally, is that they all lack vessels both in the primary as well as the secondary tissues of the stem. One can account for the absence of vessels from secondary vascular tissues in monocotyledons by the distinctive method of tracheid development, which involves intrusive extension growth from a rather short cambial derivative (Röseler, 1889); it is more difficult to account for the absence of vessels from primary vascular tissues, especially as a number of woody monocotyledons without secondary growth have stem vessels (notably the Palmae, e.g. Klotz, 1978). We still lack the integrative approach developed in later sections of this article, which would probably account for this structural phenomenon in functional and developmental terms. It seems, however, a more satisfactory approach to look at these problems from this morphogenetic and functional point of view, rather than to invoke speculative "evolutionary" explanations.

6. *Zingiberaceae* (Fig. 7). The vascular anatomy of the rhizome and aerial axis of *Alpinia speciosa* (Zingiberaceae) has been investigated using shuttle microscopy by Bell (1980a, b). Here the objective is to contrast axes which differ in the functions of the appendages (photosynthetic leaves in aerial axes versus non-assimilating protective scales in underground axes).

In the aerial axis axial continuity is lacking in the central cylinder since departing leaf traces do not form a continuing axial bundle. This overall "loss" of bundles is in part compensated by splitting of axial bundles when traced in a distal direction, but seemingly insufficient to maintain a constant vascular supply. This progressive decrease in vascular bundles may be common in determinate axes

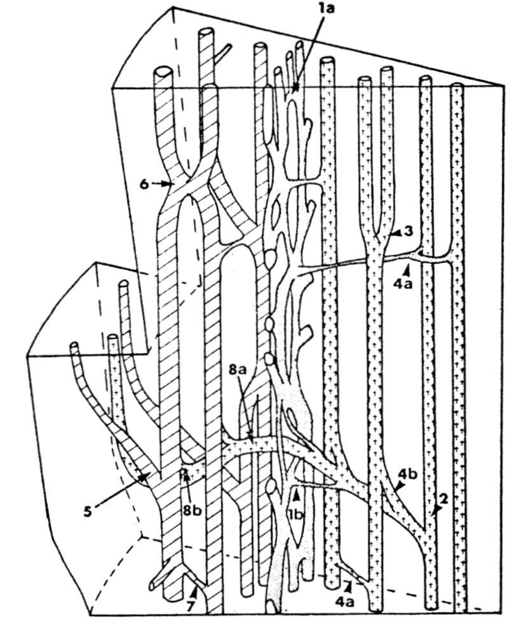

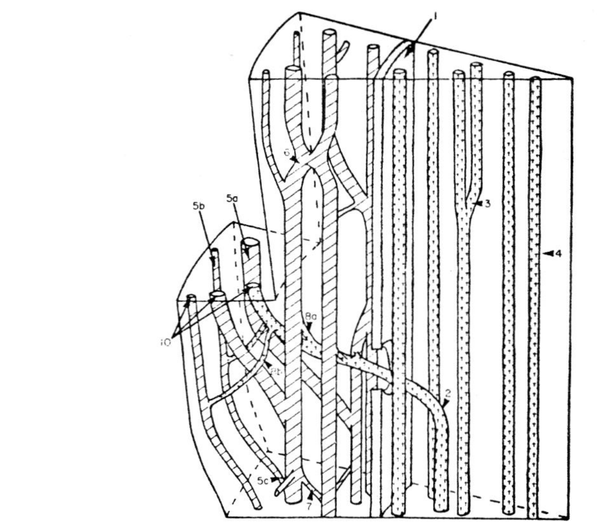

Fig. 7. Left: stem vascular system in a sector of the rhizome at the level of insertion of a scale leaf in Alpinia speciosa *(Zingiberaceae). Numbers refer to structural details described in the original papers. At position 2 is an axial bundle diverging from an outgoing leaf trace (8a) with a bridge at 4b, corresponding to the* Rhapis *principle. Right: the same analysis in a sector of the aerial stem. Note at position 2 the absence of a continuing axial bundle. Diagonal hatch: outer (cortical) system of vascular bundles; crossed: inner system; stippled: intermediate zone (after Bell, 1980a, b).*

in monocotyledons. In contrast the cortical vascular system is relatively extensive and there is anastomosing between them and outgoing leaf traces.

The rhizome vasculature is more complex. The system in the central cylinder conforms in a general way to the "*Rhapis*-type" since outgoing leaf traces develop continuing axial bundles as well as bridges. The cortical system is also elaborate, anastomosing, and with well-developed connections to the outgoing leaf traces. There is an intermediate region which partly consists of a plexus of root traces and is in direct continuity with both axial bundles and leaf traces of the central cylinder via short connecting strands. Bell in particular emphasizes the integrated nature of the vascular systems in relation to overall architecture. He especially draws attention to the effectiveness of vascular contact from one rhizome segment to another, in the sympodial branching system. Vascular continuity was demonstrated by dye injection experiments. These results emphasize that the aerial shoots can draw water (and possibly therefore nutrients) from several preceding sympodial units.

Further research is needed in this family in order to determine the extent to which a diagnostic pattern of axis vasculature can be recognized. Of particular comparative interest is the observation that rhizome and aerial axis in the palm *Rhapis* are much more like each other in their vascular pattern than the corresponding organs in the ginger. Rhizomatous axes, are not a well-developed feature of palms, but characterize most Zingiberalean families.

B. *Research in Other Areas*

1. *Vascular Cryptogams.* The cinématographic methods of analysis lends itself to the examination of the relatively simple vascular system of ferns and fern allies. It has been used by me to produce teaching films (e.g. illustrating protostelic and dictyostelic structure in a vivid manner) and has been helpful in demonstrating that branches in *Lycopodium* lack corresponding "branch gaps" in the stele and illustrates relationships between vascular pattern and phyllotaxis well. Adams (1977) examined the medullary system of the fern *Cyathea fulva* and demonstrated some regularity which had not previously been recognized in the distribution of these central bundles. Regularity is most apparent in peripheral medullary bundles and three categories of bundle were recognized, each category with a distinctive course and associated with the leaf trace system derived from the non-medullary system in a relatively constant way. Extension of these studies in the direction of comparative 3-dimensional analysis should increase our knowledge of tree fern anatomy considerably. Others of my students have used the method in morphological analysis, including Gruber (1980) in a study of the rhizome pattern of the extensively branched fern *Hypolepsis pelta* and Cox (unpublished) in a description of the anatomical contrast between stem and leaf attachment in *Lophosoria*, whose underground branched stem system is not mentioned in recent literature (e.g. Lucansky, 1982).

A rekindling of interest in vascular anatomy of ferns which takes into account functional and adaptive features is needed. At least we have relatively simple techniques for making the initial structural analyses with great precision.

2. *Gymnosperms*. Studies of the 3-dimensional distribution of resin canals in conifers by Hug (1979) in part used shuttle microscopy. In *Larix decidua* the axial and radial canal systems are continuous. Individual canals usually end in contact with another canal. The radial canals, for example, always start in contact with an axial canal. The epithelium of canals forms a continuous system of tissue interconnected with other living tissue such as axial and radial parenchyma, the cambium and phloem.

3. *Nymphaeales*. In angiosperm phylogeny a working hypothesis for the evolutionary relationship between dicotyledons and monocotyledons suggests the Nymphaeales as ancestral to monocotyledons, or at least certain of them. Part of the presumed evidence in support of this hypothesis is that members of the Nymphaeales and monocotyledons are similar in the "scattered" arrangements of stem vascular bundles, a relatively superficial observation based on examination of single transverse sections. Since we now have a clear grasp of basic structural and developmental features of monocotyledons as represented by the "*Rhapis*"-model, it seems appropriate to inquire how similar to this is the stem vasculature of water-lilies. Weidlich (1976a, b) has provided a partial answer, using a modification of the ciné method to suit the massive rhizomes of these plants. A series of axial bundles can be recognized, corresponding to well defined orthostichies, the connection of each leaf (or peduncle, which occupies a leaf position) to these axial bundles is constant and regular with 3 leaf traces derived from 3 different axial bundles, only two of which are adjacent. Weidlich concludes that in *Nymphaea* the stem vasculature is unique to angiosperms. His results show that the pattern of vascular construction of members of the order which he

analyzed could not be interpreted in terms of the *Rhapis*-principle. Similarities between monocotyledons and water-lilies in terms of stem vasculature are superficial and do not provide evidence for a phyletic relationship. Similar comparative study of the Piperales, another dicotyledonous group with "scattered" vascular bundles in the stem which superficially resembles those found in monocotyledons has not yet been undertaken.

4. *Dicotyledonous Vascular Systems.* In one sense the vascular system of dicotyledonous stems, because of the arrangement of vascular bundles in a single ring, can be said to be simpler than that of monocotyledons so that the topographic complexity of following bundles in three planes scarcely exists. Numerous illustrations exist which show the vascular system of dicotyledonous stems displayed in two-dimensions, with the vascular cylinder "unrolled" in a diagrammatic way. However, early stages of the vascular pattern are condensed and significant developmental steps occur within a few microns of the shoot apex proper. By a combination of the techniques of thin-sectioning and shuttle microscopy Larson and his co-workers have in the last decade or so considerably improved our understanding of basic processes in vascular development (see e.g. Larson, 1975 and summary in Larson, 1982). This research is reported on elsewhere in these proceedings by Dr. Larson himself so I need not elaborate. However, an important developmental feature demonstrated by Larson is the uncommitted nature of differentiating procambial strands i.e. the fact that a vascular bundle can be recognized (as a procambial strand) before the existence of the appendage which it supplies. This condition is one of the important conclusions demonstrated in our investigation of

monocotyledons, supporting the earlier observation of Priestley and his co-workers on *Tradescantia* (Scott and Priestley, 1925) and *Alstroemeria* (Priestley *et al.*, 1935; Priestley and Scott, 1937).

There still remains suggested developmental differences between dicotyledons and monocotyledons in the morphogenetic steps involved in the establishment of vascular continuity between stem and leaf. Zimmermann and Tomlinson (1972) have suggested that the process in dicotyledons is the result of an interaction between two morphogenetic centers or "poles" whereas in monocotyledons the process involves 3 such poles, the intermediate morphogenetic pole being possibly represented by the meristematic cap which stands between the axial and appendicular poles. Experimental approaches are certainly needed to test such hypotheses but would depend on precise methods of analysis if interpretable results are to be obtained.

5. *Anomalous Secondary Thickening.* In dicotyledons with so-called "anomalous" structure of either the primary or secondary vasculature, the relative simplicity of the vascular cylinder disappears. A number of investigators have used shuttle microscopy as an aid to unravelling vascular systems in such plants. For example, Zamski has investigated *Bougainvillea* (Zamski, 1980), sugar beet (Zamski and Azeakat, 1981) and *Avicennia* (Zamski, 1979). His research on *Avicennia* demonstrates the method of connection between primary and secondary tissue and the way in which discontinuous cambia, each of which functions for only a limited period of time, can produce vascular tissues which are continuous from one set of cambial derivatives to another. The vascular system is thus highly integrated and anomalous secondary thickening is not disadvantageous in terms of long-

distance transport. He also demonstrated that there was no necessary correlation between number of rings at any level and number of branches above that level (Zamski, 1981). These conclusions are supported by the few other examples where a holistic approach has been applied to vascular analysis. Using shuttle microscopy Fahn and Zimmermann (1982) have similarly demonstrated the way in which vascular integration is affected in *Atriplex* (Chenopodiaceae) which develops a thick stem by means of successive cambia. Successive cambial bands are developmentally continuous with existing bands and this leads to vascular continuity between primary and secondary vascular tissues.

6. *Floral Anatomy.* Ciné analysis lends itself to the elucidation of floral vascular systems which are frequently complex and condensed. Routinely this method is used in my research where reconstruction from serial sections is necessary, although rather little indication of this may be given in published accounts (e.g. Posluszny and Tomlinson, 1977; Tomlinson, 1982).

An extensive study of palm flowers by Uhl and Moore (e.g. Uhl, 1976; Uhl and Moore, 1971, 1977, 1980) has relied extensively on frame-by-frame ciné photography. Here the sections are commonly photographed in part-polarized light as this allows recognition of birefringent structures, notably the xylem of floral traces. The objectivity of the ciné method is particularly important here, because the observation of level of trace departure and extent of fusion of parts is important in these studies. The results obtained by this and other methods are summarized elsewhere by Dr. Uhl in these proceedings.

C. *The Palm-Stem - an Integrated View*

The previous sections present an overview of topics in which the cinématographic method has been the main contributory factor in developing our understanding. The greatest advances have come with the investigation of the palm stem and we can now present an overview of information which integrates construction, development and function. We have continued to use *Rhapis* as a "type", since we believe this serves to demonstrate basic principles which underly monocotyledonous stem construction generally.

1. *Mature Stem.* The principle of *Rhapis* stem vasculature described elsewhere (e.g. Zimmermann *et al.*, 1982; Zimmermann and Tomlinson, 1965) needs to be restated in somewhat more detail (Fig. 8). If an axial bundle is followed acropetally from a position close to the periphery of the central cylinder it migrates gradually towards the stem center. At the level of maximum radial penetration it turns abruptly towards the stem periphery and produces several derivative branches - an axial bundle, one or more bridges while the leaf trace proper finally enters the leaf base. The axial bundle maintains longitudinal continuity, the bridges provide lateral continuity. In mature stems bundles also diverge into axillary appendages; they may diverge as branches from leaf traces or connect directly in a basipetal direction with axial bundles (Tomlinson and Zimmermann, 1968; Tomlinson, 1973).

Quantitative variation in the stem vasculature follows a regular pattern. Bundles with the greatest radial displacement (major bundles) show the longest distance between successive branching at leaf insertion (leaf contact distances). Bundles with limited radial displacement (minor bundles) have short leaf contact distances. There is a continuum of intermediate bundles. Cortical bundles are short and either

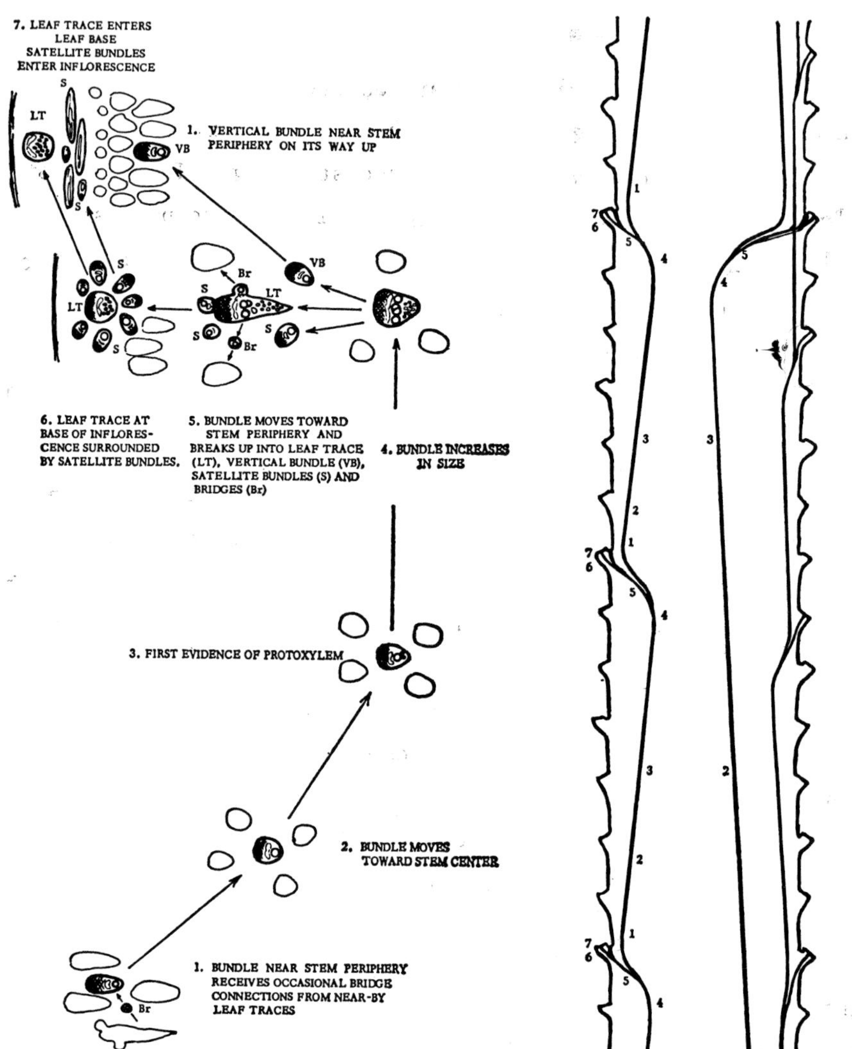

Fig. 8. Rhapis excelsa. *Axial vascular continuity in the stem represented diagrammatically:*
Right: course of bundles in one plane, a major, intermediate and minor bundle are shown (cf. Fig. 1).
Left: details of histological changes in diagrammatic transverse section in a single vascular bundle throughout one leaf contact cycle. Numbers identify corresponding levels in the two diagrams (after Zimmermann et al., 1982).

anastomose or end blindly in a basal direction. Variation in this regularity is appreciable and represents "noise" in the development process and has been described in some detail in earlier papers (e.g. Zimmermann and Tomlinson, 1968). Ciné analysis provides the needed refinement which demonstrates the repetitive nature of the pattern since all bundles in any one filmed sequence can be followed precisely. Our diagrams have represented bundles projected onto a radial plane (where in fact they are helical) and, of course, illustrate examples of only a few topographic types.

2. *Vascular Continuity* (Fig. 8). To this pattern of bundles as structural units, we must now add details of the distribution of conducting tissues which leads to the recognition of principles of hydraulic architecture in palms (Zimmermann *et al.*, 1982; Zimmermann and Sperry, 1983). Phloem is continuous between all bundle types and offers no apparent structural complexity which cannot be represented by the simple analytical diagrams. Continuity through the phloem has been demonstrated by autoradiographic techniques (Zimmermann, 1973). Xylem is more complex because one must distinguish between protoxylem and metaxylem which are not necessarily contiguous. In the proximal part of any structural cycle (i.e. when the axial bundle is in its most peripheral position) the xylem is represented (in *Rhapis*) by a single wide metaxylem vessel (or two ends of overlapping vessels). Distally the metaxylem is joined by contiguous narrow protoxylem elements, at first one or two, but progressively increasing in number. At a higher level protoxylem and metaxylem are separated by a narrow band of parenchyma; at the level of maximum penetration the metaxylem becomes elaborated and may consist of a parallel series of 3 to 4 vessels. These vessels are continuous laterally into bridges and

continuing axial bundles; only protoxylem is continuous from stem into leaf. This distribution has a simple developmental explanation, but in the mature stem it is the basis for the hydraulic bottleneck which favors axial movement of water against lateral movement into appendages (notably leaves). This functional aspect is described in detail by Zimmermann *et al.* (1982). The bottleneck results from the nature of the protoxylem continuity between stem and leaf, i.e. via numerous but narrow, tracheids (in the leaf trace) which are contiguous with one or few, but wide, vessels (in the axial bundle).

3. *Vascular Development* (Fig. 9). In describing the origin of the pattern of vascular bundles (as procambial strands) in the crown of *Rhapis* Zimmermann and Tomlinson (1967) relied on the ciné method to follow faithfully the position of individual traces connecting progressively older leaves. As bundles were followed, now in a basipetal direction from the leaf base, their location was plotted as an accurate positional representation in relation to the overall crown topography. To simplify the illustrative part of this analysis, all leaves were represented synthetically as if arranged in a single orthostichy (rather than the actual 2/5 phyllotactic spiral) and the internal helix was eliminated. This analysis established three important developmental features: - (a) axial bundles preceded in their appearance the leaf which they eventually supplied, (b) the "double-curve" of monocotyledonous vasculature, described precisely by von Mohl (1824) is a simple topographic consequence of the way in which the leaf base and crown expands following inception of individual strands, (c) axial continuity is a two-step process, involving first a connection between a leaf trace and an axial bundle generated by an umbrella-shaped meristematic cap, and a later (distal) divergence of the

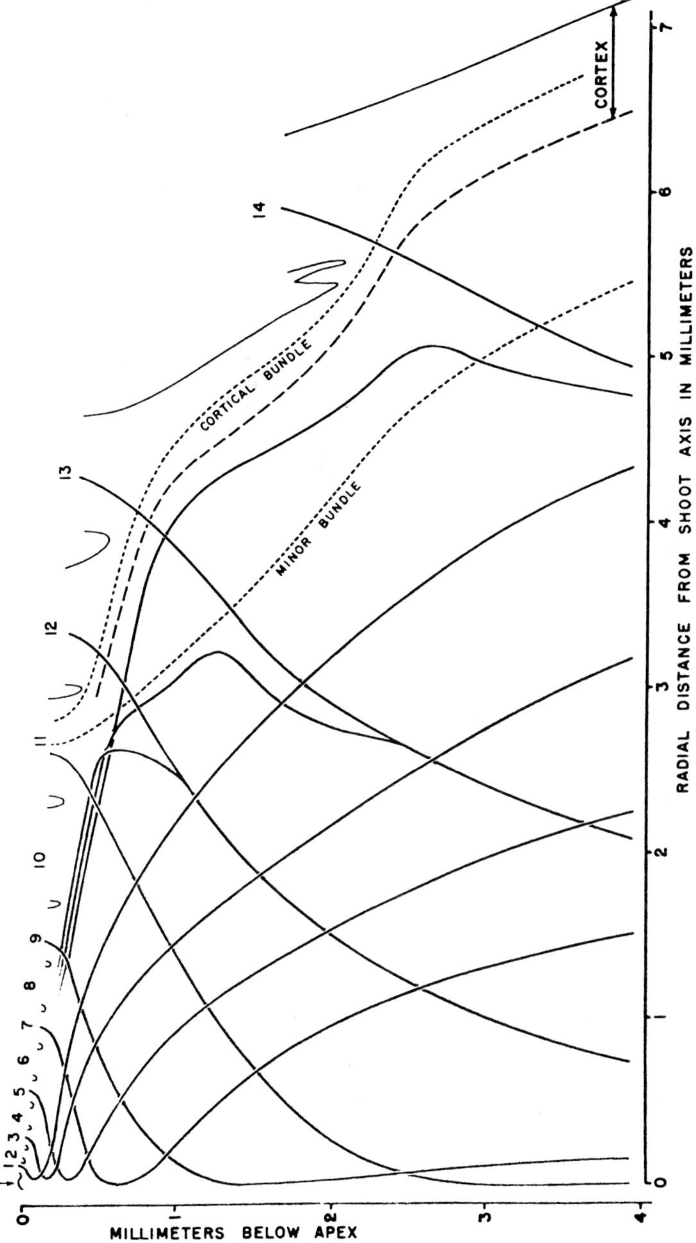

Fig. 9. Rhapis excelsa. *Course of differentiating vascular bundles in the stem apical region, represented diagrammatically in one plane. The leaf positions are rotated to occupy a single orthostichy (after Zimmermann and Tomlinson, 1967).*

previously uncommitted axial vascular bundle into a higher leaf. These processes are difficult to comprehend from static diagrams and they have been presented in a dynamic way as animated cartoon sequences in demonstration movies (Zimmermann and Mattmüller, 1982).

Quantitative variation which leads to the distinction between major, intermediate and minor bundles is also a topographic consequence of leaf trace connections being made over a long period of time (related to the long period of leaf base expansion) and increasing confinement to peripheral stem areas. Other developmental details are considered in the original research papers which have dealt with the development of vascular systems which conform to the *Rhapis* system in several genera of monocotyledons (e.g. Zimmermann and Tomlinson, 1968, 1969).

4. *Vascular Integration in Space and Time* (Fig. 10). The most recent advances have been effected by an integration of the process of initiation of the vascular system as a series of procambial strands and the inception of vascular tissues within this vascular skeleton (Tomlinson and Vincent, 1984). In making this analysis consideration has to be given to the status of each leaf which the developing trace system supplies. Each leaf primordium undergoes an initial period of growth enclosed within the crown (about 8-10 plastochrones); it then elongates rapidly, becomes visible as the central "spear leaf", after which the blade expands (within about 1 plastochrone) and the leaf is recruited into the photosynthetic apparatus of the palm as the youngest expanded leaf. The expanded leaf is then gradually displaced from a vertical to a horizontal and finally pendulous position as it senesces and is irregularly abscised. Quite marked changes in the water relations of the leaf occur as the leaf matures and these are

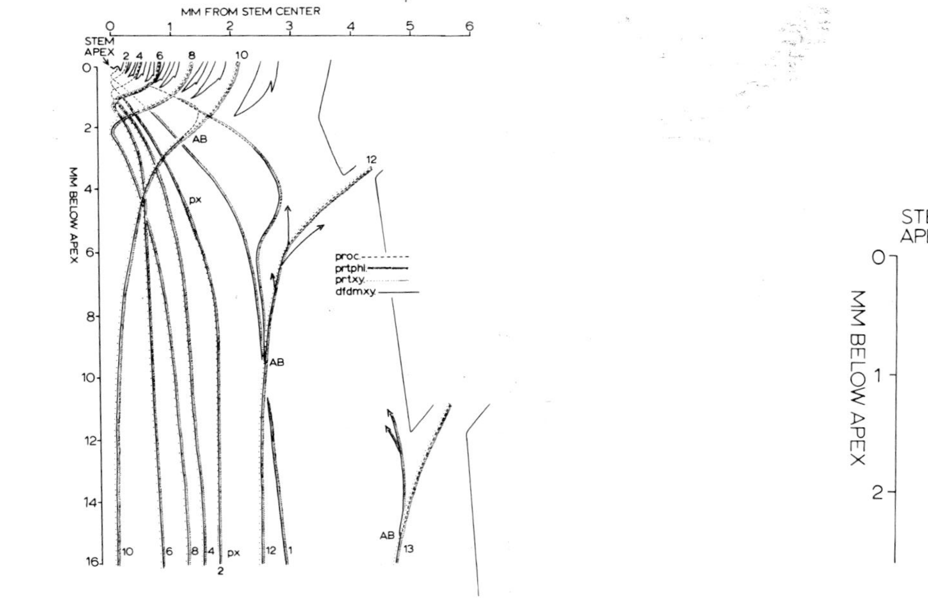

Fig. 10 (left). *Course of vascular differentiation in the crown of* Rhapis excelsa. *Position of differentiating vascular elements represented in a single major bundle to each of the numbered leaf primordia.*

Fig. 10 (right). *Details of crown region from same diagram. Protoxylem is not yet continuous into leaf 6. Arrowheads represent level of first differentiation of metaxylem.*

(After Tomlinson and Vincent, 1984; further explanation in text)

being investigated experimentally by graduate student John Sperry at Harvard Forest and at Fairchild Tropical Garden in Miami. The hydraulic status of the leaf in part is a result of the timing of the *vascular* connection it makes with the axial system (as distinct from the *morphological* connection, visible superficially as the leaf base attachment). These have been referred to respectively as the *vascular insertion* and the *morphological insertion* (J. Sperry, pers. comm.). A study of the course of differentiation of the xylem, imposed on the vascular template, shows how the process is effected.

a. *Phloem differentiation*. We can dismiss phloem development as a relatively simple process. Protophloem always differentiates in continuity with existing phloem, its extent at different levels is a consequence of the time period over which a given bundle undergoes elongation growth (one should emphasize that these topographic relations are complex in palms by virtue of the extensive primary thickening growth, with internodal elongation continuing until about plastochrone 15, i.e. to the internode below the node of insertion of the 15th leaf, which is about the fourth or fifth *visible* leaf in the crown). The distinction between protophloem and metaphloem is not an exact one; the position of obliterated protophloem elements in mature bundles is represented by sclerenchymatous elements which form part of the fibrous bundle sheath.

b. *Xylem differentiation*. Protoxylem differentiation is discontinuous. Elements appear first in association with P2 and effect continuity rapidly with the leaf in a distal direction and differentiate in a basipetal direction, initially without making contact with the functioning axial xylem system. This is because the metaxylem (which is blocked out very early and in continuity with older, functional xylem) remains radially discontinuous with the protoxylem except in the basal (older) part of each leaf cycle. These features are best appreciated with reference to the series of diagrams which represent schematically the spatial distribution of vascular tissues within a single bundle at successive plastochrones (Fig. 10).

The period over which protoxylem differentiation proceeds is an extended one since it begins in the leaf primordium at position P2 and continues until the leaf is in position P15 or P16 i.e., over a period of about 13 or 14 plastochrones. We refer here not to the differentiation of xylem within the leaf or leaf base itself, which has not been studied very extensively but to differentiation within the leaf trace system which supplies an individual leaf. The process is complicated because xylem differentiation is initiated progressively in vascular bundles as they connect to the leaf; the rate of initiation of these connections is expressed by a sigmoid growth curve. If we consider the simple condition of a single major bundle, protoxylem first appears in the upper part of the leaf trace and is discontinuous both distally and proximally (e.g. P2 in Fig. 10). Continuity into the leaf is established by P7; it must be remembered that this is still a primordium scarcely 1 cm tall. At any one level, protoxylem is continually extended, ruptured and replaced by newly differentiated elements. The extension which brings about this rupture is initially due to primary thickening growth which increases the length of the vascular bundle, it is only in quite late stages that internodal elongation influences vascular differentiation in a topographic sense. The overall result of this process, at any one level is an increase in the total number of functional xylem elements. In order to understand how xylem connection is made proximally we must first investigate metaxylem differentiaton.

Metaxylem differentiation is continuous acropetally. Two distinct phases can be recognized - (a) the initial period when the elements are blocked out during the period of vascular bundle elongation, (b) the later phase of metaxylem element maturation, when cell walls become thickened and lignified; this only occurs after vascular bundle elongation ceases. Initiation of metaxylem is recognizable in transverse section of bundles close to the shoot apex as two or more cells which vacuolate early and enlarge; they may be recognized closer to the shoot apex than the protoxylem. Metaxylem vessel initials are continuous below with a single element which represents the metaxylem in the axial portion of the bundle and which in older leaves has matured. Since differentiating metaxylem is always continuous with mature metaxylem it must differentiate continuously, acropetally into axial bundles and bridges at the level of leaf trace branching but not into the leaf trace itself beyond the level of this branching. This, of course, is the structural basis for the hydraulic constriction recognized by Zimmermann *et al.* (1982). This structural aspect is, in part, a simple topographic consequence of the appearance of differentiated tracheary elements in parts of the leaf trace which extend most but also because of the failure of tracheary elements to mature late in such regions.

5. *Functional Integration.* We now have structural evidence which will elucidate the question: when is xylem continuity and a transport pathway for water movement between stem and leaf established in relation to the age of a leaf primordium and its transpirational requirements? That this

is not a straightforward question can be appreciated from the observation that in the differentiated vascular system, in any one vascular bundle, protoxylem and metaxylem remain contiguous only at proximal levels, but at distal levels they are separated by a layer of conjunctive parenchyma across which there is no possibility of direct water transport. Vascular continuity, which establishes *vascular* insertion, can be recognized where the wide metaxylem elements mature in contact with functional protoxylem. This is only possible in late stages of protoxylem differentiation because of the topographic requirements for the contact to be maintained, i.e. not ruptured by later extensive vascular elongation. This occurs when the leaf is about position P15, i.e. when its associated internode is just about to complete its extension.

A complication is that xylem elements may continue to differentiate (or at least mature) on the inner face of the strand of protoxylem elements contiguous with the wide metaxylem elements after vascular elongation occurs. By definition these are also metaxylem elements, but they are not significantly wider than adjacent protoxylem elements and in our earlier description were not recognized. These elements presumably correspond to the narrow vessel elements recognized by Zimmermann *et al.* (1982) in macerated material.

The important physiological correlation which can be recognized once this topographic establishment of xylem is appreciated in principle, is that linkage of "leaf" and "stem" xylem coincides with the time at which a leaf becomes visible when the developing crown first begins to transpire and then dramatically increases this rate of transpiration as it expands. This dramatic rise in rate of water loss coincides more or less precisely with the establishment of the *vascular connection*, i.e. the developmental contact made between *mature* metaxylem and distally matured protoxylem.

The details of vascular development in the developing palm crown are obviously complex in relation to the large number of vascular strands which connect each individual leaf with the stem, but once developmental principles are grasped in relation to a single vascular bundle, it becomes possible to recognize that the time course of maturation of vascular tissues is designed to establish a set of functional associations *during* crown development as well as subsequently in the mature axis. The system is designed to generate the necessary continuity of xylem transport tissue exactly in proportion to the needs of the developing appendages. Clearly the system needs to be examined further in the direction of a study of the vascular connection of other lateral stem appendages (inflorescences, vegetative branches).

The very generalized nature of the vascular pattern of *Rhapis* as a "type" for monocotyledons means that the principles have wide application in understanding developmental processes in this group of plants. The very extended period of leaf development in larger monocotyledons allows the temporal and spatial separation of structural, developmental and functional processes to be appreciated. Similar considerations must apply throughout the vascular plants. Variation among them need to be correlated with the architecture of the organism in the same way as has been provided for palms.

III. LEAF TRACE, AXIAL BUNDLE, CAULINE BUNDLE

The observations presented in summary above are useful in that they allow some commentary on abstract concepts which have governed vascular analysis in plants. For example, we have used the terms "leaf trace" and "axial bundle" in our description of vascular systems in monocotyledons. Can we define these terms more precisely now that our analysis is fairly complete?

How do we distinguish leaf trace and axial bundle? The answer depends to a larger extent on whether one thinks in structural (topographic) terms, developmental terms or functional terms. In a structural context "leaf trace" would then refer to the level of maximum penetration into the stem center of each bundle entering from a leaf, "axial bundle" would be its basal continuation until it connects with another incoming leaf trace at a lower level. This disposition is easily recognized topographically, and could be useful descriptively, but has no obvious functional or developmental basis.

In a developmental context, "leaf trace" would be that part of a bundle beyond the level of divergence (branching) of the continuing "axial bundle". Developmentally this represents the level at which linkage is made between existing leaf trace and meristematic cap (as shown in the animated diagrams in Zimmermann and Mattmüller, 1982). This is topographically very variable, even within a single family, as shown by Zimmermann and Tomlinson (1972) for the palms. The linkage seems to have no clear functional significance although it is topographically easy to recognize and must have an important morphogenetic basis.

The most satisfactory definition is probably a functional one, and refers to the extent to which protoxylem is differentiated within each leaf contact cycle. The portion of the unit with protoxylem is functionally "leaf trace" in terms of appendage supply. This part of the structural cycle is the basis of the "hydraulic constriction" described by Zimmermann and Sperry (1983) in which the hydraulically significant parameter (sum of the fourth powers of the diameters of all functional water conducting elements) is abruptly reduced. This collective value is based on the concept of "vascular insertion" developed above. The "axial bundle" would then be the basal portion of each cycle in which no protoxylem is differentiated.

This concept is generalizable to all vascular plants and their examination with an appreciation of this functional characteristic in mind opens out an interesting new vista in plant anatomy. Since this feature is a basic structural prerequisite of plants which develop appendages and grow tall, comparative study could be highly illuminating.

The observation also seems to account for one general structural peculiarity of xylem development in most vascular plants i.e. the discontinuous development of the protoxylem, which is initiated in isolation from other xylem tissues. This is simply a mechanism which produces a hydraulic constriction. The structural peculiarity has been widely described but as far as I know no functional explanation for its existence has been forthcoming. In the palms, by virtue of the extended period over which vascular continuity between leaf and stem is developed, the functional benefit is particularly obvious.

Finally these investigations allow us to come to grips with the abstract notion of "cauline bundle" which has figured extensively in discussion of the "nature" of the vascular system in plants i.e. the extent to which bundles of the stem vascular system are wholly established by cauline processes of regulation and the extent to which the vascular system can be considered to consist of a set of phyllome processes of regulation. In palms, if there is a "cauline" system, it can be recognized in the metaxylem which maintains axial continuity. The "phyllome" system is the protoxylic leaf trace, as defined above, which is important in allowing the plant structurally to regulate water movement so that axial continuity is favored over appendicular continuity. This, of course, puts the concept of cauline bundle on a less abstract and more physiological basis. Morphogenetic analysis is needed to expand this hypothesis.

Clearly these several approaches to the description and interpretation of the vascular system are not mutually exclusive, they are qualified by the point of view adopted.

IV. CONCLUDING REMARKS

Despite the seemingly complexity of the vascular system of the stem of large monocotyledons we now have a very comprehensive view of the way in which it is constructed, develops and functions so as to maintain the integrity of the crown structure (via the concept of "hydraulic architecture" of Zimmermann, 1978). This is a considerable advance over early comparative descriptions of monocotyledons as having "scattered" vascular bundles or fitting their stelar pattern into some comparative scheme as an "atactostele", both of which expressions convey absolutely no information whatsoever. In some ways we have a more completely integrated view of monocotyledonous vasculature than for any other group of vascular plants. This is largely because cinématographic methods have provided the much needed objectivity. Continuing studies are needed to broaden the survey and should provide useful systematic and possibly even phylogenetic information. However, it should be appreciated that the important discoveries which have been made all relate to developmental *processes*, the structural consequences of which are not necessarily very informative in a systematic context. The distribution of secondary vascular tissues in monocotyledons illustrate this. Plants of this type of construction are not necessarily related, as their systematic distribution suggests; recognition of the innovation of secondary vascular tissues as a simple developmental extension of primary processes is facilitated by our more precise understanding of those primary processes.

Perhaps the most significant steps to be taken in the future relate to an investigation of the morphogenetic factors controlling inception of the monocotyledonous vascular system and conducting tissues. That rather dissimilar end products may be the result of rather small shifts in developmental patterns is suggested by French and Tomlinson (1983) in their study of Cyclanthaceae. They equate the characteristic "compound" vascular bundles of this family with that of the *Rhapis*-type by recognizing that developmental "tracks" remain more completely separated spatially in the first type than in the second. If we view the developing vascular "tracks" in the *Rhapis* system as less completely integrated than is at first apparent (and there is partial structural evidence for this even in "simple" bundles) the morphogenetic factors become more obviously dissociated.

Experimental approaches are obviously needed, but they can be applied rigorously and intelligently only if we have a controlled understanding of the normal pattern.

The peculiar advantage of the study of monocotyledons in these morphogenetic analyses should be evident from the previous description and available methods of analysis. It is because they have large numbers of bundles and because the developmental pattern is uniform that its modification (whether natural or manipulated) can be recognized with facility and certainty. The extended period of the developmental process and the broad topographic separation of successive phases actually facilitates study.

The broad perspectives that I have presented are simply the foothills of mountain ranges to be scaled. If some preconceptions can be eliminated we should be able to move upwards in a continued progression.

REFERENCES

Adams, D. C. (1977). Ciné analysis of the medullary bundle system in *Cyathea fulva*. *Amer. Fern. J.* *67,* 73-80.

Bell, A. D. (1980a). The vascular pattern of a rhizomatous ginger (*Alpinia speciosa* L. Zingiberaceae). I. The aerial axis and its development. *Ann. Bot. (Lond.)* n.s. *46,* 203-212.

__________. (1980b). The vascular pattern of a rhizomatous ginger (*Alpinia speciosa* L. Zingiberaceae). 2. The rhizome. *Ann. Bot. (Lond.)* n.s. *46,* 213-220.

Fahn, A. and Zimmermann, M. H. (1982). Development of the successive cambia in *Atriplex halinus* (Chenopodiaceae). *Bot. Gaz.* *143,* 353-357.

French, J. C. and Tomlinson, P. B. (1980). Preliminary observations on the vascular system in stems of certain Araceae. *In* "Petaloid Monocotyledons" (Eds. C. Brickell, D. F. Cutler and M. Gregory), pp. 105-116. Linnean Society Symposium Series, No. 8. Academic Press, London.

French, J. C. and Tomlinson, P. B. (1981a). Vascular patterns in stems of Araceae: subfamily Pothoideae. *Amer. J. Bot. 68*, 713-729.

French, J. C. and Tomlinson, P. B. (1981b). Vascular patterns in stems of Araceae: subfamily Monsteroideae. *Amer. J. Bot. 68*, 1115-1129.

French, J. C. and Tomlinson, P. B. (1981c). Vascular patterns in stems of Araceae: subfamilies Calloideae and Lasioideae. *Bot. Gaz. 142*, 366-381.

French, J. C. and Tomlinson, P. B. (1981d). Vascular patterns in stems of Araceae: subfamily Philodendroideae. *Bot. Gaz. 142*, 550-563.

French, J. C. and Tomlinson, P. B. (1983). Vascular patterns in stems of Araceae: subfamilies Colocasioideae, Aroideae and Pistioideae. *Amer. J. Bot.:* 70:756-771.

French, J. C., Clancy, K. and Tomlinson, P. B. (1983). Vascular patterns in stems of Cyclanthaceae. *Amer. J. Bot.:* in press.

French, J. C. and Tomlinson, P. B. (1984). Organization and histology of the stem vascular system in Araceae: an overview: in preparation.

Gruber, T. M. (1980). The branching pattern of *Hypolepsis repens*. *Amer. Fern. J. 71*, 41-47.

Hug, U. E. (1979). Das Harzkanalsystem im juvenilen Stammholz von *Larix decidua* Mill. Beih-Zeitschr. Schweiz. Forst. No. 61, pp. 127.

Klotz, L. H. (1978). Form of the perforation plates in the wide vessels of metaxylem in palms. *Journ. Arnold Arbor. 59*, 105-128.

Larson, P. R. (1975). Development and organization of the primary vascular system in *Populus deltoides* according to phyllotaxy. *Amer. J. Bot. 62*, 1084-1099.

__________. (1982). The concept of cambium. *In* "New Perspectives in Wood Anatomy" (Ed. P. Baas). Martinus Nijhoff/W. Junk, The Hague, Boston, London.

Lucansky, T. W. (1982). Anatomical studies of the neotropical Cyathaceae. II. *Metaxya* and *Lophosoria*. *Amer. Fern J. 72*, 19-29.

Mohl, H. von. (1824). De palmarum structura. *In* "Historia Naturalis Palmarum" (Ed. K. F. P. von Martius), 1: I-VII, 16 pls.

Posluszny, U. and Tomlinson, P. B. (1977). Morphology and development of floral shoots and organs in certain Zannichelliaceae. *Bot. Journ. Linn. Soc. 75*, 21-46.

Priestley, J. H. and Scott, L. I. (1937). Leaf venation and leaf trace in monocotyledons. *Proc. Leeds Philosophical Soc. 3*, 305-324.

Priestley, J. H., Scott, L. I. and Gillett, E. C. (1935). The development of the shoot in *Alstroemeria* and the unit of shoot growth in monocotyledons. *Ann. Bot. (Lond.) 49*, 161-179.

Röseler, P. (1889). Das Dickenwachsthum und die Entwickelungsgeschichte der secundären Gefässbündel bei baumartigen Lilien. *Jahrb. Wiss. Bot. 20*, 292-348.

Scott, L. I. and Priestley, J. H. (1925). Leaf and stem anatomy of *Tradescantia fluminensis* Vell. *Jour. Linn. Soc. Lond. (Bot.) 47*, 1-28.

Stevenson, D. W. (1980). Radial growth in *Beaucarnea recurvata*. *Amer. J. Bot. 67*, 476-489.

Stevenson, D. W. and Fisher, J. B. (1980). The developmental relationship between primary and secondary thickening growth in *Cordyline* (Agavaceae). *Bot. Gaz. 141*, 264-268.

Tieghem, P. van. (1867). Recherches sur la structure des Aroidées. *Ann. sci. nat. Bot.* Ser. 5, *6*, 72-210.

Tomlinson, P. B. (1973). The monocotyledons: their evolution and comparative biology. VIII. Branching in monocotyledons. *Quart. Rev. Biol. 48*, 458-465.

__________. (1979). Systematics and ecology of the Palmae. *Ann. Rev. Ecol. Syst. 10*, 85-107.

__________. (1982). Anatomy of Monocotyledons (Ed. C. R. Metcalfe). VII: Helobiae (Alismatidae) - including the seagrasses. Clarendon Press, Oxford. i-xiv, pp. 559, pl. 16.

Tomlinson, P. B. and Vincent, J. R. (1984). Anatomy of the palm *Rhapis excelsa*. X. Differentiation of stem conducting tissues. **J. Arnold Arb. (in press).**

Tomlinson, P. B. and Zimmermann, M. H. (1966). Anatomy of the palm *Rhapis excelsa*. II. Rhizome. *Journ. Arnold Arbor. 47*, 248-261.

Tomlinson, P. B. and Zimmermann, M. H. (1968). Anatomy of the palm *Rhapis excelsa*. VI. Root and branch insertion. *Journ. Arnold Arbor. 49*, 307-316.

Uhl, N. W. (1976). Developmental studies in *Ptychosperma* (Palmae). II. The staminate and pistillate flowers. *Amer. J. Bot. 63*, 97-109.

Uhl, N. W. and Moore, H. E. (1971). The palm gynoecium. *Amer. J. Bot. 58*, 945-992.

Uhl, N. W. and Moore, H. E. (1977). Centrifugal stamen initiation in phytelephantoid palms. *Amer. J. Bot. 64*, 1152-1161.

Uhl, N. W. and Moore, H. E. (1980). Androecial development in six polyandrous genera representing five major groups of palms. *Ann. Bot. 45*, 57-75.

Weidlich, W. H. (1976a). The organization of the vascular system in the stems of the Nymphaeaceae. I. *Nymphaea* subgenus *Castalia* and *Hydrocallis*. *Amer. J. Bot. 63*, 499-509.

__________. (1976b). The organization of the vascular system in the stems of Nymphaeaceae. II. *Nymphaea* subgenera *Anecphya, Lotos* and *Brachyceras*. *Amer. J. Bot. 63*, 1365-1379.

Zamski, E. (1979). The mode of secondary growth and the three-dimensional structure of the phloem in *Avicennia*. *Bot. Gaz. 140*, 67-76.

__________. (1980). Vascular continuity in the primary and secondary stem tissues of *Bougainvillea* (Nyctaginaceae). *Ann. Bot. 45*, 561-567.

__________. (1981). Does successive cambia differentiation in *Avicennia* depend on leaf and stem initiation? *Israel J. Bot. 30*, 57-64.

Zamski, E. and Azenkat, A. (1981). Sugarbeet vasculature. I. Cambial development and the three-dimensional structure of the vascular system. *Bot. Gaz. 142*, 334-343.

Zimmermann, M. H. (1971). Dicotyledonous wood structure (made apparent by sequential sections). Film E 1735. Encyclopaedia Cinematographica. Institut fur den Wissenschaftlichen Film. Gottingen.

__________. (1973). The monocotyledons: their evolution and comparative biology. IV. Transport problems in arborescent monocotyledons. *Quart. Rev. Biol. 48*, 314-321.

__________. (1976). The study of vascular patterns in higher plants. *In* "Transport and Transfer Processes in Plants" (Eds. I. F. Wardlaw and J. B. Passioura), pp. 221-235. Academic Press, New York.

__________. (1978). Hydraulic architecture of some diffuse-porous trees. *Can. J. Bot. 56,* 2286-2295.

Zimmermann, M. H. and Mattmüller, M. (1982). The vascular pattern in the stem of the palm *Rhapis excelsa.* I. The mature stem. II. The growing tip. Films D 1404 and D 1408. Institut fur den Wissenschaftlichen Film. Gottingen.

Zimmermann, M. H., McCue, K. F. and Sperry, J. S. (1982). Anatomy of the palm *Rhapis excelsa.* VIII. Vessel network and vessel-length distribution in the stem. *Journ. Arnold Arbor. 63,* 83-95.

Zimmermann, M. H. and Sperry, J. S. (1983). Anatomy of the palm *Rhapis excelsa.* IX. Xylem structure of the leaf meristem. *Journ. Arnold Arbor.*: in press.

Zimmermann, M. H. and Tomlinson, P. B. (1965). Anatomy of the palm *Rhapis excelsa.* I. Mature vegetative axis. *Journ. Arnold Arbor. 46,* 160-177.

Zimmermann, M. H. and Tomlinson, P. B. (1966). Analysis of complex vascular systems in plants: optical shuttle method. *Science 152,* 72-73.

Zimmermann, M. H. and Tomlinson, P. B. (1967). Anatomy of the palm *Rhapis excelsa.* IV. Vascular development in apex of vegetative aerial axis and rhizome. *Journ. Arnold Arbor. 48,* 122-142.

Zimmermann, M. H. and Tomlinson, P. B. (1968). Vascular construction and development in the aerial stem of *Pronium* (Juncaceae). *Amer. J. Bot. 55,* 1100-1109.

Zimmermann, M. H. and Tomlinson, P. B. (1969). The vascular system in the axis of *Dracaena fragrans* (Agavaceae). I. Distribution and development of primary strands. *Journ. Arnold Arbor. 50,* 370-383.

Zimmermann, M. H. and Tomlinson, P. B. (1970). The vascular system in the axis of *Dracaena fragrans* (Agavaceae). 2. Distribution and development of secondary vascular tissue. *Journ. Arnold Arbor. 51,* 478-491.

Zimmermann, M. H. and Tomlinson, P. B. (1972). The vascular system of monocotyledonous stems. *Bot. Gaz. 133,* 141-155.

Zimmermann, M. H. and Tomlinson, P. B. (1974). Vascular patterns in palm stems: variations of the *Rhapis* principle. *Journ. Arnold Arbor. 55,* 402-424.

Zimmermann, M. H., Tomlinson, P. B. and LeClaire, J. (1974). Vascular construction and development in the stems of certain Pandanaceae. *Bot. J. Linn. Soc. 68,* 21-41.

COMPARATIVE DEVELOPMENT OF VASCULAR TISSUE PATTERNS IN THE SHOOT APEX OF FERNS

Richard A. White

Department of Botany
Duke University
Durham, North Carolina

In the pertinent literature related to vascular pattern development in the ferns, two themes, at least, predominate: (1) there is a considerable cauline or stem influence in the establishment of the pattern, or (2) the primary influence which determines xylem differentiation and maturation patterns relates to the leaves and their development at the apex. A review of the literature is provided as the background for a current survey, using modern techniques, of protoxylem maturation patterns among a variety of fern species which represents a broad taxonomic and morphological diversity. There are a number of fern species in which it is clear that leaves do have a primary influence in the early establishment of the vascular pattern. In such species, provascular strands with subsequent protoxylem maturation are differentiated only in relation to leaf traces. Examples are provided from the literature (e.g. Ophioglossaceae) and from the laboratory (e.g. Cyatheaceae) which describe in detail the protoxylem strand relationship to leaf traces. Detailed interpretations of the protoxylem pattern (whatever the mature stelar pattern may be) in these species permit an analysis and derivation of the phyllotactic fraction for the species. This is a further indication of the close interrelationship between leaf development and the protoxylem pattern. In many instances this relationship is obscured by later development, the subsequent maturation of metaxylem and the establishment of a characteristic siphono-

ISBN 0-12-746620-7

stele or dictyostele. However, where there is a single main protoxylem strand in each of the meristeles of a typical highly dissected dictyostele, it is also possible to derive the phyllotactic fraction by an analysis of the meristele pattern (e.g. *Diplazium* and *Blechnum* in this study). In other species (Gleicheniaceae, Hymenophyllaceae, and Schizaeaceae) it is clear that not all protoxylem cell maturation is associated with leaves. The protoxylem cells may not constitute strands, there may be no reticulate system where strands are present, and both provascular tissue and mature protoxylem cells may occur distal to all leaf primordia. In such species, with clearly cauline (stem) vascular tissue, influences in addition to leaf development must be important. One major feature noted in several species (e.g. *Gleichenia* and *Dicranopteris*) is the association of lateral root vascularization with xylem cell maturation. Within this latter group, exemplified particularly by species with protosteles or siphonsteles, the diversity of protoxylem-metaxylem patterns is described. The general conclusion of the current research is that what appear to be relatively simple and very similar mature stelar patterns with characteristic primary xylem maturation patterns are in reality much more diverse and complex. This diversity and complexity remains to be pursued in detail. The influences of and interactions among the leaf, shoot, and root on vascular pattern establishment and xylem maturation patterns, provide a broad array of problems to be resolved by current and future researchers.

I. INTRODUCTION

Previous investigations of cauline vascular systems in the ferns focus mainly on comparisons of mature stem structure and speculation on the phylogeny of the stele and the classification of stelar types, for which there is an extensive literature (cf. most recently, Beck *et al.*, 1983; Schmid, 1983). However, it has become apparent that there are developmental and histological dissimilarities among steles of the "same type" (i.e. protosteles, siphonosteles, dictyosteles) not reflected in these classifications. The primary purpose of my current research is to investigate stelar anatomy from a more developmental perspective with emphasis on the relationship(s) between leaf initiation and the establishment of vascular patterns at the fern shoot apex. Despite numerous studies of shoot apical meristems, leaf initiation and histogenesis, and stelar patterns, relatively little detailed information exists on early differentiation of the vascular system. Studies in this laboratory incorporate a combination of approaches that should permit for the first time a detailed analysis of the interrelationship(s) among leaf initiation, phyllotaxy, and the inception of vascular tissue (stelar patterns) at the shoot apex of ferns. As this research entails the examination of a broad survey of fern species which represent a variety of stelar types the data should provide information which is potentially valuable for systematic and phylogenetic comparisons.

The history and significant literature related to vascular tissue patterns and development extends back over more than a century. As early as the mid- to late 19th century (Hanstein, 1858; DeBary, 1884) a close relationship was noted between the arrangement of leaves on the shoot and the vascular pattern which develops in the stem of _seed plants_. The different patterns of leaf traces and stem bundles in seed plant eusteles have been frequently described, and the relationships between phyllotaxy and the number of sympodial bundles in the shoot have been elaborated more recently for both gymnosperms (e.g., Namboodiri and Beck, 1968a,b) and dicotyledons (Devadas and Beck, 1971, 1972; Beck _et_ _al_. 1983).

Studies of the vascular organization of fern shoots also comprise a formidable body of literature covering over one hundred years. Following the studies of Van Tieghem and Jeffrey (Van Tieghem and Douliot, 1886; Jeffrey, 1899, 1902) broad surveys of pteridophyte stem anatomy were pursued, leading to extensive and somewhat divergent classifications of stelar types and hypothetical phylogenetic sequences. These studies emphasized the relationships of leaf traces and stem bundles, the presence or absence of pith and internal phloem, the position of protoxylem points (mesarch, etc.), and the different arrangements of leaf gaps.

A general consensus emerged from these studies that the primitive stele type in ferns (and in vascular plants in general) is a simple protostele, and that siphonosteles and dictyosteles are derived patterns (Gwynne-Vaughan, 1901, 1903, 1911; Tansley, 1907/1908). There was further agreement that vascular tissue differentiation is acropetal and that the ontogeny of species with complex dictyosteles at maturity proceeds in successive nodes from an initial protostele through a solenostelic stage before the dictyostelic pattern appears.

Beyond this, there are areas of disagreement. Some authors (e.g. Boodle, 1903; Tansley and Chick, 1901, 1903) consider the stem vasculature to be fundamentally cauline, phylogenetically (and developmentally?) independent of the leaf traces. Others (e.g. DeBary, 1884; Stevenson, 1982) view the fern stem bundle system as a composite structure of leaf traces (and root traces, in some cases). The relationship between leaf vascularization and the shoot stele is strongly emphasized by these latter workers. Schoute (1938) gives a good summary and thoughtful discussion of these views; he suggests that detailed investigations might possibly "bring more light" to the subject.

Substantial differences exist among the various descriptions of patterns of protoxylem distribution in fern steles. Protoxylem was distinguished from metaxylem using one or more of three criteria: (1) helical or spiral tracheid wall thickenings; (2) early maturation of elements; and (3) smaller cell diameters. From the numerous descriptions in the literature (e.g., Boodle, 1900, 1901a, 1903; Conard, 1908; De Bruyn, 1911; Farmer and Hill, 1902; Ford, 1902; Gwynne-Vaughan, 1903; Marsh, 1914; Tansley, 1907/1908) protoxylem and metaxylem may be distinguished in several ways: (1) longitudinal strands of helically thickened protoxylem cells occur among the metaxylem cells. These patches are variously positioned (endarch, mesarch, or exarch) and are connected with leaf traces; (2) "protoxylem" cells with other than helical wall thickenings are scattered among the metaxylem cells of the stem stele and do not form discrete strands or patches; (3) early-maturing xylem forms a "peripheral band" of scalariform-pitted protoxylem around the stem bundles. The protoxylem is not specially related to the leaf traces; (4) no protoxylem is present in the stem stele.

Where strands of helical protoxylem are present in the stem stele, they are connected with the protoxylem of leaf traces. Information extracted from the literature (DeBary, 1884; Gwynne-Vaughan, 1903; Tansley, 1907/1908; Tansley and Chick, 1903; Boodle, 1901b; Compton, 1909; Posthumus, 1924; Schoute, 1938) and from our own preliminary examinations of several species indicates that several different patterns may exist among leptosporangiate ferns: (1) protoxylem strands are absent from the stem stele and present only in leaf trace or petiole bundle; (2) protoxylem strands are decurrent below the leaf trace insertion and end blindly in the stem stele; (3) protoxylem forms a continuous longitudinal system of strands that extends the length of the shoot and branch to connect with leaf trace strands; (4) interconnected protoxylem strands form a reticulate system in the stem; (5) additional cauline protoxylem strands not related to leaf traces are present. Because many studies determined protoxylem/metaxylem patterns from serial sections of mature tissues, some caution is necessary in accepting descriptions of shoots with discontinuous protoxylem or with protoxylem absent. It is possible that some early-maturing xylem has been obliterated or overlooked due to the absence of typical small cell size or wall patterns.

The frequent restriction of "helical protoxylem" to leaf traces in some "primitive" ferns (e.g., Schizaeaceae) and the observed relationship between stem protoxylem and leaf traces in other species led to statements that "differentiation of stem protoxylem is in close relation to the leaf trace and that spiral protoxylems were a comparatively late character to appear in ferns" (e.g. Boodle, 1901a; Gwynne-Vaughan, 1901).

Various patterns of leaf trace/protoxylem relationships in the mature stem were routinely described but there were few attempts beyond basic description of stem patterns to relate the stem vascular patterns to leaf arrangement. As a result, however, elegant and usually accurate descriptions were presented of mature stem bundles fusing and separating with leaf traces departing in regular patterns. Schematic diagrams of the longitudinal patterns of bundles and traces through several nodes are abundant in the literature. No clear connection was made between phyllotaxy and the complexity of the dictyostele patterns which were described. This is important later (below), but, for example, where phyllotaxy was described and a diagram of the stem accompanied the description, a clear relationship can now be seen (see below) between the denominator of the phyllotactic fraction (i.e., the number of leaf orthostichies on the shoot) and the number of individual meristeles in the stem. Stems with 2/5 phyllotaxy have five major bundles in cross section and one protoxylem point in each bundle (e.g. Anemia, Mohria, Boodle, 1901a). Osmunda regalis, with a phyllotaxy of 5/13, has 13 meristeles in the stem cross section and each has been reported to have a single protoxylem point. There is disagreement in the literature on this latter point and recent preliminary review of two species of Osmunda (in this laboratory) disputes this.

Stem-leaf relationships and the protoxylem association with leaf traces in species with polycyclic siphonosteles have not been clearly described. Such classically studied species as Saccoloma, Matonia, and members of the Marattiaceae were described in great detail, but conclusions concerning leaf trace patterns remain varied (e.g. Compton, 1909; DeBary, 1884; Wigglesworth, 1902; Tansley, 1907/1908; Tansley and Lulham, 1905).

Comparative anatomy of ferns followed these early descriptions up through Bower's time (Bower, 1923, 1926) and remained primarily descriptive (e.g. Ogura, 1972) until a new emphasis on development was highlighted by Wardlaw, his co-workers, and some current workers (for a review see White, 1971, 1979). But even this survey of the most current literature finds a lack of information and considerable disagreement concerning the basic developmental patterns which occur at the fern shoot apex. For example, relatively recent literature still indicates that fern leaves arise either as a group or cluster of cells at the shoot apex (cf. Wardlaw, 1957) or as a single initial cell (cf. Bierhorst, 1977). It may well be that both patterns do occur, but the matter has not been resolved. The standard literature ascribes a major role to the leaves as having the primary influence on vascular pattern development (e.g. Esau, 1965; Wardlaw, 1944, 1945, 1957; Wetmore and Wardlaw, 1951; review in White, 1971, 1979), but recent data suggest strongly that the reverse may be true: the shoot vascular pattern may influence the sites of leaf initiation at the shoot apex (see Larson 1983, for a review).

This research on *Populus* (e.g. Larson, 1975, 1976, 1979) indicates that there are precocious procambial strands in the shoot apex which extend distal to the last observable (most distal, or youngest) leaf primordium. These strands develop before a visible primordium can be identified at the shoot apex and they are said to participate in the determination of the leaf sites. This is a complete reversal of much current thought in which the leaves arise (by whatever process) and control the pattern(s) of procambial differentiation and xylem maturation. A review of the literature indicates that precocious strands similar to those recently described for *Populus* (Larson 1975) have been described previously for gymnosperms (e.g. Gunckel and Wetmore, 1946a, b; Sterling, 1945a, b, 1947) and angiosperms (e.g. dicotyledons, Tucker, 1962; monocotyledons, Maze, 1977). However, relatively little has been made of these reports in modifying the generally accepted ideas of leaf initiation and vascular pattern development. Currently there are no data for ferns related to precocious strands or the presence of procambium distal to the last leaf primoridum on the shoot apex. However there is a recent description of precocious vascularization of adventitious buds on Osmunda primordia (Kuehnert and Larson, 1983). Among other pteridophytes distal procambium was described earlier for Lycopodium (Freeburg and Wetmore, 1962), but a more recent study contradicts that finding (Stevenson, 1976). The most recent studies in fern shoots did not pursue this point (e.g. Stevenson, 1980; White and Lucansky, 1984; White and Weidlich, 1984).

What is required to resolve some of these questions and to clarify what does occur at the shoot apex are detailed, systematic examinations of leaf initiation, early tissue differentiation and vascular pattern development at the fern shoot apex. Despite a careful review of the old literature, and a survey of data currently available, very few generalizations are possible at this time about these topics. As Esau indicates with reference to vascular differentiation in ferns "...the variability and complexity of this system in the ferns does not justify generalization in the absence of extensive comparative studies" (Esau, 1965, p. 52). This same statement holds true for leaf origin and the relationship between leaf primordia and the development of patterns in the vascular tissue.

Ferns have been considered to retain some cauline vascular tissues as distinct from foliar traces (see White, 1971, for review), but many consider the highly dissected dictyostele to represent a pattern influenced most strongly by leaf vascularization (e.g. White and Lucansky, 1984; White and Weidlich, 1974, 1984), and details of the relationships between protoxylem in the shoot steles and leaf traces and phyllotaxy remain obscure in most fern species.

The relationship of the number of sympodial bundles and leaf phyllotaxy in the eusteles of seed plants has been described in the literature, and elaborated upon recently for both gymnosperms (e.g. Namboodiri and Beck, 1968a, b) and dicotyledons (Devadas and Beck, 1971, 1972; Beck *et al.*, 1983). The observed similar numerical correspondence of protoxylem strands and leaf orthostichies in ferns (see below) raises the possibility that in some details of vascular organization and development some ferns may be more closely comparable to seed plants than has been hitherto supposed. A recent review, however, still suggests that there is a "fundamental distinction between the seed plant and the filicalean-stele" (Beck *et al.*, 1983). More developmental and comparative data are needed for the ferns if this question is to be resolved.

For the ferns, these more recent data raise the issue that appears frequently in earlier literature: the developmental relationship between leaves and the mature shoot vascular system, especially the association of leaf primordia with the development of the protoxylem system. Our analyses to date indicate continuous acropetal differentiation of procambium and xylem in shoots of a variety of fern species; no disjunct xylem maturation has been noted. A recent report, however, suggests that both procambium and xylem differentiates bidirectionally in the leaf base of *Botrychium* (Stevenson, 1980) and bidirectional xylem maturation has been described in young leaf primorida in Dryopteris (Kuehnert and Larson, 1983).

The data (see below) for both the protoxylem system in young tree fern sporophytes and for the relationship between meristele number and leaf phyllotaxy in Blechnum and Diplazium tempt one to propose the possibility that protoxylem points (actually strands) may, at least in species with highly dissected dictyosteles, provide a "template" for the mature stelar pattern (i.e., may represent the positions of developmental centers within the vascular cylinder). This has been suggested specifically in the earlier literature (Gwynne-Vaughan, 1903; Tansley, 1907/1908; Schoute, 1938) and a variation on the theme was elaborated as a means for the origin of the dictyostele and eustele (Jeffrey, 1902). The varied descriptions in the literature cited above provide reasons for caution, however, since some species are said to lack protoxylem in the stem and others may have more than one major protoxylem strand in each meristele (e.g. Chau, 1981). For other species with well-developed protoxylem strands, the unresolved conflicts and unanswered questions in the literature require more detailed analyses (e.g., Gleichenia). Detailed examination is especially needed for ferns with complex, polycyclic, or dissected steles such as those found in Pteridium, Thyrsopteris, Saccoloma, or Matonia, in order to clarify the vascular relationships between leaves and shoots in these plants.

Several genera of ferns exhibit patterns in protoxylem strands that seem to differ significantly from the numerical relationship described above. Preliminary studies of species of Dennstaedtia, Hypolepis, and

Gleichenia reveal well-developed longitudinal protoxylem strands in their long stem internodes, yet the shoots have only a distichous leaf arrangement. The number of protoxylem strands immediately below the shoot apex greatly exceeds the number of leaf primorida at the apex. Part of the protoxylem system in these species is apparently not connected with leaf trace vascularization and may represent a wholly cauline component of the vascular system. Much more adequately detailed information on shoot vascular systems of these and other ferns is required before clearer morphological generalizations or systematically useful comparisons can be made.

Based on more recent studies (referred to above and elaborated upon below) and a review of the literature, current research should center clearly on the detailed examination and description of the relationship(s) among leaf origin, phyllotaxy, protoxylem differentiation and maturation, and the establishment of the mature stelar pattern. Techniques are available that allow the examination of each of these topics in the same apex, thus assuring the accuracy of the data. The apex may be dissected and stained and the pattern(s) of leaf origin (phyllotaxy) may be observed through epi-illumination techniques. This same apex may then be processed for paraffin or thin-section embedding and sectioning and carefully oriented serial transverse or longitudinal sections may then be analyzed for vascular tissue differentiation patterns. These basic techniques were used to obtain the following data on selected species of ferns.

II. CURRENT DATA

A. Eusporangiate Ferns

Species of the Ophioglossaceae have become the focus of several recent studies of vascular anatomy (e.g. *Botrychium*, Chau, 1981; Stevenson, 1980; *Ophioglossum*, Chau, 1981; Webb, 1975, 1981).

Although complex in presentation, by interpretation of recent reports, a relationship can be noted between leaf phyllotaxy and gap arrangement in the stele for both *Ophioglossum petiolatum* (Webb, 1975) and *O. crotalophoroides* (see Figs. 2, 3, 19 & 20, Webb, 1981). No detailed descriptions of vascular ontogeny at the shoot apex were provided though reference was made to protoxylem in *O. crotalophoroides*: (1) it is on "the inner side" of each bundle (endarch bundles), (2) it closes the gaps, and (3) it is continuous throughout the plant body (Webb, 1981). The conclusion based upon serial cross sections and three-dimensional reconstruction of the stele was that at least in *O. crotalophoroides* the stele is *not* a leaf trace complex (ấlà Campbell, 1921) but that leaf traces and root traces are appended to a stem stele. One of the earliest studies of eusporangiate ferns to examine carefully the development of provascular tissue and protoxylem patterns in particular at the fern shoot apex was done on *Botrychium multifidum* (Stevenson, 1980). This substantial work provides data for *Botrychium* comparable to that developed more recently for *B. dissectum* and several *Ophioglossum* species (Chau, 1981). In *B. multifidum*, with a characteristic mature phyllotaxy of 2/5, there are five major sympodia in the stem stele. Sequential leaves (and their associated leaf traces) are seen to be inserted in five orthostichies (Fig. 1). In the primary vascular

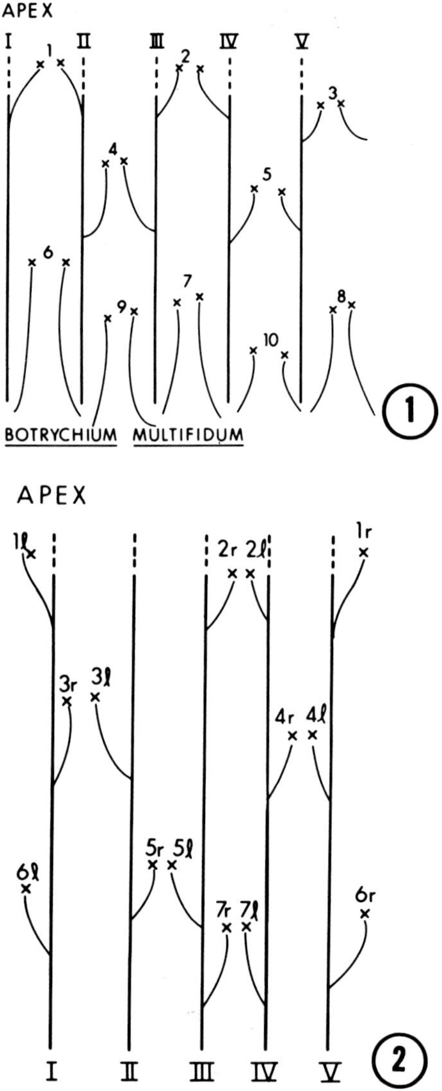
APEX
I
II
III
IV
V
1
2
3
4
5
6
7
8
9
10
1
BOTRYCHIUM MULTIFIDUM
APEX
1ℓ
2r 2ℓ
1r
3r 3ℓ
4r 4ℓ
5r 5ℓ
6ℓ
7r 7ℓ
6r
I
II
III
IV
V
2
O. ENGELMANNII

system, the stem sympodia are joined by two protoxylem strands per leaf trace. The subsequent development of a fascicular cambium in Botrychium rapidly obscures the organization of the primary (proto- and when present, metaxylem) strands. Analyses of serial sections from the shoot apex indicate the absence of a cauline procambium system, and the presence at the most distal regions of procambial strands related only to the two youngest leaf primordia. In like manner, no true (classical) leaf gaps occur in the primary network (Webb, 1981). Rather, characteristic gaps become isolated from each other following secondary xylem development. Stevenson concludes, in contrast to Webb's interpretation of Ophioglossum, that Campbell's interpretation (1921) of the stelar tissue as an anastomosing of leaf and root traces is correct (Stevenson, 1980). In the xylem, there appears to be no metaxylem, but rather secondary cambial activity forms a continuous cylinder, except across leaf gaps, and thus provides the basis for the classical interpretation of the Botrychium stele. It is clear that there is true cambial activity and secondary xylem in Botrychium (Chau, 1981; Stevenson, 1980). It is equally clear that the establishment of the procambial system, and maturation of protoxylem is related to leaf trace differentiation and maturation.

FIGURES 1-2. Fig. 1. Diagram of the protoxylem strands which compose the primary vascular pattern of B. multifidum. Modified from Stevenson, 1980. (---provascular strands; - mature protoxylem. Fig. 2. Diagram of protoxylem strands which compose the vascular pattern of Ophioglossum engelmannii. Modified from Chau, 1981.

The cellular details of differentiation and maturation at the shoot apex of *Botrychium dissectum* (Chau, 1981) parallel those described for *B. multifidum* (Stevenson, 1980).

A similar investigation into primary vascular patterns was undertaken of a number of species of *Ophioglossum* (*O. engelmanii*, *O. vulgatum* var. *pseudopodum*, *O. crotalophoroides*, *O. lusitanicum* var. *californicum*, *O. palmatum*, *O. pendulum* (Chau, 1981). The results parallel those described for *Botrychium multifidum* in that five non-anastomosing sympodia with two traces to each leaf were described. These five sympodia, similar to the situation in *Botrychium*, reflect the five leaf orthostichies which characterize the species (Fig. 2).

B. Osmundaceae

Osmunda, with its unusual vascular pattern(s) and leaf trace system, has been examined and reexamined over the past century. Many of the early observations and conclusions have been confirmed, some have been corrected, and still others remain to be resolved (see Tansley, 1907/1908; Posthumus, 1924; Miller, 1971). An association of leaf phyllotaxy and protoxylem strand differentiation and maturation referred to in earlier works is confirmed and further clarified in our recent work with *Osmunda regalis* and *O. cinnamomea*. Typical helical protoxylem is found in the leaf trace and continues for several nodes in the stem, basipetal to the insertion of the leaf trace to the stele. Early reports indicate acropetal differentiation of provascular tissue and protoxylem, but a recent analysis in this laboratory questions this. An analysis of serial sections which begins basal to the leaf trace departure and proceeds distally provides the following data. Leaf trace formation follows a regular pattern. In the meristeles proximal to the trace departure, the protoxylem strand is central (mesarch xylem maturation). A few parenchyma cells appear internal to the protoxylem, and thus the pattern becomes endarch. The bundle appears to "split open," i.e. become horseshoe shaped (in sequential sections). The outer, central portion of this xylem bundle, departs as a leaf trace (including the protoxylem point) and two (lateral) xylem bundles remain in the stem. There is thus a single large leaf trace to each leaf. The two lateral xylem bundles which remain in the shoot lack protoxylem points. In any given cross section, therefore, there will be two kinds of meristeles: those with and those without protoxylem points. The protoxylem strand in the trace can be traced upward through the cortex in the leaf trace, and downward in the stem bundle where it eventually ends blindly. There are no interconnections among any of the protoxylem strands associated with the separate leaves. A continuous dictyostele-type stelar pattern characterizes the metaxylem, while the protoxylem strands, one per leaf trace, are discontinuous and end abruptly several nodes below the departure of each leaf trace.

At the shoot meristem, internal to the peripheral zone of the apical meristem, which gives rise to the protoderm and cortex, is the "incipient vascular tissue" (Steeves, 1963) or procambium (Imaichi, 1977). This constitutes a zone of small, dense, narrowly elongate cells which are continuous with cells of the more distinct provascular tissue below. At a point (± 200-300 µm) proximal to the apical dome, reliable interpretation could not be made of individual bundles because the medullary rays could not be distinguished from immature xylem at this level. Analyses of serial sections through the youngest nodes at the apex indicate that disjunct protoxylem patches occur at the leaf-trace stele junction, and that protoxylem maturation from this site is bidirectional; acropetal into the trace and leaf and basipetal down into the shoot meristele. At the site of the youngest visible leaf primordium provascular tissue traverses the leaf trace from the most distal incipient vascular tissue. In some sections, by careful analysis of the provascular strands, the site of the initiation of the next leaf (I_1) could be determined. Although the external shoot surface is not yet affected by this primordium, a connection between the incipient vascular tissue and this new leaf site could be noted and suggests that the initiation of provascular strands may precede the physical induction of the leaf primordium from the surface cells. Detailed transmission electron microscope studies remain to be done to confirm these initial observations.

This most recent review of vascular tissue development at the shoot apex in Osmunda provides data related to several points of current interest:

(1) A single protoxylem strand occurs in each leaf base and extends some distance down the stem meristele before ending blindly.

(2) The protoxylem pattern is one of discontinuous strands; no interconnecting of the strands occurs.

(3) The discontinuity of the strands and the pattern of leaf trace departure result in meristeles in the shoot cross section which lack protoxylem points.

(4) In addition to meristeles which lack protoxylem points, others may have two protoxylem points by virtue of the number of internodes through which the protoxylem strand (associated with each leaf) extends proximal to its insertion on the shoot meristele before ending.

(5) Preliminary data indicate that provascular strands occur distal to the level of the youngest visible leaf primordium and may well precede the actual development of the next leaf primordium and indicate its site.

C. Cyatheaceae

Detailed studies have been reported recently on the mature stem anatomy of species from a broad survey of genera in the Cyatheaceae (e.g. Lucansky, 1974a, b; Lucansky and White, 1974, 1976). These studies described mature stelar patterns with a particular emphasis on nodal anatomy. As a continuation of such studies, attention was then focused on the establishment of the vascular pattern at the shoot apex. Eight species representing five genera of the Cyatheaceae (sensu, Tryon 1970) were examined: *Alsophila salvinii* Hook, *Nephelea mexicana* (Schlect. and Cham.) Tryon, *Trichipteris mexicana* (Mart.) Tryon, *T. stipularis* (Christ) Tryon, *Cyathea multiflora* Sm., *C. maxonii* Maxon, *C. divergens* Kze., and *Cnemedaria mutica* (Christ) Tryon. A full report of the comparative anatomy of young sporophytes of these species will appear elsewhere (White and Lucansky, 1984). Suffice it to say that the basic pattern at the shoot tip, which is described below, is similar among all the species examined.

Examination of serial sequential sections beginning at the base of the shoot of young sporophytes (14-17 leaves) and proceeding toward the shoot apex reveals a very close association of the protoxylem strands in the shoot with leaf origin and development. Proximal to the fourth or fifth leaf primordium on the young shoot, the mature metaxylem of the shoot stele obscures the details of the protoxylem patterns. Once distal to mature metaxylem, however, it is possible to trace the complex pattern which characterizes the protoxylem system in the shoots of these species. This is a reticulate protoxylem pattern with leaf trace protoxylem strands connecting the shoot network with the leaf primordia.

Proceeding from the shoot tip down to the level at which a fully mature metaxylem pattern occurs, initial vascular tissue development is as follows. P_1 is without mature xylem, but provascular tissue can be distinguished. Mature protoxylem (xylem cells with characteristic secondary wall patterns) ends several hundred μm proximal to the opening of a gap in the stele associated with P_2. Strands of mature protoxylem cells with secondarily thickened walls are continuous from the stele below P_3 and extend into the first of two traces which depart from each side of the gap to that leaf. Only provascular strands extend into the second trace on either side of the gap. The next leaf in basipetal order, P_4, is characterized by four leaf traces, each of which has protoxylem cells with secondarily thickened walls. Each trace departs from a xylem strand in the shoot stele on either side of the gap. These strands in the shoot are characterized by cells with helical secondary wall thickenings, extend distally beyond the departure of the leaf traces to P_4 several hundred μm, and end blindly. Above this point, no xylem cells with secondarily thickened walls occur in the provascular tissue, associated with

P_4. Characteristically, three traces characterize the leaf vascular supply at P_5 and all of these traces have protoxylem cells with secondarily thickened walls. Cells with similarly thickened walls constitute the strands in the shoot vascular cylinder from which these leaf traces depart. Distal to the departure of the three leaf traces to P_5, the lignified xylem cells, which compose the shoot strands, extend for only a short distance on either side of the gap. Above this level at P_5, provascular tissue occurs in which xylem cells lack secondary wall thickening. Proximal to the gap associated with P_5, the tissues of the shoot stele are fully mature, the xylem being characterized by typical mature proto- and metaxylem cells in a mesarch pattern. Within the xylem of this mature stele, the individual protoxylem vascular strands associated with particular leaves are difficult to distinguish clearly by virtue of the maturation of the numerous surrounding metaxylem cells which comprise the remainder of the xylem in the siphonostele.

A composite analysis of serial sections from this mature siphonostele level (e.g. at P_6) in the shoot and toward the shoot tip reveals a rapid decrease in the number of mature tracheids (easily distinguished with the use of polarized-light microscopy). Those mature cells which remain constitute strands of cells which relate specifically to the vascularization of the leaves. In more distal regions of the shoot, these cells become restricted to specific strands which eventually form the individual traces to the leaves. No mature protoxylem cells occur in the shoot distal to those which constitute the first pair of leaf traces to P_3 and the strands which end blindly proximal to the gap at P_2. The detailed analysis of the vascular tissue distal to P_6 in Cyathea indicates interconnections among the protoxylem strands all of which are associated with leaf vascularization and constitute a reticulate system. Analysis of this system indicates that the protoxylem strand pattern reflects accurately the phyllotaxis of the plant which is 2/5 at this stage of development. In these sporophytes beginning at a particular leaf, two rotations around the shoot are necessary to arrive at the leaf site which is directly below (or above) the initial leaf. In so doing, five leaves are traversed. In the stele cross section at this level (above mature metaxylem) the number of protoxylem strands in the shoot system is also equal to the denominator of the phyllotactic fraction: five. The number of meristeles in the stele cross section at this level (e.g. distal to P_5) is two or three.

In the tree ferns, protoxylem strands occur in the leaf traces and are continuous in the shoot. At the shoot tip, a closed reticulate pattern of protoxylem strands occurs distal to mature metaxylem. Analysis of the details of differentiation and maturation indicates continuous acropetal differentiation of provascular tissue, and in this study no convincing evidence was seen of disjunct xylem maturation. The pattern of protoxylem maturation is clearly associated with leaf development and with the interconnecting of leaf-related protoxylem strands which constitute the reticulate protoxylem network. An examination of this reticulate pattern of strands reveals a direct reflection of the phyllotactic arrangement. The number of major protoxylem strands in cross section at this level (e.g. 5) reflects the denominator of the phyllotactic fraction. Such an association is not true of the number of meristeles seen in cross section of the mature siphonostele (e.g. 3), because several protoxylem strands may occur in a meristele and thus obscure the direct association of the protoxylem with leaf vascularization.

D. Polypodiaceae (s.l.)

The characteristic stelar pattern of the more "advanced" species of ferns is a highly dissected dictyostele. As an extension of earlier work on the mature stelar patterns of tree ferns, a detailed analysis was made of the three-dimensional vascular patterns which occur in species outside of the Cyatheaceae, but which were also typically tree-like in their growth form.

Two species of each of the genera *Diplazium* and *Blechnum* were examined through the use of surface movies of sequentially sectioned mature stems following the technique of Zimmerman and Tomlinson (1972, 1974). Details of the studies of the reticulate system of the meristeles which constitute the dictyostele of these species will appear elsewhere (White and Weidlich, 1984). In one species of *Diplazium* (*Diplazium expansum*) there are eight meristeles and eight leaf gaps at all levels. Leaves are borne opposite leaf gaps. Successive leaves in the genetic spiral are separated from each other in cross section by three meristeles. At the base of a particular node a meristele divides into two separate meristeles thus opening up a leaf gap. The leaf traces depart from the meristeles radially and enter the leaf base distally. Distal to the departure of a particular leaf from the stem, the two meristeles which border the leaf gap fuse and thus close the gap. Distal to the level where a particular leaf gap closes, the leaf gap for the next leaf in the orthostichy opens by the splitting of the same meristele. Continued splitting and rejoining of a particular meristele to open and close leaf gaps and the association of a leaf with each of these gaps throughout the longitudinal course of the stem generates a row, or orthostichy, of leaves on the stem surface. The continued splitting and rejoining of the eight meristeles to open and close leaf gaps in *D. expansum* throughout their longitudinal course, and the association of a leaf with each of these gaps generates eight orthostichies of leaves on the stem surface. Successive leaves of an orthostichy are separated from each other by eight members of the genetic spiral and three turns about the axis, i.e. there is a 3/8 phyllotaxis. The number of meristeles in the stem equals the number of orthostichies present on the stem surface which also equals the denominator of the phyllotactic fraction and the number of leaves which separates successive longitudinal members of an orthostichy.

The stems of the *Blechnum* species are more complex than the pattern seen in *Diplazium*. *Blechnum discolor* has a highly dissected dictyostele which consists of 13 interconnected meristeles and 13 leaf gaps. Successive members of the genetic spiral are separated from each other (as seen in cross section) by five meristeles. As in *Diplazium* gaps related to each leaf are formed by the division of a meristele into two separate meristeles. Distal to leaf trace departure, gap closure occurs by the fusion of the two meristeles which border the leaf gap. The single meristele formed by this fusion separates again at a higher level to form the leaf gap associated with the next distal leaf of same orthostichy.

The continued splitting and rejoining of the 13 meristeles to open and close leaf gaps throughout the longitudinal course of the stem, and the association of a leaf with each of these gaps generates 13 orthostichies of leaves on the stem surface. Successive members of a particular orthostichy are separated from each other by 13 members of the genetic spiral and five turns about the axis. Thus the phyllotactic fraction of 5/13 may be determined both from the stem vascular system and from the surface features of the stem. The number of meristeles in the stem equals the number of orthostichies present on the stem surface which also equals the denominator of the phyllotactic fraction and the number of leaves which separates successive longitudinal members of an orthostichy.

The stem vascular system of *Blechnum meridense* is a highly dissected dictyostele also with closely overlapping leaf gaps. The phyllotactic fraction is 5/13 and as in the other species of *Diplazium* and *Blechnum* this can be determined both by an analysis of the stem vascular system and from the surface features (leaf scars) of the stem.

The preceding descriptions of the vascular patterns in *Diplazium* and *Blechnum* are based on analyses of mature stem anatomy and meristele configurations in mature stems. The relationship of meristele number to phyllotaxy parallels that described earlier for the other fern species. However, those patterns were based upon analyses of *protoxylem* strands associated with leaf development and vascularization. A careful examination of the meristeles in *Diplazium* and *Blechnum* reveals a single main protoxylem strand in each mesarch bundle. This single protoxylem strand is exclusive of the branches from the protoxylem strand which are with the two leaf traces characteristic of each leaf gap.

In Diplazium and Blechnum, therefore, there is a one-to-one relationship between the number of protoxylem strands and the number of meristeles: each meristele has a single main protoxylem strand. Thus the earlier described relationship between protoxylem strands and leaf phyllotaxy in the Cyatheaceous species is maintained in these more dissected dictyostelic species. In these latter species, however, the relationship is also expressed by the number and organization of the mature meristeles by virtue of the consonance of protoxylem strand number with the number of meristeles.

At the other end of the spectrum of stelar diversity are those patterns, frequently described for various species in the literature, which include protoxylem strands that are not associated with leaf traces or leaf primordia. These may be referred to as cauline strands. In addition to this feature, which is interesting in view of the preceding data which stress a protoxylem strand/leaf trace relationship, many of these latter species are also characterized by very long internodes. Distal to the most recent leaf primordium that can be identified

FIGURES 3-4. Fig. 3. Diagram of protoxylem strands which compose the vascular pattern of Gleichenia bifida. Mature protoxylem strands extend distal to the most distal leaf primordium. Fig. 4. Diagram of protoxylem strands which compose the vascular pattern of Dicranopteris pectinata. Of the twelve protoxylem strands in the siphonostele only two or three appear to be directly associated with the leaf traces to leaf primordia at any given node. (-- immature protoxylem strands).

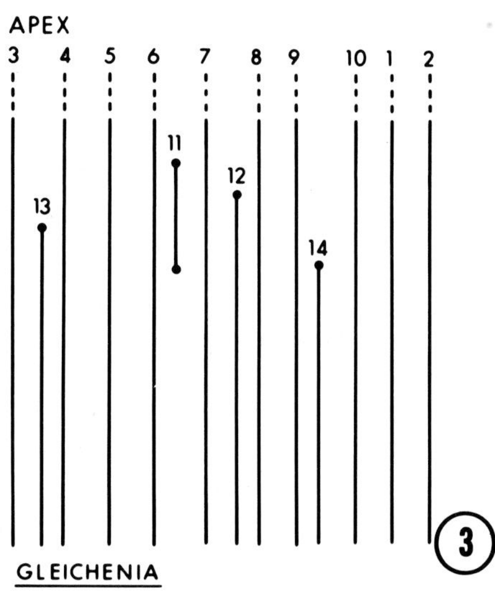

APEX
3 4 5 6 7 8 9 10 1 2
11
12
13
14
GLEICHENIA
3

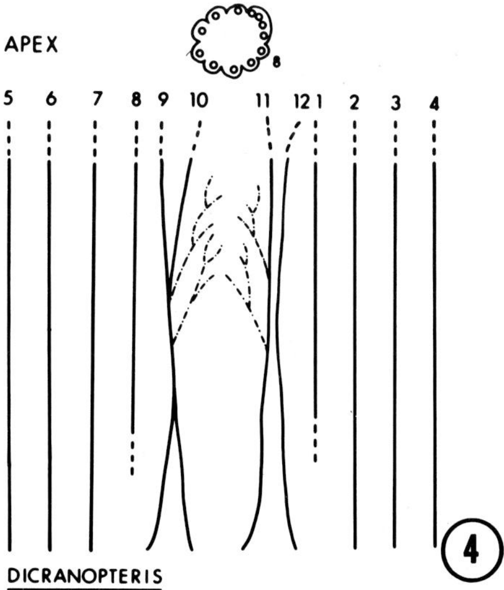

APEX
1
8
5 6 7 8 9 10 11 12 1 2 3 4
DICRANOPTERIS
4

morphologically, there are characteristic patterns of xylem differentiation and maturation. In the absence of leaf primordia, the presence of presumed leaf-associated protoxylem strands and totally cauline (not leaf trace or root trace-associated) strands raises anew the question of the relationship between leaf primordium development and the establishment of vascular patterns in the shoot apex. There appear to be then two major modes of protoxylem differentiation in some species of ferns: (1) leaf related and (2) not related to leaf development.

It is clear from a developmental analysis of protosteles, for example, that there is a great diversity of protoxylem differentiation patterns in this classic stelar pattern which in part reflects this foliar vs. cauline protoxylem distinction.

E. Gleicheniaceae

Gleichenia. Several species have been surveyed and based on the association of leaf primordia at the shoot apex and the development of protoxylem strands, it is clear that of the several to numerous protoxylem strands characteristic of a particular species, only a few are associated with each leaf primordium. The remainder of the protoxylem strands differentiate acropetally, remain distinct and do not interconnect (Fig. 3). These latter

FIGURES 5-6. Fig. 5. Cross section of solenostele of *Dicranopteris pectinata*. Arrows indicate protoxylem strands (x.s.). X160. Fig. 6. Cellular detail of protoxylem points (arrows). Xylem maturation pattern is mesarch. (X400).

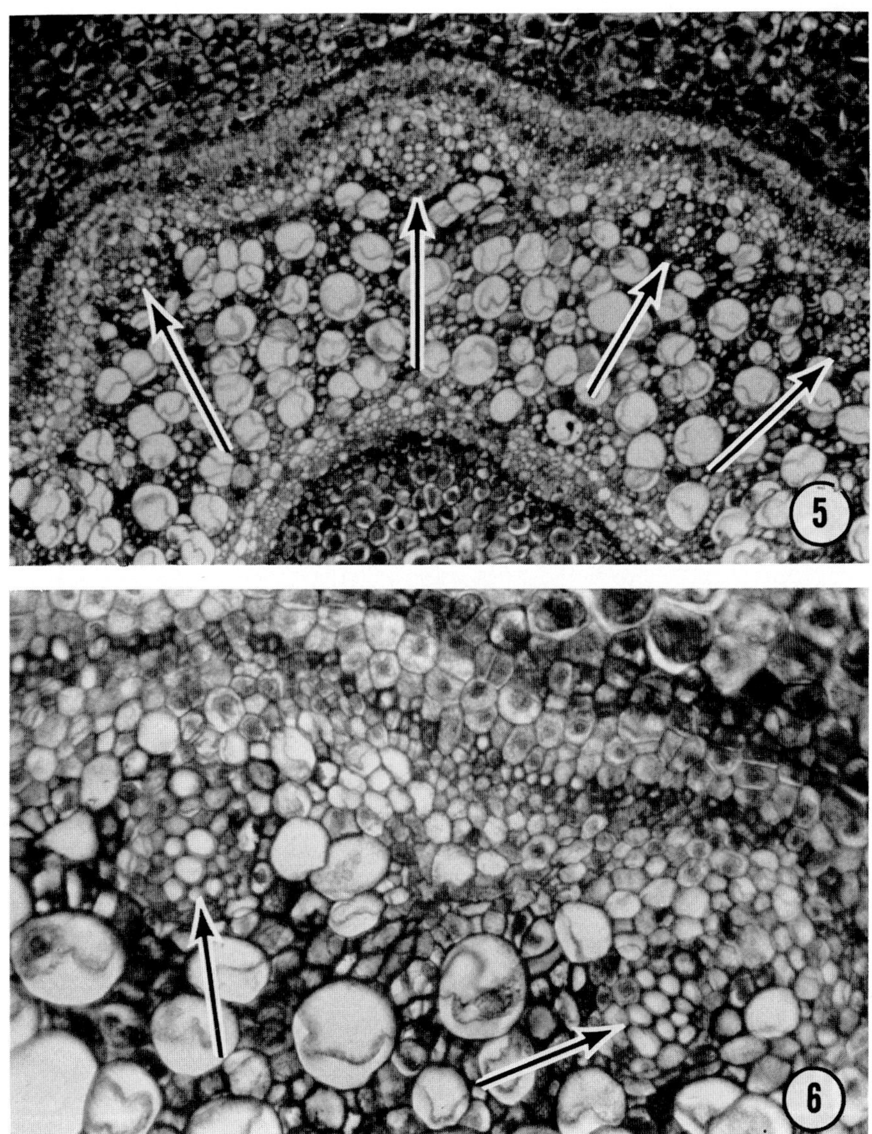
5
6

strands are basically cauline in their development and are in no way clearly associated with leaf trace vascularization.

Dicranopteris. This genus, a close relative of *Gleichenia* is characterized by a solenostele. An analysis of the longitudinal traverse of the protoxylem strands embedded in the metaxylem reveals a pattern very similar to *Gleichenia* (Fig. 4). Of the 12 protoxylem strands in *Dicranopteris pectinata* (Figs. 5,6), only four are associated with leaf trace vascularization at any particular node. The other strands continue over long internodes as independent strands, not interconnected with each other, and associated occasionally only with root traces. In both *Gleichenia* and *Dicranopteris*, there is essentially a dorsiventral procumbent rhizome, with leaf primordia in a distichous arrangement on the upper or dorsal surface (Figs. 7,8). The protoxylem strands in the ventral portion of the stele remain distinct from those strands associated with leaf vascularization and distinct

FIGURES 7-8. Surface view of shoot tip of *Gleichenia bifida*. Shoot apex (a) has no leaf primordia. Sections reveal a complete complement of 12-14 mature protoxylem strands distal to the last-formed leaf primordium (l). X20). Fig. 8. Surface view of shoot tip of *Dicranopteris pectinata*. This species is characterized by long internodes and lateral buds. a, shoot apex; b, bud; l, youngest leaf primordium.

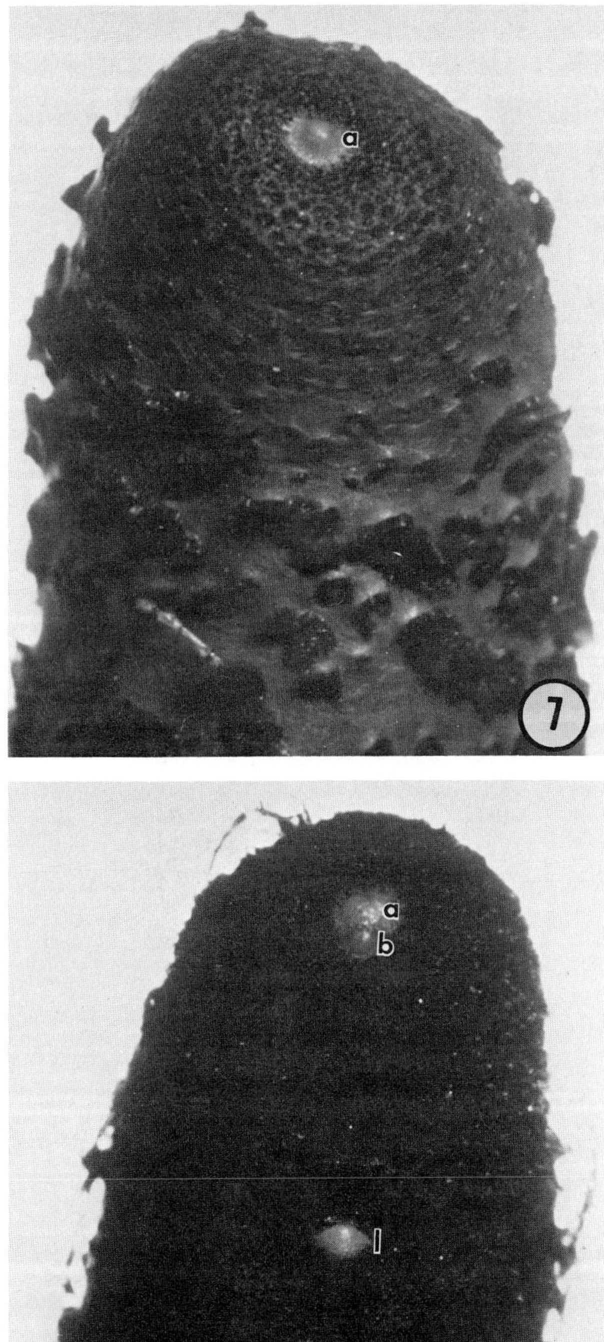
a
7
a
b
l
8

from each other (over the distance examined in this preliminary survey) (Fig. 4). The internodal distance between leaves is great for both genera (Figs. 7,8) and clearly distal to any observable leaf primordia at the apex protoxylem strands are distinguishable and characterized by tracheids with helical wall thickenings. A similar situation occurs in the cauline bundles. Thus pattern and maturation of protoxylem strands in typical Gleicheniaceae appear more independent of leaf primordia than is the leaf/protoxylem strand relationship in the previously described species of Blechnum, Diplazium, and the several species of Cyatheaceae.

F. Hymenophyllaceae

The Hymenophyllaceae have recently become the object of renewed interest (e.g. Hébant-Maurie, 1973). The characteristic protostele is quite different in development from that found in the Gleicheniaceae. Preliminary analysis of sequential sections of Vandenboschia (Fig. 9) reveals that at the shoot apex, there is a cauline protoxylem strand, distal to any recognizable leaf primordia, which is characterized by typical secondary wall thickenings. Where leaf primordia occur on the rhizome, separate leaf trace protoxylem strands differentiate and mature. In following the two protoxylem strands, cauline and foliar, proximal from the shoot apex, it is clear that these strands remain separate for quite a long distance (Fig. 9) and traverse several nodes before becoming contiguous with each other. Analyses are currently

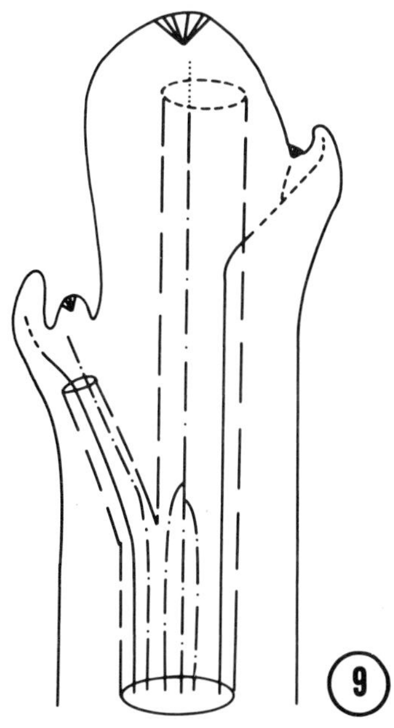

FIGURE 9. Vandenboschia sp. Diagram of immature protoxylem strands (---), mature protoxylem strands associated with leaf traces (---) and cauline vascular patterns (-.-.-). The cauline strand, with mature protoxylem cells, extends distal to leaf primordia and remains unconnected to the leaf traces for some distance down the shoot tip.

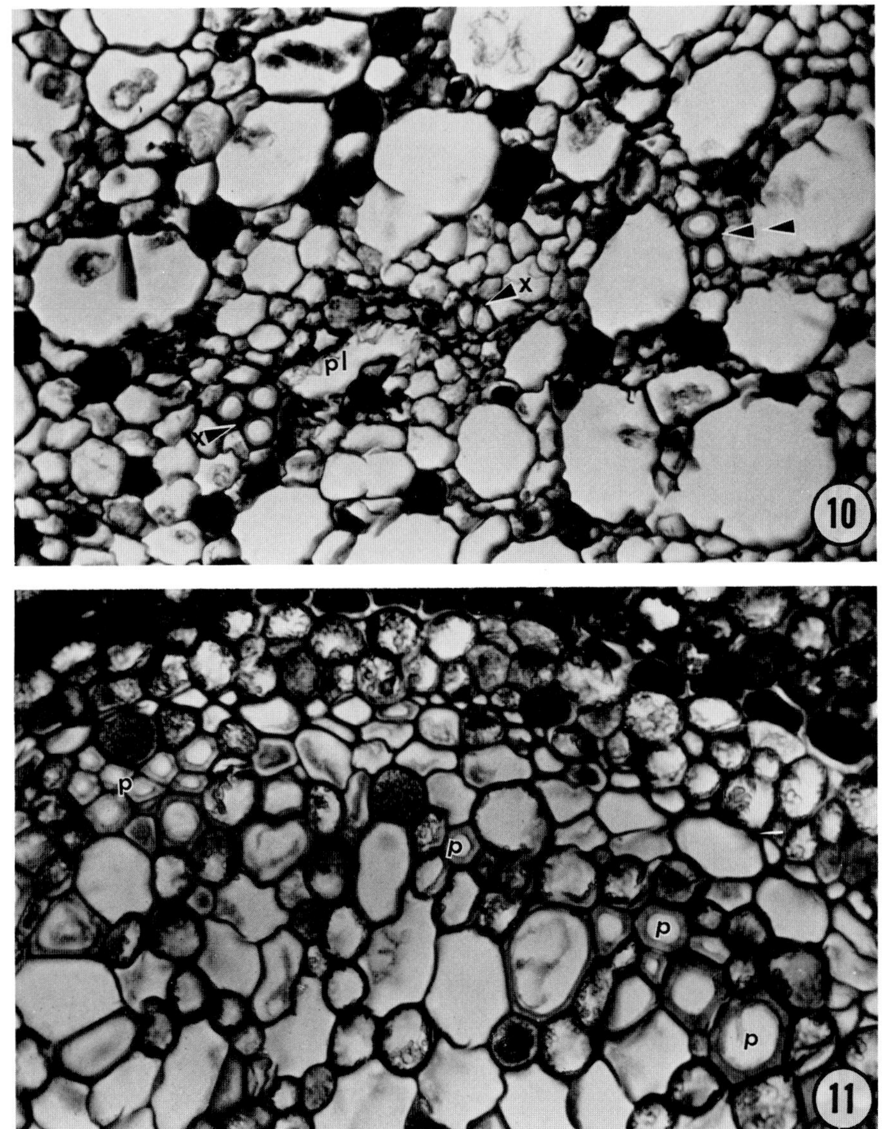
x
pl
x
10
p
p
p
p
11

underway of long internodes in an attempt to determine in detail whether the cauline and leaf trace protoxylem strands do indeed interconnect regularly. The extension of cells and the growth of the internodes quickly crushes and in many cases obscures the basic protoxylem pattern(s) which occur in the mature protostele pattern which characterizes Vandenboschia (Fig. 10).

G. Schizaeaceae

Lygodium exemplifies another basic protoxylem pattern which occurs in protostelic ferns. In this species no protoxylem strands are observable in the stem, and thus no longitudinal pattern of protoxylem is discernable (Fig. 11). The basic developmental pattern may be described as a diffuse protoxylem system. Sequential serial sections reveal no clear association of the maturing protoxylem cells with leaf trace connections to the stem stele. The earliest maturing cells are external to those which mature later and therefore an exarch xylem maturation pattern

FIGURES 10-11. Vandenboschia. Cellular detail of the central portion of the protostele. Double arrows indicate the leaf trace. Arrows marked x indicate cauline protoxylem. A distinctive protoxylem lacuna (pl) occupies the central portion of the cauline protoxylem strand (X400). Fig. 11. Lygodium japonicum. Detail of the stele indicating the diffuse peripheral distribution of mature protoxylem cells (p) among the metaxylem cells of the stele. Protoxylem tracheids have scalariform wall thickenings. (X400).

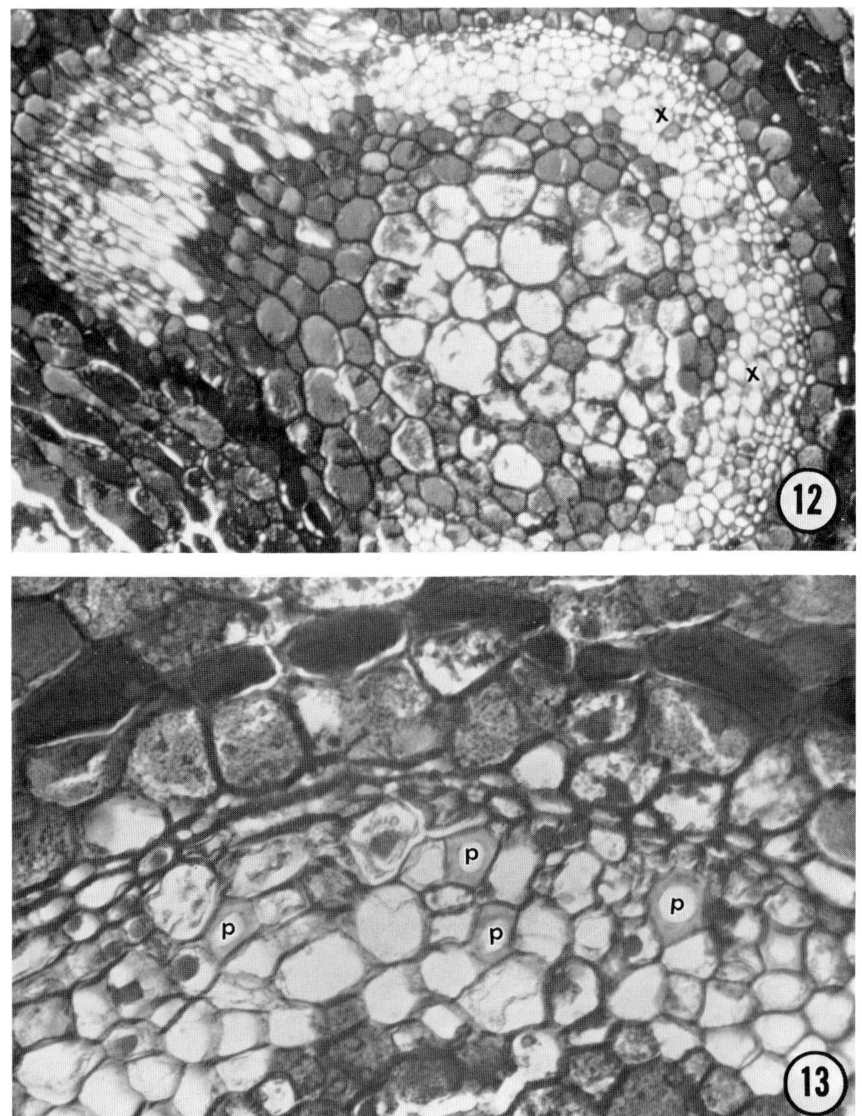
x
x
12
p
p
p
p
13

exists (Fig. 11). Schizaea, also characterized by a stele with diffuse protoxylem cell maturation (Fig. 12), differs from Lygodium, however, in that the earliest maturing cells (Fig. 13) are not clearly exarch but are just internal to the outermost metaxylem cells. The maturation of these diffusely arranged protoxylem cells is not associated with leaf vascularization. More detailed analyses of both Lygodium and Schizaea patterns are necessary. It is clear, however, that protoxylem cell maturation in these species is not as closely linked to leaf development or vascularization as has been described previously for other fern species in this study.

From the preceding survey of a few fern species with simple steles, it is apparent that the patterns of maturation of protoxylem cells even within simple steles are variable. In Gleichenia (protostele, Fig. 15) and Dicranopteris which has a simple solenostele (Figs. 5,6), there are helically thickened xylem cells that form strands of protoxylem in the stele. A portion of these strands is associated with leaf vascularization but a significant number of strands differentiates and matures independent of leaf traces. In Vandenboschia it is possible to identify below the shoot apex a protoxylem strand, more or less central to the protostele which extends in the shoot distal

FIGURES 12-13. Fig. 12. Schizaea dichotoma. A cross section of the stele indicating xylem pattern (x) X160. Fig. 13. Schizaea. Cellular detail of the stele indicating the diffuse occurrence of mature protoxylem cells (p) among the metaxylem cells (X400).

to any discernable leaf primordium (Fig. 9). At each node, an additional protoxylem strand develops which is associated with leaf trace vascularization. In *Vandenboschia*, the cauline and the foliar systems retain their identity in the protostele for several nodes. Procambium and protoxylem differentiation and maturation is acropetal in both systems, and mature protoxylem cells occur in the cambium strand, distal to any observable leaf primordia. Finally, in a separate series of ferns with relatively simple steles, *Lygodium* and *Schizaea* are both characterized by another different protoxylem maturation pattern. In these species protoxylem is defined by earliest maturation (the cells have usually scalariform secondarily thickened walls) rather than by characteristic

FIGURES 14-15. Fig. 14. *Dryopteris aristata*. Surface view of the shoot apical dome (a) and sequentially older leaf primordia. Such a close arrangement of leaves is characteristic of many fern species with little internode elongation and is associated with a dictyostele. An analysis of the protoxylem strand network of such species reveals a direct association of these strands with leaf vascularization. (X25) Fig. 15. *Gleichenia bifida*. Cellular detail of protoxylem strands (p) surrounded by metaxylem cells. In contrast to ferns with short internodes, and densely packed leaves, the protoxylem strands in this species do not form a reticulate network and all of the strands are not associated with leaf traces: some appear strictly cauline for long distances in the shoot while others may be associated with root vascularization.

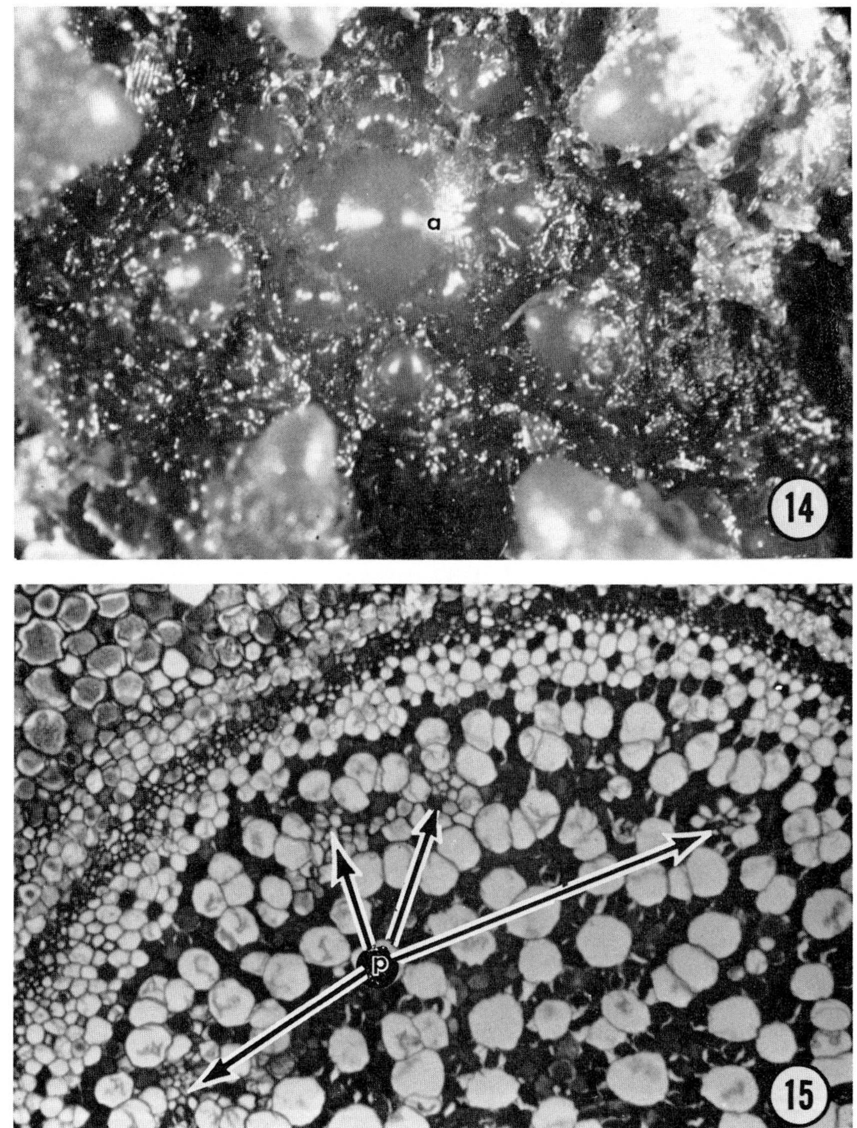
a
14
p
15

helical thickenings, the maturation is diffuse among cells in the periphery of the stele and there is no apparent association between protoxylem cell maturation and leaf vascularization (Figs. 11,12,13).

Finally, in contrast to several previously described species (e.g. species of Cyatheaceae, Blechnum, Diplazium) and others with similar leaf arrangement, among these species with simple steles are representatives with very long internodes (e.g. Figs. 7,8), and leaf arrangements quite different from the regular radial symmetry characteristic of many ferns (e.g. Fig. 14). Careful examination of the shoots of species with long internodes reveals no primordia to be present at apices which are literally centimeters distal to the youngest visible leaf primordium (cf. Figs. 7,8). Yet in these extended distal "internodes", characteristic protoxylem patterns can be described, including strands with mature protoxylem cells

FIGURES 16-17. Fig. 16. Dicranopteris pectinata. Surface view of a clearing of the shoot tip revealing the apex (a) and the first leaf primordium (l). Provascular tissue and mature protoxylem strands extend distal to the last leaf primordium and all do not appear to be associated with leaf vascularization. (X160). Fig. 17. Vandenboschia sp. Surface view of the shoot tip revealing the apex (a) and the first leaf primordium (l). Close to the apex a separate cauline protoxylem strand in the protoxylem extends distal to the leaf trace and is composed of mature protoxylem cells. (X100)

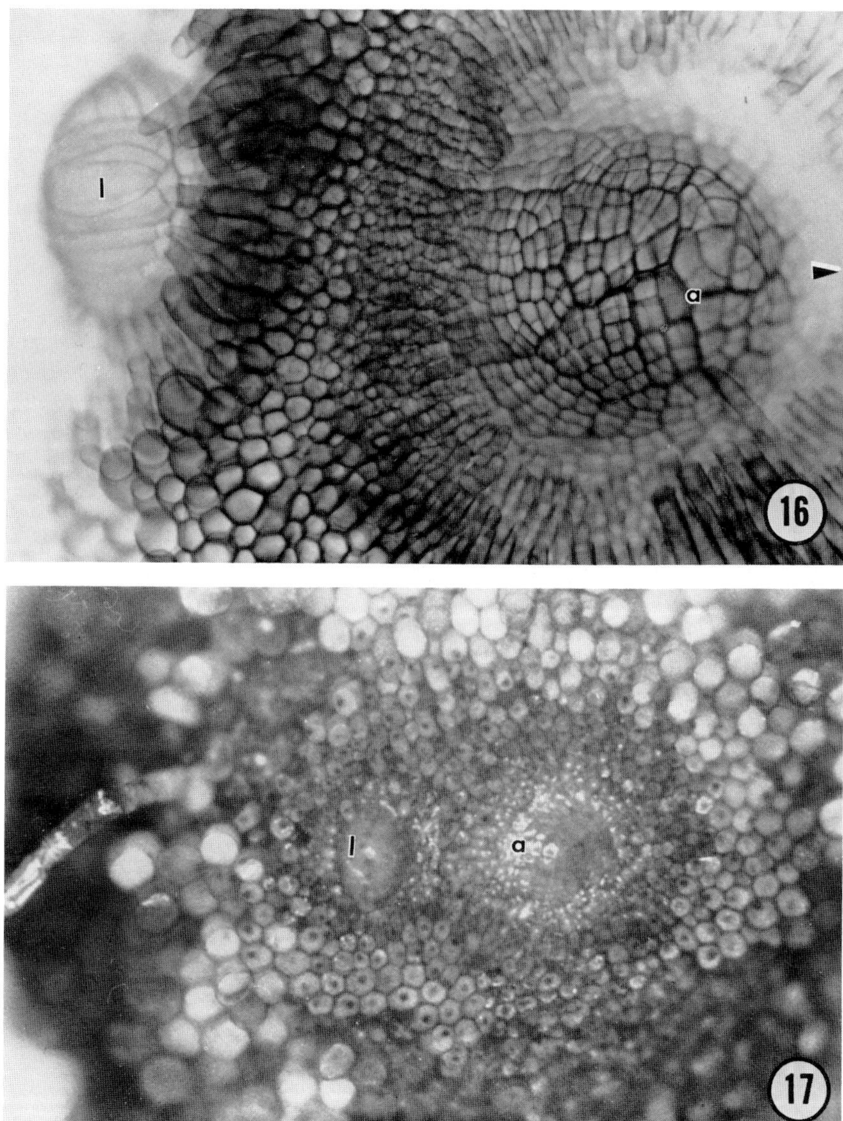
l
a
16
l
a
17

(Fig. 15). In the absence of distal leaf primordia an explanation other than the usual (leaf determination of protoxylem strand differentiation) is necessary. And of course, those strands typically on the ventral side of the stele, which are never associated with leaf primordia (which are basically restricted to the dorsal rhizome surface) (Figs. 16-19) must also develop under influences which have not yet been adequately explained. Procumbent rhizomes, exemplified here for example by *Lygodium*, *Schizaea*, *Dicranopteris* and *Vandenboschia*, have clear dorsiventral rhizomes, with leaves in a distichous arrangement on the dorsal surface (e.g. Figs. 16,17,18,19). Preliminary observations suggest that maturation of protoxylem cells in the so-called cauline strands on the ventral side of the stele may well be associated with adventitious root development. Considerably more detailed work needs to be done to clarify this, however.

FIGURES 18-19. Fig. 18. *Schizaea dichotoma*. Surface view of the shoot tip revealing the apex (a) and the first leaf primordium (l). Protoxylem cells mature among the metaxylem cells in the stele independent of the leaf primordium trace (X100). Fig. 19. *Lygodium japonicum*. Surface view of the shoot tip indicating the shoot apex (a) and the distichus arrangement of the leaves (l). Protoxylem maturation is diffuse among the metaxylem cells in the stele and appears to be independent of leaf vascularization. Arrow at right indicates the ventral side of the shoot. Leaf primordia form in two rows close to the mid-dorsal line of the rhizone. (X100).

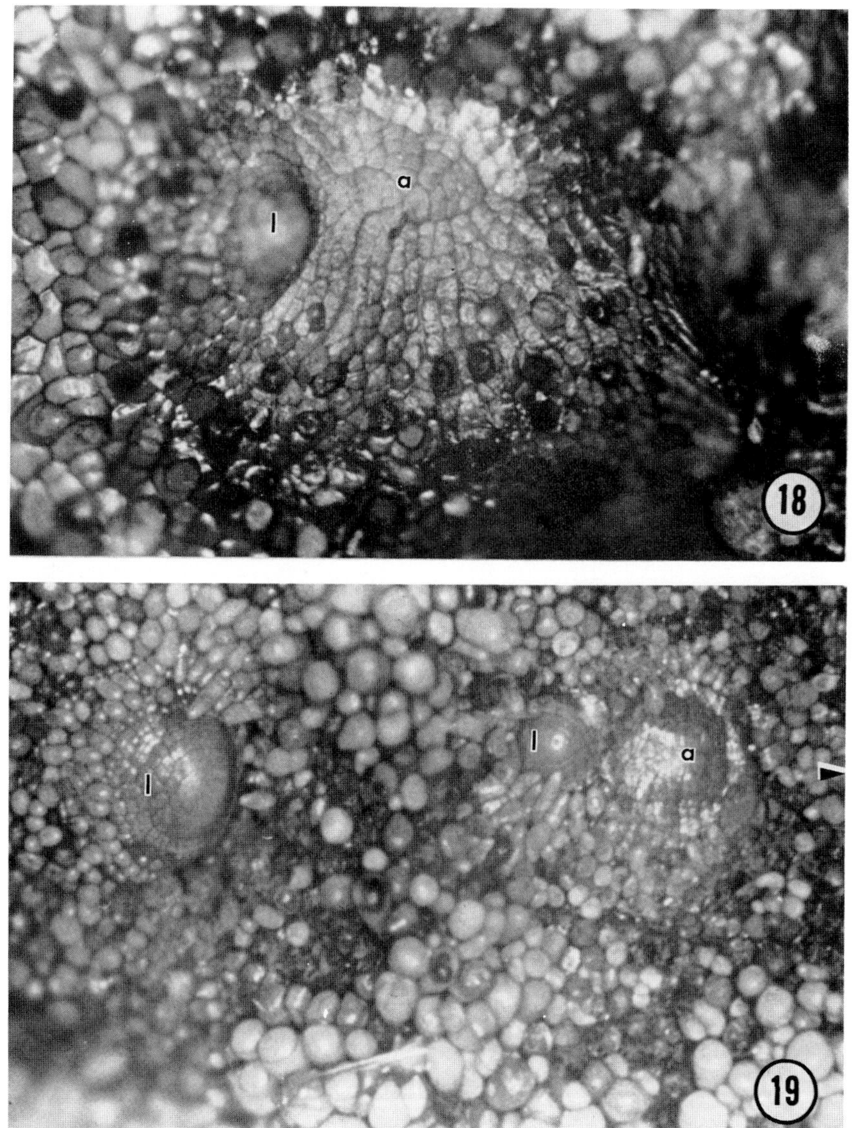
a
l
18
l
l
a
19

CONCLUSION

Many of the species described in this report have been described in previous studies. With a recent renewed interest in details of differentiation at the shoot apex, however, a new perspective can be applied both to the earlier data and to those presented here. In the literature on the primary vascular systems of seed plants, the intimate association of the vascular pattern (particularly the xylem pattern) and leaf development has been described and clarified in great detail (see for example Beck _et_ _al_., 1983).

It is possible by careful analysis of the three-dimensional primary xylem pattern in seed plants to determine the leaf phyllotaxy and phyllotactic fraction from the primary vascular pattern. The clear association, therefore, of leaves with the pattern is established. In addition, histologically, differentiation of the procambium and maturation of the protoxylem has been described as intimately associated with the origination and early development of leaf primordia. Classically, the leaves are said to determine the vascular pattern. Recently there has been a shift of emphasis by some workers in the interpretation of the pattern (see Larson, 1983), but no one questions the intimate association of leaves and the establishment of the primary vascular pattern.

That association of leaf vascularization and the primary vascular pattern has not been previously so clearly made with regard to vascular tissue differentiation and the development of characteristic fern stelar patterns. Descriptive anatomy of leaf gaps and traces has predominated in the fern literature. Detailed analyses of reports in the literature and of original preparations indicate that among some ferns there is an interrelationship of leaf primordia origination and development with the establishment of the protoxylem pattern which is basically the same as that seen in seed plants. It is possible among some ferns following careful analysis of the protoxylem patterns to determine the phyllotactic fraction from the pattern. Hence the association of leaves with protoxylem strands, their differentiation and maturation is established for these species. In those species in which the major protoxylem strands are equal in number to the number of meristeles, for example in some highly dissected dictyostelic ferns, an analysis of the mature primary vascular pattern provides an accurate determination of the phyllotactic fraction for those species. It would appear in the fern species included in the preceding two categories the association of leaves with the vascular pattern parallels that described for seed plants.

However, the situation among the ferns is not so homogeneous as might be implied by the preceding discussion. In some species, including those with protosteles as well as others with very complex and highly dissected steles, there are protoxylem patterns (and meristeles) which are not clearly associated with leaves. In this report, this situation was exemplified by species with supposedly simple steles: *Gleichenia*, *Dicranopteris*, *Vandenboschia*, *Lygodium*, and *Schizaea*. In *Gleichenia*, *Dicranopteris*, and *Vandenboschia* in addition to protoxylem which is associated with leaves, there is protoxylem which is not associated with leaves. This latter form has been referred to as cauline (proto-) xylem. No detailed studies are available which describe the developmental basis for cauline xylem maturation, but in many species it appears to be associated with the development of adventitious roots. In *Lygodium* and *Schizaea* another pattern can be distinguished. In these species no helically thickened protoxylem cells occur in the stem stele. The protoxylem is defined in these species not by cell morphology but by the fact that the cells mature first among all the xylem cells of the stele. No clear pattern of differentiation could be determined (i.e. protoxylem maturation is diffuse), and it is clear by the examination of numerous replicates that the earliest maturing cells in such species are not clearly associated with leaf trace development.

A broader array of fern species which represents a wider diversity of stelar types remains to be analyzed from this current perspective. At least three major patterns of vascular pattern establishment in the shoots of ferns can however be described: (1) *protoxylem* patterns are associated closely with leaves, and the shoot phyllotactic arrangement can be determined by careful analysis of these patterns independent from whatever mature stelar pattern characterizes the species; (2) in species characterized by highly dissected dictyosteles, where there is a major protoxylem strand associated with each meristele, the mature vascular system is directly related to phyllotaxy, and (3) in numerous species, there are protoxylem patterns and meristeles which differentiate and mature apparently independent from any leaf association (so-called cauline bundles). Any major scheme which would attempt to generalize the situation in the ferns beyond this would be premature at present. The variation in details of stelar pattern differentiation and development is great among the ferns, and requires a much broader sampling than has been accomplished thus far. The association of the leaf and stem to constitute the shoot of the vascular plant and the interrelationship of the leaves to the establishment of vascular tissue patterns is likely to be so fundamental that similarities at this level among major groups of vascular plants might be expected. In view of the similarities in procambium and protoxylem association with leaves among seed plants, and between seed plants and some ferns, it appears unjustified at present to conclude that the vascular patterns of seed plants and ferns are fundamentally different. Protoxylem and protoxylem patterns in particular appear to be clearly closely associated with leaf vascularization in both groups. The development of metaxylem and the cauline strands in ferns provide major problems in need of further detailed analysis. Further detailed studies of fern vascular patterns distal to leaf primordia may provide additional useful data related to precocious strands and their potential role in determining the sites of leaf primordia. And if there are indeed totally cauline vascular bundles (meristeles), the control of cell differentiation and the establishment and coordination of these patterns with leaf trace-related patterns constitute major areas of investigation which remain to be pursued.

ACKNOWLEDGMENTS

I wish to express my grateful appreciation to Melvin D. Turner for his technical assistance and for very helpful comments and discussion throughout the course of research reflected in this chapter. Portions of the work reported here were supported by NSF grant DEB-77-14648.

REFERENCES

Beck, C.B., R. Schmid and G.W. Rothwell. 1983. Stelar morphology and the primary vascular system of seed plants. Bot. Rev. 48(4): 691-815.

Bierhorst, D.W. 1977. On the stem apex, leaf initiation and early leaf ontogeny in Filicalean ferns. Amer. J. Bot. 64: 125-152.

Boodle, L.A. 1900. Comparative anatomy of the Hymenophyllaceae, Schizaeaceae and Gleicheniaceae. II. On the anatomy of the Hymenophyllaceae. Ann. Bot. 14: 455-496.

_____. 1901a. Comparative anatomy of the Hymenophyllaceae, Schizaeaceae and Gleicheniaceae. II. On the anatomy of the Schizaeaceae. Ann. Bot. 15: 359-421.

_____. 1901b. Comparative anatomy of the Hymenophyllaceae, Schizaeaceae and Gleicheniaceae. III. On the anatomy of the Gleicheniaceae. Ann. Bot. 15: 703-747.

_____. 1903. Comparative anatomy of the Hymenophyllaceae, Schizaeaceae and Gleicheniaceae. Ann. Bot. 17: 511-537.

Bower, F.O. 1923, 1926. The Ferns (Filicales) Vol. 1, 2. Cambridge University Press, Cambridge.

Campbell, D.H. 1921. The eusporangiate ferns and the stelar theory. Amer. J. Bot. 8: 303-314.

Chau, R. 1981. Vascular morphology of the Ophioglossaceae. Ph.D. dissertation. The University of Michigan, Ann Arbor.

Compton, R.H. 1909. The anatomy of *Matonia sarmentosa*. Baker. New Phytol. 8: 299-310.

Conard, H.S. 1908. The Structure and Life History of the Hay-Scented Fern. Carnegie Inst. of Washington. 56 pp. + plates.

DeBary, A. 1884. Comparative Anatomy of the Vegetative Organs of the Phanerogams and Ferns. Clarendon Press. Oxford.

De Bruyn, H. 1911. The ontogenetic development of the stele in two species of _Dipteris_. _Ann_. _Bot_. 25: 761-772.

Devadas, C. and C.B. Beck. 1971. Development and morphology of stelar components in the stems of some members of the Leguminosae and Rosaceae. _Amer_. _J_. _Bot_. 58: 432-446.

Devadas, C. and C. B. Beck. 1972. Comparative morphology of the primary vascular systems in some species of Rosaceae and Leguminosae. _Amer_. _J_. _Bot_. 59: 557-567.

Esau, K. 1965. Vascular differentiation in plants. Holt Rinehart and Winston. New York.

Farmer, J. B. and T. G. Hill. 1902. On the arrangement and structure of the vascular strands in _Angiopteris evecta_, and some other Marattiaceae. _Ann_. _Bot_. 16: 371-402.

Ford, S.O. 1902. The anatomy of _Ceratopteris thalictroides_. _Ann_. _Bot_. 16: 95-121.

Freeberg, J.A. and R.H. Wetmore. 1967. The Lycopsida - a study in development. _Phytomorphology_ 17: 78-91.

Gunckel, J. E. and R. H. Wetmore. 1946a. Studies of development in long shoots and short shoots of _Ginkgo_ _biloba_ L. I. The origin and pattern of development of the cortex, pith and procambium. _Amer_. _J_. _Bot_. 33: 285-295.

_____. 1946b. Studies of development in long shoots and short shoots of *Ginkgo biloba* L. II. Phyllotaxis and the organization of the primary vascular system, primary phloem and primary xylem. *Amer. J. Bot.* 33: 532-543.

Gwynne-Vaughan, D. T. 1901. Observations on the anatomy of solenostelic ferns. I. *Loxsoma*. *Ann. Bot.* 15: 71-98.

_____. 1903. Observations on the anatomy of solenostelic ferns. Part II. *Ann. Bot.* 17: 689-742.

_____. 1911. Some remarks on the anatomy of the Osmundaceae. *Ann. Bot.* 25: 525-536.

Hanstein, J. 1858. Über den Zusammenhang der Blattstellung mit dem Bau des dicotylen Holzringes. *Jahrb. Wiss. Bot.* I: 233-283.

Hebant-Maurie, R. 1973. Fonctionnement apical et ramification chez quelques fougères du genre *Trichomanes* L. (Hymenophyllacees). *Adansonia* 13: 495-526.

Imaichi, R. 1977. Anatomical study of the shoot apex of *Osmunda japonica* Thunb. Bot. Mag., Tokyo 90: 129-141.

Jeffrey, E.C. 1899. The morphology of the central cylinder in the angiosperms. *Trans. Roy Canad. Inst.* 6: 599-636.

Jeffrey, E. C. 1902. The structure and development of the stem in the Pteridophyta and Gymnosperms. *Proc. Roy. Soc.* London Ser. B, 195: 119-146.

Kuehnert, C.C. and P.R. Larson. 1983. Development and organization of the primary vascular system in the Phase II leaf and bud of *Osmunda cinnamomea* L. *Bot. Gaz.* 144(3): in press.

Larson, P. R. 1975. Development and organization of the primary vascular system in *Populus deltoides* according to phyllotaxy. *Amer. J. Bot.* 63: 1332-1348.

Larson, P. R. 1976. Procambium vs. cambium and protoxylem vs. metaxylem in *Populus deltoides*. *Amer. J. Bot.* 63: 1332-1348.

Larson, P. R. 1979. Establishment of the vascular system in seedlings of *Populus deltoides* Bartr. *Amer. J. Bot.* 66: 452-462.

Larson, P. R. 1983. Primary vascularization and the siting of primordia. pp. 25-51 *in* the Growth and Functioning of Leaves. J. E. Dale and F.L. Milthorpe (eds). Cambridge University Press, London.

Lucansky, T. W. 1974a. Comparative studies of the nodal and vascular anatomy in the neotropical Cyatheaceae. I. *Metaxya* and *Lophosonia*. *Amer. J. Bot.* 61: 464-471.

Lucansky, T. W. 1974b. Comparative studies of the nodal and vascular anatomy in the noetropical Cyatheaceae. II. Squamate genera. *Amer. J. Bot.* 61: 472-480.

Lucansky, T. W. and R. A. White. 1974. Comparative studies of the nodal and vascular anatomy in the neotropical Cyatheaceae III. Nodal and petiole patterns; summary and conclusions. *Amer. J. Bot.* 61: 618-628.

Lucansky, T. W. and R. A. White. 1976. Comparative ontogenetic studies in young sporophytes of the ferns. I. A primitive and an advanced taxon. *Amer. J. Bot.* 63: 463-472.

Marsh, A. S. 1914. The anatomy of some xeriphilous species of *Cheilanthes* and *Pellaea*. *Ann. Bot.* 28: 671-684.

Maze, J. 1977. The vascular system of the inflorescence axis of Andropogon gerardii (Gramineae) and its bearing on concepts of monocotyledon vascular tissue. Amer. J. Bot. 64: 504-515.

Miller, C. N., Jr. 1971. Evolution of the fern family Osmundaceae based on anatomical studies. Contr. Mus. Paleontol. Univ. Michigan 21: 139-203.

Namboodiri, K.K. and C.B. Beck. 1968a. A comparative study of the primary vascular system of conifers. I. Genera with helical phyllotaxis. Amer. J. Bot. 55: 447-457.

Namboodiri, K. K. and C. B. Beck. 1968b. A comparative study of the primary vascular system of conifers. II. Genera with opposite and whorled phyllotaxis. Amer. J. Bot. 55: 458-463.

Ogura, Y. 1972. Comparative anatomy of the vegetative organs of the pteridophytes. Handbuch der Pflanzenanatomie Ed. 2, B. VII. Teil 3. Gebrüder Borntraeger, Berlin.

Posthumus, O. 1924. On some principles of stelar morphology. Rec. Trav. Bot. Neerl. 21: 111-296.

Schmid, R. 1977. Tracheary secondary wall patterns and the definition of protoxylem and metaxylem I.A.W.A. Bull. s.n. 1: 7-9.

Schmid, R. 1983. The terminology and classification of steles: historical perspective and the outlines of a system. Bot. Rev. 48(4): 817-931.

Schoute, J.C. 1938. Anatomy. in Manual of Pteridology. F. Verdoorn Ed. pp. 65-104. Nijhoff. The Hague.

Steeves, T. A. 1963. Morphogenetic Studies on Osmunda cinnamomea. L. The shoot apex. J. Indian Bot. Soc. 42: 225-236.

Sterling, C. 1945a. Growth and vascular development in the shoot apex of Sequoia sempervirens (Lamb.) Endl. I. Structure and morphology of the apex. Amer. J. Bot. 32: 118-126.

Sterling, C. 1945b. Growth and vascular development in the shoot apex of Sequoia sempervirens Endl. II. Vascular development in relation to phyllotaxis. Amer. J. Bot. 32: 380-386.

Sterling, C. 1947. Organization of the shoot of Pseudotsuga taxifolia (Lamb.) Britt. II. Vascularization. Amer. J. Bot. 34: 272-280.

Stevenson, D. W. 1976. Observations on phyllotaxis, stelar morphology, the shoot apex and gemmae of Lycopodium lucidulum. Michaux (Lycopodiaceae). Bot. J. Linn. Soc. 72: 81-100.

Stevenson, D.W. 1980. Ontogeny of the vascular system of Botrychium multifidum (S.G. Gmelin) Rupr. (Ophioglossaceae) and its bearing on stelar theories. Bot. J. Linn. Soc. 80: 41-52.

Stevenson, D.W. 1982. An interpretation of Filicinean vascular systems. p. 74 (Abstr.) Bot. Soc. of Amer. Misc. Publ. 162.

Tansley, A.G. 1907/1908. Lectures on the evolution of the Filicinean vascular system. New Phytol. 6: 26-39, 54-68, 110-120, 136-147, 148-155, 188-203, 219-238, 254-269; 7: 1-16, 29-40.

Tansley, M. A. and E. Chick. 1901. Notes on the conducting tissue-system in Bryophyta. Ann. Bot. 15: 1-38.

_____. 1903. On the structure of Schizaea malaccana. Ann. Bot. 17: 493-510.

Tansley, A.G. and R.B.J. Lulham. 1905. A study of the vascular system of _Matonia pectinata_. _Ann. Bot._ 19: 474-517.

Tryon, R. 1970. The classification of the Cyatheaceae. _Contr. Gray. Herb._ 200: 1-53.

Tucker, S.C. 1962. Ontogeny and phyllotaxis of the terminal vegetatives hoot of _Michelia fuscata_. _Amer. J. Bot._ 49: 722-737.

Van Tiegham, P. and H. Douliot. 1886. Sur la polystelie. _Ann. Sci. Nat. Bot._, ser. 7, 3: 275-322.

Wardlaw, C.W. 1944. Experimental and analytical studies of Pteridophytes. III. Stelar morphology: the initial differentiation of vascular tissue. _Ann. Bot._ 8: 173-188.

Wardlaw, C.W. 1945. Experimetnal and analytical studies of Pteridophytes V. Stelar morphology: The development of the vascular system. _Ann. Bot._ 9: 217-233.

Wardlaw, C.W. 1957. On the organization and reactivity of the shoot apex in vascular plants. _Amer. J. Bot._ 44: 176-185.

Webb, E. 1975. Stem anatomy and phyllotaxis in _Ophioglossum petiolatum_. _Amer. Fern. J._ 65: 87-94.

_____. 1981. Stem anatomy, phyllotaxy, and stem protoxylem tracheids in several species of _Ophioglossum_. I. _O. petiolatum_ and _O. crotalophoroides_. _Bot. Gaz._ 142: 597-608.

Wetmore, R. H. and C.W. Wardlaw. 1951. Experimental morphogenesis in vascular plants. _Ann. Rev. Pl. Physiol._ 2: 269-292.

White, R.A. 1971. Experimental and developmental studies of the fern sporophytes. _Bot. Rev._ 37: 509-540.

White, R.A. 1979. Experimetnal investigations of fern sporophyte development. pp. 505-549. In The Experimental Biology of Ferns. A.F. Dyer Ed., Academic Press, London.

White, R.A. and T.W. Lucansky. 1984. Comparative ontogenetic studies in young sporophytes of tree ferns with special emphasis on vascular pattern development. Amer. J. Bot. (in press).

White, R.A. and W. Weidlich. 1974. The relationship between stem anatomy and growth habit in tree ferns (Cyatheaceae) and other ferns with erect stems. Amer. J. Bot. 6: 39-40.

White, R.A. and W. Weidlich. 1984. Organization of the vascular system in the stems of *Blechnum* and *Diplazium*. Bot. Gaz. (in press).

Wigglesworth, G. 1902. Notes on the rhizome of *Matonia pectinata* R. Br. New Phytol 1: 157-160.

Zimmerman, M.H. and P.B. Tomlinson. 1972. The Vascular system of Monocotyledonous stems. Bot. Gaz. 133: 141-155.

Zimmerman, M.H. and P.B. Tomlinson. 1974. Vascular patterns in palm stems: variations of the *Rhapis* principle. J. Arnold Arbor. 55: 402-424.

THE ROLE OF SUBSIDIARY TRACE BUNDLES IN STEM AND LEAF DEVELOPMENT OF THE DICOTYLEDONEAE

Philip R. Larson

North Central Forest Experiment Station
U.S. Department of Agriculture, Forest Service
Rhinelander, Wisconsin

Subsidiary bundles are subdivisions of a leaf trace that do not arise as independent divergencies from a parent trace. They normally originate in the node and develop either basipetally in the stem, acropetally in the leaf, or both. The literature describing different nodal types is examined, and it is noted that multiple bundles exit the stem through a common gap, particularly in many unilacunar and trilacunar species. These extra-trace bundles are considered subsidiaries, and it is suggested that many of them might differentiate basipetally in the stem from a nodal initiation site. Acropetal subsidiary bundles originate by subdivision of trace bundles in the node where they are also reoriented and mixed before passing through the petiole and redistributed as veins in the lamina. Acropetal and basipetal subsidiary bundles must develop in synchrony to maintain functional continuity. Basipetal subsidiary bundles presumably integrate the vasculature of the stem and contribute to vascular redundancy; acropetal bundles perform similar functions in the lamina. The question is raised whether subsidiary bundles might perform a regulatory role in leaf development.

ISBN 0-12-746620-7

The vascular systems of higher plants are both highly complex and extremely efficient conductive systems. The ways in which vascular systems function have been extensively examined and we are now well informed about the essential processes by which water and minerals are transported from absorbing roots and assimilates are translocated from photosynthesizing leaves. Interestingly, however, we are less well informed as to how the structural system develops and how it is organized to perform these complex functions. Such information is particularly sparse for woody plants and trees that undergo secondary stem thickening.

To perform its varied functions, the vascular system must develop in stages with each advancement in structural complexity leading to greater functional diversity. Obviously, an adult plant is structurally more complex and functionally more demanding than a seedling of the same species. Nonetheless, each new leaf, whether juvenile or adult, must progress through a given series of ontogenetic stages before it matures, and its vasculature must be developmentally coordinated and functionally integrated with that of other leaves and essential parts of the plant. The vascular system must also be highly redundant in its interconnections to ensure survival of the leaf, and occasionally of the entire plant, under adverse circumstances. Vascular redundancy is achieved in a variety of ways, but in every case the nodal region, with its upward and downward extensions in leaf and stem, respectively, is the focal point for vascular interchange. This vascular continuum between stem and leaf will be the subject of this treatise. My objective is to examine how nodal vascular systems _might_ develop to meet the functional requirements of broad classes of plants and what features these systems _might_ have in common. Thus, my discussion will be somewhat speculative, in the belief that speculation of what "might be" often incites further study of the more intriguing question of "what is". For convenience, I will confine the discussion to three parts: (I) stem and node vasculature, (II) leaf vasculature, and (III) the stem-node-leaf continuum. In the latter section, I will speculate on the functional roles subsidiary bundles might perform in both the stem and leaf.

I. STEM AND NODE VASCULATURE

A. *The Procambial System*

Despite the nodal diversity exhibited by dicotyledonous plants, their vascular systems originate and develop in remarkably similar ways. The vascular system in all plants is established during embryonic development and thereafter it is perpetuated as a plant enlarges (Larson, 1982a). The vascular system is perpetuated by procambial strands that progress acropetally and continuously in the apical shoot. In most species that have been critically examined, a single procambial strand--the central or median strand--develops acropetally to initially serve a new primordium. The central strand always progresses toward the new primordial site several plastochrons in advance of its associated lateral strands, when present. Because the central strand is precocious and makes first contact with a new primordium, it has been referred to as the perpetuating member of the procambial system (Larson, 1982a).

By definition, a procambial strand becomes a procambial trace, or leaf trace, when it establishes contact with the primordium it serves. A new central strand most commonly originates as a divergence from a parent central trace. In some species, new central strands are believed to originate as divergencies from parent lateral traces, but such origins have not been proven. Nonetheless, with few exceptions, the central trace is the dominant trace serving a leaf. Because all leaf traces develop acropetally from their points of origin, I will describe them as either departing or exiting the stem vascular cylinder and entering the nodal region. I will define subsidiary bundles as subdivisions of a leaf trace that do not arise as independent divergencies from a parent trace in the stem. These bundles normally originate in the node and develop either basipetally in the stem, acropetally in the leaf, or both. The actual boundaries of the node are ill-defined (Isebrands and Larson, 1977b), although the node has been defined in broad terms (Mitra and Majumdar, 1952; Howard, 1970, 1974; Dickison, 1975). For purposes of discussion I will use the terms nodal region or nodal gap with reference to the stem and leaf base with reference to the leaf. These terms apply to the approximate lower and upper bounds, respectively, of the commonly accepted interpretation of the node.

B. *Leaf Gaps or Nodal Lacunae*

The central trace and the lateral traces, when present, depart the vascular cylinder via leaf gaps or lacunae. Leaf gaps are formed by the acropetal closing of procambial strands above a departing leaf trace (Majumdar, 1942; Larson and Pizzolato, 1977, Fig. 2). Species are often classified by the number of lacunae that occur at each node, the most common types being unilacunar, trilacunar, and multilacunar (Sinnott, 1914; Sinnott and Bailey, 1914). Although most species have odd numbers of lacunae, bilacunar (Kato, 1966; Sehgal and Paliwal, 1974), tetralacunar (Watari, 1934; Kato,

1967; Hilger, 1978), hexalacunar, and greater (Millington and Gunckel, 1950; Tucker, 1962; Kato, 1967; Sugiyama, 1972, 1974) have been reported. Moreover, many modifications of the basic nodal types have been recognized (Howard, 1974). One modification occurs when two traces with independent origins in the vascular cylinder are associated with a unilacunar node, the unilacunar, two-trace node (Bailey and Swamy, 1948; Marsden and Bailey, 1955). Another variant is the so-called "split-lateral" in which the outermost lateral traces of opposite leaves in a decussate phyllotaxy share a common gap (Howard, 1970). These variants develop in contrasting ways. In the former, two independent traces merge before exiting the single gap of one leaf, whereas in the latter, a single trace bifurcates before exiting a lateral gap that serves two opposite leaves. Most studies of nodal vasculature have been conducted on species exhibiting secondary stem thickening. Consequently, the nodal condition in herbaceous species is sparsely documented and less well known (Howard, 1970).

Attempts have been made to correlate the number of leaf gaps with morphological characteristics. For example, stipules are best developed in families with trilacunar nodes and leaves that are either lobed or with dentate, serrate or glandular margins. The strongest tendency toward reducing and eliminating stipules occurs in families with unilacunar nodes and leaves with entire margins (Sinnott and Bailey, 1914; Dormer, 1944; Bailey, 1956). Species with multilacunar nodes most often have sheathing leaf bases and their leaves are usually alternate, seldom opposite (Sinnott and Bailey, 1914); multilacunar nodes are found in few families of the Dicotyledoneae (Howard, 1970; Dickison, 1975). Among woody species, those with trilacunar nodes predominate in temperate regions and those with unilacunar nodes in tropical regions. The palmately lobed leaf is largely confined to woody species of temperate regions (Bailey, 1956).

Howard (1970) commented on the fact that "our knowledge of differentiation within the procambial tissue leading to the development of one or more traces per leaf is still scanty". The multiple traces in trilacunar and multilacunar species are presumed to originate from different parent traces situated on different sympodia in the phyllotactic array. Trace origins have been examined in many trilacunar species, and little doubt remains that each of the three traces originates independently and develops acropetally in the stem. However, it is still an open question whether the numerous traces reported in some multilacunar species are all true leaf traces that have independent origins and develop acropetally. Few detailed anatomical studies have been conducted on trace origins in these species because of the inherent technical difficulties involved and the objectives of the investigators. By far, most investigations on nodal vasculature have been phylogenetic, not developmental. My interests, to the contrary, are developmental. I will therefore attempt to extract from the literature certain features common to the recognized nodal types that might be useful in relating stem and leaf vasculatures.

C. *Variations in Nodal Vasculature*

1. VARIATION in NUMBER of GAPS per NODE. The number of gaps that occur at each node varies greatly both among plants of a species and among nodes on a plant. Such variation may be attributed either to ontogeny or to environmentally induced growth changes that occur during ontogeny. With regard to ontogeny, the number of leaf gaps usually increases with plant size and age. The cotyledons of most dicotyledonous species, including those with palmately or pseudopalmately veined cotyledons, are served by two traces that depart through a single gap (Bailey, 1956). This nodal pattern has been referred to as a "double-trace" (Thomas, 1907). The cotyledonary double-trace node is of embryonic origin and it therefore differs from the adult unilacunar, two-trace node described by Marsden and Bailey (1955). Structurally, however, they are similar because two traces with independent origins exit the stem cylinder through a common gap. Origin of the double cotyledonary traces and organization of the subsequent leaf-trace system have been described in detail for *Populus deltoides* seedlings (Larson, 1979). This basic cotyledonary pattern is modified in various ways among different families (Money, Bailey and Swamy, 1950); and even in those families with relatively simple cotyledonary nodes, deviations from the typical pattern have been frequently reported (Shimaji, 1957; Kato, 1966). In the Juglandaceae, Stone (1970) observed that those species exhibiting epigeal germination had the typical unilacunar, two-trace node, whereas those exhibiting hypogeal germination had nodal diversity. In the latter group, the number of cotyledonary gaps was directly related to seed and seedling size (Conde and Stone, 1970). The cotyledonary nodal structure of many multilacunar families is variable--some species apparently possess trilacunar rather than unilacunar cotyledonary nodes (Sugiyama, 1976a).

Adult leaves often differ greatly in nodal structure from that of their post-cotyledonary seedling leaves (Kato, 1966; Post, 1958; Sugiyama, 1976a; Hilger, 1978). Most species with trilacunar adult nodes have unilacunar seedling nodes (Bailey, 1956; Philipson and Philipson, 1968), but species with multilacunar adult nodes may have either unilacunar (Canright, 1955; Kato, 1966) or trilacunar (Swamy, 1949; Canright, 1955) seedling nodes. Ontogenetically, the unilacunar nodal condition is the most stable, trilacunar relatively stable, and multilacunar the most variable (Watari, 1939). For example, species with trilacunar nodes seldom revert to the unilacunar state (Watari, 1939; Bailey, 1956). However, Watari (1939) observed many cases in the Saxifragaceae in which vigorous leaves of trilacunar species became multilacunar and, conversely, weak leaves of multilacunar species retrogressed to the trilacunar state.

The nodal structure of multilacunar species varies not only ontogenetically but also among plants of a species and among adult leaves of a plant (Bailey and Nast, 1944; Millington and Gunckel, 1950; Canright, 1955). The latter variability is undoubtedly environmentally induced because the number of lateral traces has been observed to increase in vigorous leaves (Swamy, 1949; Post, 1958; Benzing, 1967a; Neubauer, 1979a) and to decrease in less-vigorous leaves (Watari, 1934, 1939). Thus variability may result from either the addition, fusion, or loss of lateral traces (Sugiyama, 1976a, b). Kato (1966, 1967) documented many instances of apparently random nodal variation on lateral branches of multilacunar species that he attributed to environment. For example, either 5, 7 or 9 lacunae were observed in ten branches of five plants of *Rubus sieboldi*. The nodal types on one branch occurred in the sequence 5, 5, 7, 7, 5, 7, 9 and 5, and on another branch in the sequence 7, 7, 9, 5, 5 and 5. Nine-lacunae nodes were rare; the outermost traces at these nodes were weak and seldom persisted in the petiole. In *Michelia*, Tucker (1962) noted that the vascular supply of each leaf primordium consisted of seven to twelve procambial strands that diverged from points around the vascular cylinder. However, during development the largest strands exited the cylinder via three to five separate gaps, while later-differentiating, weaker ones shared gaps with larger strands in most nodes.

2. VARIATION in NUMBER of BUNDLES per GAP. A leaf gap is not necessarily limited to a single trace or bundle. Two traces may share a common gap as in the "double-trace" (Thomas, 1907) or the "split-lateral" (Howard, 1970). Many other observations have been made of several traces or multiple-trace bundles departing through a common gap both in cotyledons (Kato, 1966; Sugiyama, 1976b) and in adult leaves (Watari, 1934, 1939; Benzing, 1967a; Kato, 1967). Such reports are most frequent in species with unilacunar (Sinnott, 1914; Philipson and Philipson, 1968; Dickison, 1975; Neubauer, 1981) and trilacunar (Howard, 1970; Seghal and Paliwal, 1974; Neubauer, 1979) nodes. For example, in *Euptelea polyandra* five to nine small bundles may depart through the unilacunar gap (Benzing, 1967a), and in *Quercus* sp. as many as five bundles may depart through the trilacunar central gap (Smith, 1937). The central trace of trilacunar species often consists of multiple bundles (Watari, 1934; Tucker, 1962; Benzing, 1967a; Philipson and Philipson, 1968; Howard, 1970). Thus, the number of leaf gaps may vary within some multilacunar species, and the number of trace bundles departing through a gap may vary among plants of a species and among nodes of a plant.

D. Basipetal Subsidiary Bundles

1. BASIPETAL BUNDLES in POPULUS. Not all vascular bundles serving a leaf develop acropetally. In *Populus deltoides*, basipetal subsidiary bundles are also produced (Larson, 1975, 1976a, 1982b). These subsidiary bundles are initiated at the base of each primordium, and they develop basipetally in the stem using the original procambial trace, the one that

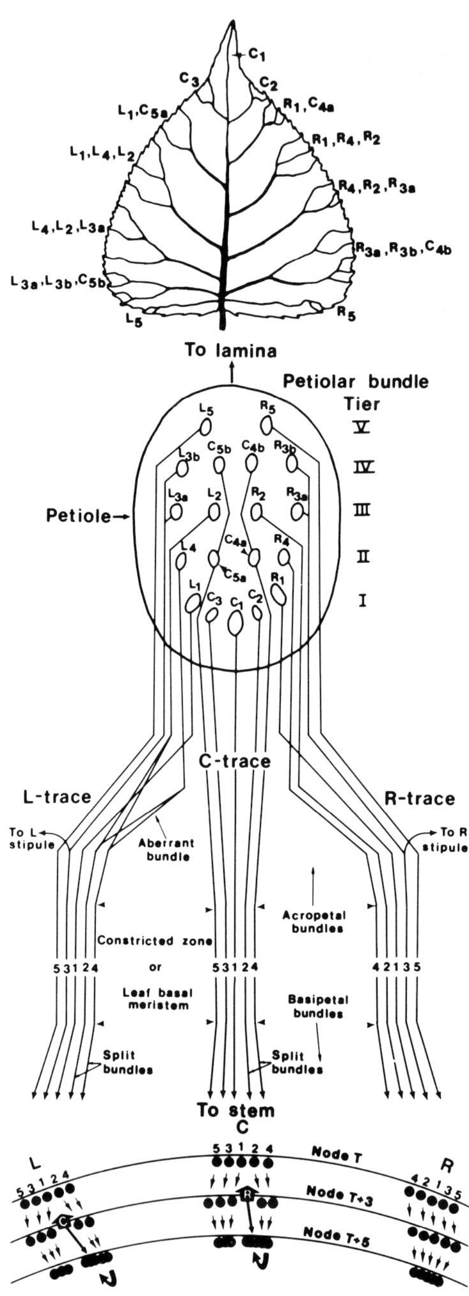

To lamina
Petiolar bundle
Tier
Petiole
C-trace
L-trace
R-trace
To L stipule
To R stipule
Aberrant bundle
Acropetal bundles
Constricted zone
or
Leaf basal meristem
Basipetal bundles
Split bundles
To stem
Node T
Node T+3
Node T+5

developed acropetally, as a template in their descent (Fig. 1). The central trace may produce four to eight and each lateral trace two to six basipetal subsidiary bundles in addition to the original acropetal traces. Thus, the vascular cylinder in a Populus stem is formed by an acropetally developing system of procambial strands and a basipetally developing system of subsidiary bundles consisting of both xylem and phloem.

Basipetally developing subsidiary bundles have not been specifically described for other species. However, I doubt that such bundles are unique to Populus, and I suggest that they may be present but undocumented in certain other species for two main reasons. First, few investigators of nodal vasculature have been concerned with the directionality of trace bundle development because their objectives were different. Secondly, most investigations of nodal vasculature have been conducted on mature plants with secondary development; apparently, this is the customary procedure (Bailey,

Figure 1. The stem-node-leaf vascular continuum in Populus deltoides. During development, the original procambial strand of the central leaf trace (bundle C_1) establishes first contact with the new primordium followed by the original procambial strands of the lateral leaf traces (bundles L_1 and R_1); these procambial strands develop acropetally and continuously in the stem. Subsidiary trace bundles (2 thru 5) are initiated later from the leaf basal meristem or nodal constricted zone. They differentiate acropetally in the node and basipetally in the stem, using the original procambial bundles as templates in their ascent and descent, respectively. The acropetal bundles are mixed in the node to form petiolar bundles and these bundles are redistributed in the lamina as lateral veins; the vein pattern shown will vary among leaves. The basipetal bundles of the C- and L-traces are split by the exiting R- and C-traces, respectively, three nodes below in the stem. The split bundles usually merge with the trace that split them, and the remaining bundles merge to form a single trace bundle.

1956; Post, 1958; Howard, 1970). It is impossible to determine either the true origin or the direction of development of traces and trace bundles from microsections of secondary material in which vascular patterns are already established. Accurate reconstructions from clearings of primary tissues are also extremely difficult because clearings do not reveal procambial trace bundles. They are even more difficult to obtain from herbarium specimens. The only way that both the origin and the direction of bundle differentiation can be recognized with certainty is by detailed developmental studies of the apex and primary region of the shoot where these events first occur.

Although investigators have not specifically described subsidiary bundles in other species, they have either commented on or alluded to multiple-bundle traces consisting of several to many bundles exiting through a common gap as noted earlier. Moreover, many photomicrographs and camera lucida drawings in the literature show multiple bundles in the nodal gaps. As I will discuss later, it is highly probable that these extra-trace bundles are subsidiaries and it is conceivable that many of them develop basipetally.

2. *BASIPETAL VESSEL GROUPS*. Frequent reference also has been made in the literature to the fact that vessels are initiated in proximity to nodes and that they differentiate basipetally in the stem (Esau, 1954, 1965; Höster, 1964). In *Populus*, differentiating primary vessels use established leaf traces and subsidiary bundles as templates in their descent (Larson, 1980a, 1982b). Primary vessels in most species are arranged in discrete radial tiers (Esau, 1954), and occasionally the tiers themselves are arranged in relatively distinct vessel groups. Both proto- and metaxylem vessels begin differentiating in the vessel group or tier associated with

the median bundle of each leaf trace and further vessel differentiation proceeds laterally (Esau, 1945; Ashworth, 1963; Benzing, 1967a; Devadas and Beck, 1971), just as it does in the subsidiary trace bundles of Populus (Larson, 1976a). Apparently, the environment also influences the number of groups or tiers of vessels in a leaf trace because they have been found to be more numerous in larger leaves (Lehmberg, 1924). Thus, vessel groups, each with a phloem counterpart, may also be the unrecognized equivalents of basipetal subsidiary bundles.

3. BASIPETAL BUNDLE SPLITS. An interesting and highly relevant consequence of basipetal bundle differentiation in both Populus deltoides (Larson, 1975, 1976a) and Populus grandidentata (Larson, 1976b) are the so-called "bundle-splits". In their descent, one or more of the outermost subsidiary bundles are split from the main trace when they encounter an exiting trace three nodes below (Fig. 1). For example, basipetally developing subsidiary bundles of the left and central traces of the leaf at node T are split by the exiting central and right traces, respectively, of the leaf at node T+3. The split bundles eventually merge in lower internodes either with the trace that split them or with the adjacent trace. These bundle splits occur during procambial development (Larson, 1975) and they are perpetuated in both the xylem (Larson, 1980a; Meicenheimer and Larson, 1983) and the phloem (Vogelmann, Larson and Dickson, 1982).

Alexandrov and Alexandrova (1929) described trace splits in the trilacunar species of Helianthus that were almost identical to those of Populus. As each leaf trace entered the vascular cylinder, it split the trace of a leaf situated three nodes higher on the stem. According to the authors, this meant that the lower trace departed the stem cylinder before the higher bundles were formed, thus implying basipetal differentiation of the trace bundles. They noted that the vascular cylinder must somehow separate to accommodate these new bundles. Smith (1937) also described trace splits in trilacunar species of Acer and Quercus and in a heptalacunar species of Platanus. The central leaf trace in each of these species, consisting of three to five bundles, was split by an exiting trace three nodes below. However, Smith did not comment on the direction of trace development. All traces in the multilacunar species Ricinus communis appeared to continue through two to three nodes before being blocked by exiting traces from lower leaves (Reynolds, 1942). At this lower level, the traces diverged with some passing horizontally inward to the septum and others being either divided or deflected to the side of the exiting leaf traces. Schnettker (1977) described and illustrated what appeared to be trace bundle splits in Clematis.

4. *TRACE FORKINGS*. When leaf traces are followed downward in the stem, configurations similar to bundle splits are often referred to as trace or bundle forkings without specific reference to the directionality of development (Thoday, 1922; Lehmberg, 1924; White, 1955). However, trace forkings observed basipetally must be distinguished from trace fusions observed acropetally and also from basipetal bundle splits. The apparent basipetal forking of a leaf trace above a gap has usually been shown to result from the fusion of two acropetally developing traces. According to Bailey (1956), the leaf trace exiting the median gap of many dicotyledons frequently forks above the next subtending leaf on the orthostichy. But when examined developmentally, each fork can be related at lower levels to two independent parts of the eustele. Thus, the two traces tend to merge acropetally before entering a

common gap as found in Austrobaileya (Bailey and Swamy, 1948) and Ascarina (Balfour and Philipson, 1962).

Other studies have revealed similar developmental patterns. For example, Jensen (1968) observed that in some species of the Crassulaceae the central trace consisted of two strands from different sympodia that joined before exiting the unilacunar node as a single trace. However, at the apex one strand always established contact with the new primordium before the other. Girolami (1953) noted that the single trace of Linum branched to form half-traces at its basal end, but the longer of the two branches was always more prominent during development. Although he described the system basipetally, acropetal development and trace fusion were implied.

Trace forkings and fusions usually can be demonstrated as the same phenomenon observed in opposite directions. Although similar in configuration to bundle splits in mature stems, they are different developmentally. Trace fusions occur during acropetal development of the procambial strand that first establishes contact with a new primordium, whereas bundle splits occur during basipetal development of subsidiary bundles that differentiate after primordium establishment. Many reports of trace forkings were based on either clearings or the examination of mature tissues. Thus, some reported forkings may in fact be bundle splits, particularly in those studies in which neither the presence of multiple trace bundles nor the direction of bundle development were considered.

5. *MULTIPLE BUNDLES in the STEM*. As noted earlier, multiple bundles per trace are encountered more often and are more numerous in unilacunar and trilacunar than in multilacunar species. When present in the latter, the multiple bundles are usually confined to the central trace or the

central trace and its nearest major laterals. Although often referred to as multiple traces, it is unlikely that all these bundles are true leaf traces, each departing from a separate parent trace in the vascular cylinder. It is more likely that they are subdivisions of a trace, in which case they should be considered subsidiary bundles in all nodal systems. But do they develop acropetally or basipetally? Many investigators have commented on the fact that established leaf traces, but not developing procambial strands, decrease in size basipetally (Alexandrov and Alexandrova, 1929; Pulawska, 1965; Devadas and Beck, 1971; Larson, 1976a). This downward tapering of leaf traces is particularly evident in the primary stems of woody species and in those species that will later undergo secondary thickening; it is less pronounced, if present at all, in herbaceous species (Devadas and Beck, 1971). Thoday (1922) described this process of leaf trace development in _Helianthus_ as a grand period of transverse growth corresponding to that of the leaf served.

Two alternative interpretations can be offered to account for the multiple bundles and taper of the leaf trace. First, the trace may dichotomize at some level in the stem and the resultant subsidiaries develop acropetally to exit the node as multiple bundles. Trace dichotomies are particularly common in cotyledons which, almost without exception, exhibit parallel or palmate venation (Bailey, 1956). Such dichotomies can occur in nodes of any age either below, at, or above the level of the leaf gap. When they occur below, they appear as either pseudotraces or multiple bundles within a gap (Kato, 1966; Sehgal and Paliwal, 1974). Pseudotraces have been described for the bud scales of _Populus_ that also exhibit parallel venation (Larson and Richards, 1981, Fig. 14). Because pseudotraces exit the vascular cylinder via independent gaps, they can easily be misinterpreted as multilacunar nodes.

Secondly, multiple bundles may be subsidiary bundles that differentiate within the node and develop basipetally in the stem. For example, the number of subsidiary bundles present, consequently the degree to which a leaf trace tapers or narrows downward, is apparently related to leaf vigor. Multiple bundles are more frequent in traces serving vigorous leaves and less frequent in those serving weak leaves (Watari, 1934, 1939; Swamy, 1949; Kato, 1966). Such variability also occurs in the numbers and sizes of basipetally differentiating vessel groups discussed earlier. The pattern is also analogous to the one found in *Populus* where the number of subsidiary bundles per leaf trace increases during the growing season as leaf size increases (Larson, 1980b) and declines with the onset of dormancy when leaf size decreases (Goffinet and Larson, 1981). In leaves that abort just before the bud scales are initiated, subsidiary bundles are either few and small or nonexistent both on terminal shoots (Goffinet and Larson, 1982) and lateral branches (Richards and Larson, 1981). Neither interpretation can be substantiated at this time because of the lack of verifiable information on the origin of multiple-trace bundles in most species. However, available evidence suggests that both interpretations may be correct depending on the family and species group.

II. LEAF VASCULATURE

Vascularization of the leaf, like that of the stem, can be shown to develop acropetally and continuously from the leaf base. The acropetally developing procambial strand that first establishes contact with a new primordium to become a leaf trace continues its acropetal development as the primordium enlarges. Invariably, the central trace develops first and its acropetal extension in the primordium initiates the incipient midvein in all nodal types (Hara, 1962; Lersten, 1962; Tucker, 1962; Ramji, 1967; Lieu and Sattler, 1976). Even in those unilacunar species in which two independent traces merge before entering the gap, one of the two traces always establishes first contact with the primordium (Jensen, 1968).

In most leaves with a midvein, particularly those displaying pronounced bilateral symmetry, the central trace not only initiates the midvein but its main derivative continues developing as a midvein to the tip of the mature lamina (Deinega, 1898; Bailey and Nast, 1944; Herbst, 1972; Sugiyama, 1972; Fournioux and Bessis, 1977). Pinnate leaves are apparently vascularized in a similar way (Deinega, 1898; Lersten, 1972; Hickey and Wolfe, 1975; Kaplan, 1980). However, some asymmetric leaves are vascularized differently. Although the central trace may initiate an incipient midvein in the new primordium, it often fails to perpetuate as a midvein because of subsequent branching or ramification (cf. Lieu and Sattler, 1976). Clearly then, the central trace and its derivative bundles play a crucial role in both establishing first contact with a new primordium and initiating its vasculature.

Lateral traces, when present, also develop acropetally in the primordium but later than the central trace, often several plastochrons later (Majumdar, 1942; Shushan and Johnson, 1955; Höster and Zimmermann, 1961; Tucker, 1962; Denne, 1966). The lateral traces develop adjacent to, but usually independent of, the central trace. Thus, lateral traces contribute to the midvein complex until they either diverge or branch to form secondary lamina veins. The position at which a lateral trace diverges as a secondary vein depends on both the time of formation and position of the lateral trace relative to the central trace in the node; e.g., the first-formed lateral traces adjacent to the central trace in the node extend farthest in the midvein complex before diverging.

Subsidiary bundles often develop similarly. Following the subdivision of leaf traces in the nodal region, the resultant bundles develop acropetally adjacent to, but independent of, other midvein bundles before diverging. For example, the unilacunar trace of *Linum* is tripartite in the node. The median bundle traverses the entire lamina but the subsidiary bundles diverge short of the lamina tip (Girolami, 1953). In this way, the midvein of many species becomes increasingly less complex toward the lamina apex.

Insufficient evidence is available for most species to determine the true origins of secondary veins during embryonic leaf development. Some authors either assume or imply that secondary veins simply branch from the midvein (Meinhardt, 1976; Franck, 1979). One can easily gain this impression when examining either clearings or microsections of mature or near-mature leaves. As a leaf matures, interfascicular activity and secondary development not only obscure the independence of bundles in the midvein complex but also create cross-linkages among these bundles. The midvein complex then becomes a composite midrib from which secondary veins appear to branch. But it should be noted that leaf traces, although related to leaf veins, can only be directly equated with veins during the earliest stages of primordium development when they are clearly recognizable throughout their extent.

That part of the node lying between the levels at which leaf traces depart the stem vascular cylinder and enter the leaf or petiole base is one of extreme vascular diversity. Within this region, the vasculatures of many species are reoriented in a multitude of ways by the proliferation of new bundles and frequently by the fusion of bundles both within and among leaf trace derivatives. Consequently, the leaf base has been referred to as the "primary nerving center" (Croizat, 1940), the "leaf basal meristem" (Larson, 1976a), and the "lower vascular plexus" (Neubauer, 1972, 1979a, b). Although reorientation of vasculature at the leaf base appears chaotic and random, it is highly ordered and apparently conforms to a basic species or generic pattern. The number of bundles proliferating at this level in the node is again a function of leaf vigor, increasing in vigorous leaves and decreasing in weaker and smaller leaves (Kato, 1966; Sehgal and Paliwal, 1974; Neubauer, 1979a).

The leaf base is obviously a major center for bundle mixing, or for proliferating and reorienting trace bundles from the stem. The lamina base, on the contrary, serves primarily as a center for redistributing or dispersing petiolar bundles as lamina veins. The latter applies particularly to palmately compound and palmately veined leaves or leaves with asymmetrical venation patterns lacking a true midvein (Watari, 1934, 1936; Foster and Arnott, 1960). In leaves with a well-defined midvein, the petiolar vasculature often continues unaltered in the lamina but with decreasing complexity as lateral veins diverge.

Ashworth (1963) referred to the lamina base as the "region of maximal vascularity" in _Ilex_ because the vasculature decreased progressively from this point to the lamina apex. The lamina base, or petiole top, serves as a secondary center for bundle mixing in some species. That is, petiolar bundles are not only subdivided but also reorganized before being distributed as lamina veins. However, as a rule, little vascular reorientation occurs at the lamina base. Similarly, little reorientation occurs within the petiole middle or the slender part of the petiole, and this region is therefore preferred for taxonomic and diagnostic studies (Howard, 1974). Petiole elongation is a relatively late event in leaf development. Consequently, the vasculatures of the leaf and lamina bases are essentially established by the time the petiole is intercalated between them. In plants with sessile leaves, the leaf and lamina bases are of course congruent.

It appears that vascularization of most dicotyledonous leaves conforms to a general pattern. Leaf traces exit the node and enter the leaf base where they are subdivided and mixed in various ways. The reoriented bundles traverse the main part of the petiole with little or no further branching, although adjacent bundles may merge or anastomose. At the lamina base, the petiolar bundles are redistributed as veins, either as a series of primary veins or as divergencies from a midvein. Modifications of this general pattern naturally occur among the many leaf types in the Dicotyledoneae. Yet, the essential point is that leaf vascularization has been found to proceed acropetally and continuously from the leaf base in most every species critically examined. The branching of secondary, tertiary, and higher order veins has also been found to proceed acropetally and continuously, even in those leaves in which the veins recurve downward at the leaf margin (Höster and Zimmermann, 1961). All minor venation in the lamina also develops in continuity with lower order veins, and no evidence has been found of discontinuities in minor vein differentiation (Foster, 1952; Pray, 1955, 1963; Hara, 1962; Höster, 1964; Lersten, 1965; Ramji, 1967; Herbst, 1972).

III. THE STEM-NODE-LEAF CONTINUUM

A. *Bidirectional Subsidiary Bundles*

Persuasive evidence has been presented herein that the primordial leaf trace system of most dicotyledonous species develops acropetally in the stem, exits via nodal gaps, and continues in the leaf. Evidence also indicates that the leaf traces are subdivided into smaller bundles, usually at some level within the node but occasionally at the lamina base. Invariably, subdivision of the leaf traces results in an increase in vascular area distal to the node. To accommodate this increase, a simultaneous and equivalent increase in vascular area must occur in the leaf trace components of the stem. Precisely how these leaf traces divide and how the resultant bundles maintain functional continuity between stem and leaf have received little attention. Maintenance of functional continuity obviously requires that vascular development of stem, node, and leaf be closely synchronized, which in turn requires an initiation site or center of vascular development. Both direct and indirect evidence implicates the node as the most probable initiation site.

Direct evidence has been obtained from developmental studies of *Populus deltoides*. Subsidiary bundles differentiate both basipetally in the stem and acropetally in the petiole from the leaf basal meristem. That is, the original procambial bundle (i.e., strand) of each leaf trace develops acropetally and continuously, but the subsequently formed subsidiary bundles differentiate bidirectionally. A part of the nodal transition region is referred to as the "constricted zone" (Larson and Isebrands, 1978) because at this level each of the three traces appears to form a single, unified trace (Fig. 1). Nonetheless, the component bundles can be recognized as independent entities in carefully prepared microsections (Isebrands and Larson, 1977a, b). Downward from the constricted zone, the component bundles continue for several internodes before either separating as bundle splits or merging to form a single trace bundle. Upward from the constricted zone, the bundles diverge and are mixed before entering the petiole in conformance with the basic species pattern. Thus, it was suggested that the subsidiary bundles extending both basipetally and acropetally in *Populus deltoides* were produced by a "leaf basal meristem" (Larson, 1976a).

Several lines of indirect evidence presented earlier suggest that similar events may occur at a nodal initiation site in many other species. For example, at some time after primordium establishment the leaf traces begin subdividing in the node before diverging in the petiole and/or lamina. Also, leaf traces with multiple bundles or vessel groups usually taper basipetally in the stem. If these bundles were examined developmentally, many of them would undoubtedly qualify as subsidiaries. In this case, each acropetal subsidiary would require a basipetal counterpart to maintain functional continuity, and differentiation would occur bidirectionally from a nodal initiation site. The alternative would be a unidirectional trace system that developed acropetally from stem to leaf with local dichotomies and fusions in transit. At present, little published evidence supports such a system.

B. *Concluding Questions*

The foregoing discussion raises three important questions that I would like to address. First, what function might basipetal subsidiary bundles perform? I suggest that they serve to integrate the vascular system and provide vascular redundancy among leaves on the stem. Lateral bundles of trilacunar species provide a degree of redundancy because each of the three traces serving a leaf arises from a different parent trace situated on a different sympodium. Similarly, the unilacunar, two-trace system provides a degree of redundancy. However, the vasculatures of many unilacunar and trilacunar species are considered open systems (Balfour and Philipson, 1962; Devadas and Beck, 1971), primarily those that undergo secondary thickening (Dormer, 1945); i.e., leaf traces arising from different sympodia are not interconnected in the stem. Thus, basipetally developing subsidiary bundles would increase not only vascular trace area but also redundancy of the vasculature serving each leaf. The latter would apply particularly in the case of basipetal bundle splits that create new trace unions among sympodia. That such unions occur has been demonstrated in *Populus*. ^{14}Carbon is readily translocated via bundle splits in the phloem (Vogelmann *et al*., 1982) and histological dyes are similarly transported in the xylem (Meicenheimer and Larson, 1983). Subsidiary bundles resulting from trace dichotomies that develop acropetally could produce neither bundle splits nor equivalent increases in vascular redundancy.

Secondly, what relation might exist between basipetal bundles and the numerous lateral traces in multilacunar species? Little information is available regarding the precise origins and directions of development of the lateral traces in multilacunar species. Indications are that they develop acropetally as true leaf traces, each from a different parent trace in the stem (Watari, 1934; Millington and Gunckel, 1950; Tucker, 1962; Sugiyama, 1972, 1974). But might these additional traces in multilacunar species bear a functional relation to the multiple-trace bundles of unilacunar and trilacunar species. The number of lateral traces is often variable among plants and among nodes on a plant in multilacunar species (Watari, 1934, 1939; Kato, 1966, 1967). They increase numerically in more vigorous leaves, and the outermost laterals are often first to be either weakened or deleted as leaf vigor declines (Watari, 1934, 1939; Kato, 1966, 1967; Benzing, 1967b; Sehgal and Paliwal, 1974; Hilger, 1978; Neubauer, 1978). These weak, outermost traces often terminate blindly at some point in the stem, node or petiole. In contrast, the nodal vasculature of unilacunar and trilacunar species is remarkably stable; lateral traces in trilacunar species are rarely absent and seldom end blindly. However, the number of bundles per trace or gap is more variable in these species and it often increases with increasing leaf vigor. Consequently, a common feature apparently exists among all these broad classes of plants; namely, that vascular redundancy can be accommodated to conform with leaf vigor, in multilacunar plants by varying the number of leaf traces and in unilacunar and trilacunar species by varying the number of subsidiary bundles in the traces.

Finally, what role might the acropetal subsidiary bundles perform in lamina development? One point often overlooked in descriptive studies is the progressive development of leaf vasculature. Not only does lamina vascularization proceed developmentally in the sequence central trace → lateral traces → subsidiary bundles, but also petiole and midvein vascularization develops dorsally to ventrally in the same sequence. Consequently, the degree to which trace bundles proliferate in the node and ramify in the lamina may increase vascular redundancy in the leaf just as the basipetal subsidiary bundles presumably do in the stem.

One might also ask whether the vasculature performs a

regulatory role in leaf development. The ramifying but orderly system of leaf veins has led past workers to suggest such a role (cf., Foster, 1952). The usual response has ranged from inconclusive to negative. However, past consideration of this question has been confined to the lamina. The fact that the entire vascular system develops acropetally and continuously from the stem through the node and lamina to the finest of the minor veins was not previously considered. Although this question is beyond the scope of the present paper, it is pertinent to the extent that subsidiary bundles may be involved not only in lamina vascularization but also in the determination of lamina development.

The questions I have raised are not entirely academic. They may have less pertinence to how plant vascular systems develop than to how they function. Although the subsidiary bundle system may be a normal consequence of leaf ontogeny, its formation represents a degree of autonomy for the leaf in regulating it vasculature. This autonomy is attested to by the extent to which nodal vasculature is altered by the environment. Thus, what begins as a relatively simple vascular system during the earliest stages of primary development, becomes a highly complex but efficient functional system between stem and leaf at maturity.

REFERENCES

Alexandrov, W. G. and Alexandrova, O. G. (1929). Über die Struktur verschiedener Abschnitte ein und desselben Bündels und den Bau von Bündeln verschiedener Internodien des Sonnenblumenstengels. Planta 8, 465-486.

Ashworth, R. P. (1963). Investigations into the midvein anatomy and ontogeny of certain species of the genus

Ilex. J. Elisha Mitchell Sci. Soc. 73, 126-138.

Bailey, I. W. (1956). Nodal anatomy in retrospect. J. Arnold Arbor. 37, 269-287.

Bailey, I. W. and Nast, C. G. (1944). The comparative morphology of the Winteraceae. IV. Anatomy of the node and vascularization of the leaf. J. Arnold Arbor. 25, 215-221.

Bailey, I. W. and Swamy, B. G. L. (1948). The morphology and relationships of Austrobaileya. J. Arnold Arbor. 30, 211-226.

Balfour, E. E. and Philipson, W. R. (1962). The development of the primary vascular system of certain dicotyledons. Phytomorphology 12, 110-143.

Benzing, D. H. (1967a). Developmental patterns in stem primary xylem of woody Ranales. I. Species with unilacunar nodes. Amer. J. Bot. 54, 813-820.

_____. (1967b). II. Species with trilacunar and multilacunar nodes. Amer. J. Bot. 54, 813-820.

Canright, J. E. (1955). The comparative morphology and relationships of the Magnoliaceae. IV. Wood and nodal anatomy. J. Arnold Arbor. 36, 119-140.

Conde, L. F. and Stone, D. E. (1970). Seedling morphology in the Juglandaceae, the cotyledonary node. J. Arnold Arbor. 51, 463-477.

Croizat, L. (1940). A comment on current notions conerning the leaf and budscale of the Angiosperms. Lingnan Sci. J. 19, 49-66.

Deinega, V. (1898). Beiträge zur Kenntniss der Entwickelungsgeschichtes des Blattes und der Anlage der Gefässbündel. Flora 85, 439-498.

Denne, M. P. (1966). Leaf development in Trifolium repens. Bot. Gaz. 127, 202-210.

Devadas, C. and Beck, C. B. (1971). Development and

morphology of stelar components in the stems of some members of the Leguminosae and Rosaceae. Amer. J. Bot. 58, 432-446.

Dickison, W. C. (1975). The bases of angiosperm phylogeny: vegetative anatomy. Ann. Missouri Bot. Gard. 62, 590-620.

Dormer, K. J. (1944). Some examples of correlation between stipules and lateral leaf traces. New Phytol. 43, 151-153.

_____. (1945). An investigation of the taxonomic value of shoot structure in angiosperms with especial reference to Leguminosae. Ann. Bot. 9, 141-153.

Esau, K. (1945). Vascularization of the vegetative shoots of Helianthus and Sambucus. Amer. J. Bot. 32, 18-29.

_____. (1954). Primary vascular differentiation in plants. Biol. Rev. 29, 46-86.

_____. (1965). Vascular differentiation in plants. Holt, Rinehart & Winston, New York.

Foster, A. S. (1952). Foliar venation in angiosperms from an ontogenetic standpoint. Amer. J. Bot. 39, 752-766.

Foster, A. S. and Arnott, H. J. (1960). Morphology and dichotomous vasculature of the leaf of Kingdonia uniflora. Amer. J. Bot. 47, 684-698.

Fournioux, J. C. and Bessis, R. (1977). Principales etapès de l'histogenése vasculaire dans les nervures de la feuille de vigne (Vitis vinifera L.). Rev. gèn. Bot. 84, 377-395.

Franck, D. H. (1979). Development of vein pattern in leaves of Ostrya virginiana (Betulaceae). Bot. Gaz. 140, 77-83.

Girolami, G. (1953). Relation between phyllotaxis and primary vascular organization in Linum. Amer. J. Bot. 40, 618-625.

Goffinet, M. C. and Larson, P. R. (1981). Structural changes in Populus deltoides terminal buds and in the vascular transition zone of the stems during dormancy induction. Amer. J. Bot. 68, 118-129.

_____ and _____. (1982). Xylary union between the new shoot

and old stem during terminal bud break in *Populus deltoides*. *Amer. J. Bot.* 69, 432-446.

Hara, N. (1962). Histogenesis of foliar venation of *Daphne pseudo-mezereum* A. Gray. *Bot. Mag., Tokyo* 75, 107-113.

Herbst, D. (1972). Ontogeny of foliar venation in *Euphorbia forbesii*. *Amer. J. Bot.* 59, 843-850.

Hickey, L. J. and Wolfe, J. A. (1975). The bases of angiosperm phylogeny: vegetative morphology. *Ann. Missouri Bot. Gard.* 62, 538-589.

Hilger, H. H. (1978). Der multilakunäre Knoten einiger *Melianthus*- und *Greyia*-Arten im Vergleich mit anderen Knotentypen. *Flora* 167, 165-176.

Höster, H. R. (1964). Weitere Untersuchungen zur Protoxylem-Differenzierung. *Zeit. Bot.* 52, 29-45.

Höster, H. R. and Zimmermann, W. (1961). Die Entwicklung des Leitbündelsystems im Keimblatt von *Pulsatilla vulgaris* mit besonderer Berücksichtigung der Protoxylem-Differenzierung. *Planta* 56, 71-96.

Howard, R. A. (1970). Some observations on the nodes of woody plants with special reference to the problem of the 'split-lateral' versus the 'common gap'. *Bot. J. Linn. Soc.* (Suppl.) 63, 195-214.

_____. (1974). The stem-node-leaf continuum of the Dicotyledoneae. *J. Arnold Arbor.* 55, 125-181.

Isebrands, J. G. and Larson, P. R. (1977a). Organization and ontogeny of the vascular system in the petiole of eastern cottonwood. *Amer. J. Bot.* 64, 65-77.

_____ and _____. (1977b). Vascular anatomy of the nodal region in eastern cottonwood. *Amer. J. Bot.* 64, 1066-1077.

Jensen, L. C. W. (1968). Primary stem vascular patterns in three subfamilies of the Crassulaceae. *Amer. J. Bot.* 55, 553-563.

Kaplan, D. R. (1980). Heteroblastic leaf development in Acacia. Morphological and morphogenetic implications. Cellule 73, 135-203.

Kato, N. (1966). On the variation of nodal types in woody plants. (1). J. Jap. Bot. 41, 101-106.

_____. (1967). On the variation of nodal types in woody plants. (2). J. Jap. Bot. 42, 161-168.

Larson, P. R. (1975). Development and organization of the primary vascular system in Populus deltoides according to phyllotaxy. Amer. J. Bot. 62, 1084-1099.

_____. (1976a). Procambium vs. cambium and protoxylem vs. metaxylem in Populus deltoides seedlings. Amer. J. Bot. 63, 1332-1348.

_____. (1976b). Development and organization of the secondary vessel system in Populus grandidentata. Amer. J. Bot. 63, 369-381.

_____. (1979). Establishment of the vascular system in seedlings of Populus deltoides Bartr. Amer. J. Bot. 66, 452-462.

_____. (1980a). Control of vascularization by developing leaves. In "Control of Shoot Growth in Trees" (Ed. C. H. A. Little), pp. 157-172. Proc. IUFRO Workshop, Fredericton, N.B.

_____. (1980b). Interrelations between phyllotaxis, leaf development and the primary-secondary vascular transition in Populus deltoides. Ann. Bot. 46, 757-769.

_____. (1982a). The concept of cambium. In "New Perspectives in Wood Anatomy" (Ed. P. Baas), pp. 85-121. Nijhoff/Junk, The Hague.

_____. (1982b). Primary vascularization and the siting of primordia. In "Growth and Functioning of Leaves" (Ed. F. L.

Milthorpe and J. E. Dale), pp. 25-51. Cambridge Univ. Press, Cambridge.

Larson, P. R. and Isebrands, J. G. (1978). Functional significance of the nodal constricted zone in Populus deltoides. Can. J. Bot. 56, 801-804.

Larson, P. R. and Pizzolato, T. D. (1977). Axillary bud development in Populus deltoides. I. Origin and early development. Amer. J. Bot. 64, 835-848.

Larson, P. R. and Richards, J. H. (1981). Lateral branch vascularization: its circularity and its relation to anisophylly. Can. J. Bot. 59, 2577-2591.

Lehmberg, K. (1924). Zur Kenntnis des Baues und der Entwicklung der wasserleitenden Bahnen bei der Sonnenblume (Helianthus annuus). Beih. Bot. Centr. 40, 183-236.

Lersten, N. R. (1962). The pattern of differentiation of procambium and protoxylem in the leaflets of Trifolium. Amer. J. Bot. 49, 661.

_____. (1965). Histogenesis of venation in Trifolium wormskioldii (Leguminosae). Amer. J. Bot. 52, 767-774.

Lieu, S. M. and Sattler, R. (1976). Leaf development in Begonia hispida var. cucullifera with special reference to vascular organization. Can. J. Bot. 54, 2108-2121.

Majumdar, G. P. (1942). The organization of the shoot in Heracleum in the light of development. Ann. Bot. 6, 49-81.

Marsden, M. and Bailey, I. W. (1955). A fourth type of nodal anatomy in dicotyledons, illustrated by Clerodendron trichotomum Thunb. J. Arnold Arbor. 36, 1-51.

Meicenheimer, R. D. and Larson P. R. (1983). Empirical methods for xylogenesis in Populus deltoides. Ann. Bot. 51, (In press).

Meinhardt, H. (1976). Morphogenesis of lines and nets. Differentiation 6, 117-123.

Millington, W. F. and Gunckel, J. E. (1950). Structure and development of the vegetative shoot tip of Liriodendron tulipifera. Amer. J. Bot. 37, 326-335.

Mitra, G. C. and Majumdar, G. P. (1952). The leaf-base and the internode--their true morphology. The Paleobotanist 1, 351-367.

Money, L. L., Bailey, I. W. and Swamy, B. G. L. (1950). The morphology and relationships of the Monimiaceae. J. Arnold Arbor. 31, 372-404.

Neubauer, H. F. (1972). Über den Bau der Blattstiele von Pelargonien. Österr. Bot. Zeit. 120, 391-412.

_____. (1978). Über Knotenbau und Vaskularisation von Blattgrund und Blattstiel bei einigen Cornaceae und einigen ihnen nahestehenden Arten, sowie über Knotenbau im allgemeinen. Bot. Jahrb. 99, 410-424.

_____. (1979a). On nodal anatomy and petiolar vascularization of some Valerianaceae and Dipsacaceae. Phytomorphology 28, 431-436.

_____. (1979b). Knotenbau, Blattgrund- und Achselknospenvaskularisation bei Begonien. Flora 168, 329-343.

_____. (1981). Der Knotenbau einiger Rubiaceae. Plant Syst. Evol. 139, 103-111.

Philipson, W. R. and Philipson, M. N. (1968). Diverse nodal types in Rhododendron. J. Arnold Arbor. 49, 193-224.

Post, D. M. (1958). Studies in Gentianaceae. I. Nodal anatomy of Frasera and Swertia perennis. Bot. Gaz. 120, 1-14.

Pray, T. R. (1955). Foliar venation of angiosperms. II. Histogenesis of the venation of Liriodendron. Amer. J. Bot. 42, 18-27.

_____. (1963). Origin of vein endings in angiosperm leaves. Phytomorphology 13, 60-81.

Pulawska, Z. (1965). Correlations in the development of the leaves and leaf tissues in the shoot of Actinidia arguta Planch. Acta Soc. Bot. Polon. 34, 697-712.

Ramji, M. V. (1967). Morphology and ontogeny of the foliar venation of Calophyllum inophyllum L. Austral. J. Bot. 15, 437-443.

Reynolds, M. E. (1942). Development of the node in Ricinus communis. Bot. Gaz. 104, 167-170.

Richards, J. H. and Larson, P. R. (1981). Morphology and development of Populus deltoides branches in different environments. Bot. Gaz. 142, 382-393.

Schnettker, M. (1977). Die Verteilung der leitenden Strukturen in der Achse von Clematis vitalba (Ranunculaceae). Plant Syst. Evol. 127, 87-102.

Sehgal, L. and Paliwal, G. S. (1974). Studies on the leaf anatomy of Euphorbia. III. The node. Bot. J. Linn. Soc. 69, 37-43.

Shimaji, K. (1957). On the behaviour of the vascular strands at the cotyledonary and the first node of the genus Quercus. Bull. Tokyo Univ. For. No. 53, 125-133.

Shushan, S. and Johnson, M. A. (1955). The shoot apex and leaf of Dianthus caryophyllus L. Bull. Torrey Bot. Club 82, 262-283.

Sinnott, E. W. (1914). Investigations on the phylogeny of the angiosperms. I. The anatomy of the node as an aid in the classification of angiospersm. Amer. J. Bot. 1, 303-322.

Sinnott, E. W. and Bailey, I. W. (1914). Investigations on the phylogeny of angiosperms. 3. Nodal anatomy and the morphology of stipules. Amer. J. Bot. 1, 441-453.

Smith, E. P. (1937). Nodal anatomy of some common trees. Trans. & Proc. Edinburgh Bot. Soc. 32, 260-277.

Smithson, E. (1954). Development of winged cork in Ulmus x

hollandica Mill. Proc. Leeds Phil. Soc. 6, 211-220.

Stone, D. E. (1970). Evolution of cotyledonary and nodal vasculature in the Juglandaceae. Amer. J. Bot. 57, 1219-1225.

Sugiyama, M. (1972). A vascular system of "node to leaf" in Magnolia virginiana L. J. Jap. Bot. 47, 313-320.

_____. (1974). A vascular system of "node to leaf" in Michelia champaca L. J. Jap. Bot. 49, 250-256.

_____. (1976a). Comparative studies of the vascular system of node-leaf continuum in woody Ranales. I. Diversity in successive nodes of first-year plants of Magnolia virginiana L. Bot. Mag., Tokyo 89, 33-43.

_____, (1976b). II. Node-leaf vascular system of Eupomatia laurina R. Br. J. Jap. Bot. 51, 169-174.

Swamy, B. G. L. (1949). Further contributions to the morphology of the Degeneriaceae. J. Arnold Arbor. 30, 10-38.

Thoday, D. (1922). On the organization of growth and differentiation in the stem of the sunflower. Ann. Bot. 36, 489-510.

Thomas, E. N. (1907). A theory of the double leaf-trace founded on seedling structure. New Phytol. 6, 77-91.

Tucker, S. O. (1962). Ontogeny and phyllotaxis of the terminal vegetative shoots of Michelia fuscata. Amer. J. Bot. 49, 722-737.

Vogelmann, T. C., Larson, P. R. and Dickson, R. E. (1982). Translocation pathways in the petioles and stem between source and sink leaves of Populus deltoides Bartr. ex Marsh. Planta 156, 345-358.

Watari, S. (1934). Anatomical studies on some Leguminous leaves with special reference to the vascular system in petioles and rachises. J. Fac. Sci. Tokyo Univ. Sect. III 4, 225-365.

_____. (1936). Anatomical studies on the vascular system in the petioles of some species of Acer, with notes on the external morphological features. J. Jap. Fac. Sci. Tokyo Univ. Sect. III 5, 1-73.

_____. (1939). Anatomical studies on the leaves of some Saxifragaceous plants, with special reference to the vascular system. J. Fac. Sci. Tokyo Univ. Sect. III 5, 195-316.

White, D. J. B. (1955). The architecture of the stem apex and the origin and development of the axillary buds in seedlings of Acer pseudoplatanus L. Ann. Bot. 19, 437-449.

COMPARATIVE STRUCTURE OF PHLOEM

Ray F. Evert

Department of Botany
University of Wisconsin
Madison, Wisconsin

During maturation, the protoplasts of sieve elements in all groups of vascular plants undergo a selective autophagy and the plasmodesmata in their common walls are enlarged to sieve-area pores, resulting in a greater degree of continuity between contiguous protoplasts. At maturity, the sieve elements lack nuclei and vacuoles, and the surviving protoplasmic components are entirely parietal in distribution. The sieve elements in all taxa are associated with parenchymatous elements, and together these two cell types constitute interdependent physiological units. The phloem of angiosperms, with its sieve-tube members and their ontogenetically related companion cells, is the most highly specialized among vascular plants. The open sieve-plate pores in their end walls provide a high degree of protoplast continuity between contiguous sieve-tube members. The sieve-tube members of dicotyledons and some monocotyledons contain a proteinaceous substance called P-protein, the function of which remains to be determined. Moreover, use of this protein

ISBN 0-12-746620-7

category is in need of further scrutiny. The sieve cells in gymnosperm phloem typically are associated with albuminous cells, the counterparts of the angiospermous companion cell. The mature gymnospermous sieve cell contains a vast and complex network of tubular endoplasmic reticulum (ER), which forms massive aggregates opposite the sieve areas and is continuous from sieve cell to sieve cell via the sieve-area pores. With the exception of the sieve elements of lycopods, the sieve-element protoplasts of vascular cryptogams are characterized by the presence of refractive spherules, the function and exact chemical composition of which remain to be determined. The lycopod sieve elements are characterized by unoccluded sieve-area pores. By contrast, the sieve-area pores in *Psilotum* and eu- and protoleptosporangiate ferns are occluded with membranes, which are structurally distinct from the parietal ER of the cell lumen. The sieve-area pores in root and mid-internodal regions of aerial stems of *Equisetum* are traversed by large numbers of membranes, which are continuous with the parietal ER of the cell lumen. In leptosporangiate ferns, the sieve-area pores contain variable numbers of ER membranes, which also are continuous with those of the parietal network. Although specialized parenchyma cells analogous to companion cells and albuminous cells are lacking in vascular cryptogams, the sieve elements in this diverse group of vascular plants have numerous cytoplasmic connections with their neighboring parenchyma cells. Information on translocation velocities in selected vascular cryptogams would contribute greatly to an understanding of sieve-element structure in relation to function.

During the evolutionary development of land plants from their aquatic ancestors, complex tissues and mechanisms gradually evolved that made possible the development of the large plants with which we are most familiar. Among these tissues and mechanisms are those associated with the long-distance movement of water and dissolved solutes from one part of the plant body to another. Concomitant with the development of root, stem, leaves and reproductive structures by the evolving land plants was the development of tissues specialized for conducting water and organic food materials between the spatially separated plant parts. Known today as xylem and phloem, respectively, these tissues are integrated both structurally and functionally and together form a continuous vascular system throughout virtually all parts of the plant.

Both the xylem and phloem may be classified, ontogenetically, into primary and secondary tissues. The primary vascular tissues are derived from the procambium and the secondary from the vascular cambium. Like the meristem from which they are derived, the secondary vascular tissues are organized into axial and radial (ray) systems. The rays are continuous through the vascular cambium from the secondary xylem to the secondary phloem.

The basic components of the phloem are the sieve elements, several kinds of parenchyma cells, fibers and sclereids. The principal conducting cell of the phloem is the sieve element, which received its name from the presence in its walls of sieve areas — aggregates of sieve-like perforations. Although the phloem varies in structure in different plants and in different parts of the same plant, the unifying feature of the various types of phloem is the presence of the sieve element.

Esau (1969) has provided a comprehensive review of the literature on structure and development of phloem from its beginnings in the late 17th century to the late 1960s. More recent reviews of literature on various aspects of phloem structure and development can be found in the proceedings of two international conferences on phloem transport (Aronoff et al., 1975; Eschrich and Lorenzen, 1980) and in a volume of the "Encyclopedia of Plant Physiology" devoted to the same topic (Zimmermann and Milburn, 1975), as well as in two contributions to the *Annual Review of Plant Physiology* (Evert, 1977; Cronshaw, 1981). The main aim of the present discussion is not, therefore, to provide an exhaustive literature review but rather an overview of phloem structure in the major groups of vascular plants, with emphasis on the sieve element. Only the more recent literature will be cited here.

I. ANGIOSPERM PHLOEM

The phloem of angiosperms is regarded as the most highly evolved among vascular plants. With few exceptions (*Austrobaileya scandens*, Srivastava, 1970), its sieve elements are sieve-tube members. By definition, sieve-tube members are sieve elements in which some of the sieve areas bear larger pores than others and are localized on the walls to form sieve plates (Esau, 1969). Typically, the sieve plates occur on the end walls, and the sieve-tube members are arranged end-on-end to form sieve tubes (FIGS. 1 and 2). In addition to the presence of sieve-tube members, a distinctive feature of angiosperm phloem is the presence of companion cells, specialized parenchyma cells closely related to the sieve-tube members both ontogenetically and functionally.

A. *The Sieve-Tube Member*

Requisite to an understanding of mature sieve-element structure is a thorough knowledge of sieve-element development. At the early stages of development, the young sieve-tube member contains all of the components characteristic of young plant cells. As the sieve-tube member differentiates, however, it undergoes profound change. During the final stages of maturation, a selective autophagy results in disorganization and/or disappearance of many of its cellular components, including nucleus, ribosomes, dictyosomes, microtubules, and tonoplast, or vacuolar membrane. At maturity the sieve-tube member retains a plasmalemma, mitochondria, plastids, and endoplasmic reticulum (ER), all of which occupy a parietal position within the cell (FIGS. 3 and 4). In addition, the mature sieve-tube members of most dicotyledons and some monocotyledons contain some form of proteinaceous inclusion, the P-protein. During development of the sieve-tube member, changes also occur to the cell walls, most notable being the differentiation of the sieve plates.

1. *The Protoplast.* We shall consider first the structural changes that occur to the protoplasmic components of the sieve-tube member during the course of differentiation.

a. *Nucleus.* Degeneration of the nucleus during the final stages of maturation has long been recognized as one of the peculiar features of the sieve-element protoplast and has been thoroughly documented with developmental studies at the ultrastructural level (Esau, 1969). In sieve-tube members, the nucleus may disappear completely (Walsh and Popovich, 1977; Hoefert, 1980; Singh, 1980; Danilova and Telepova,

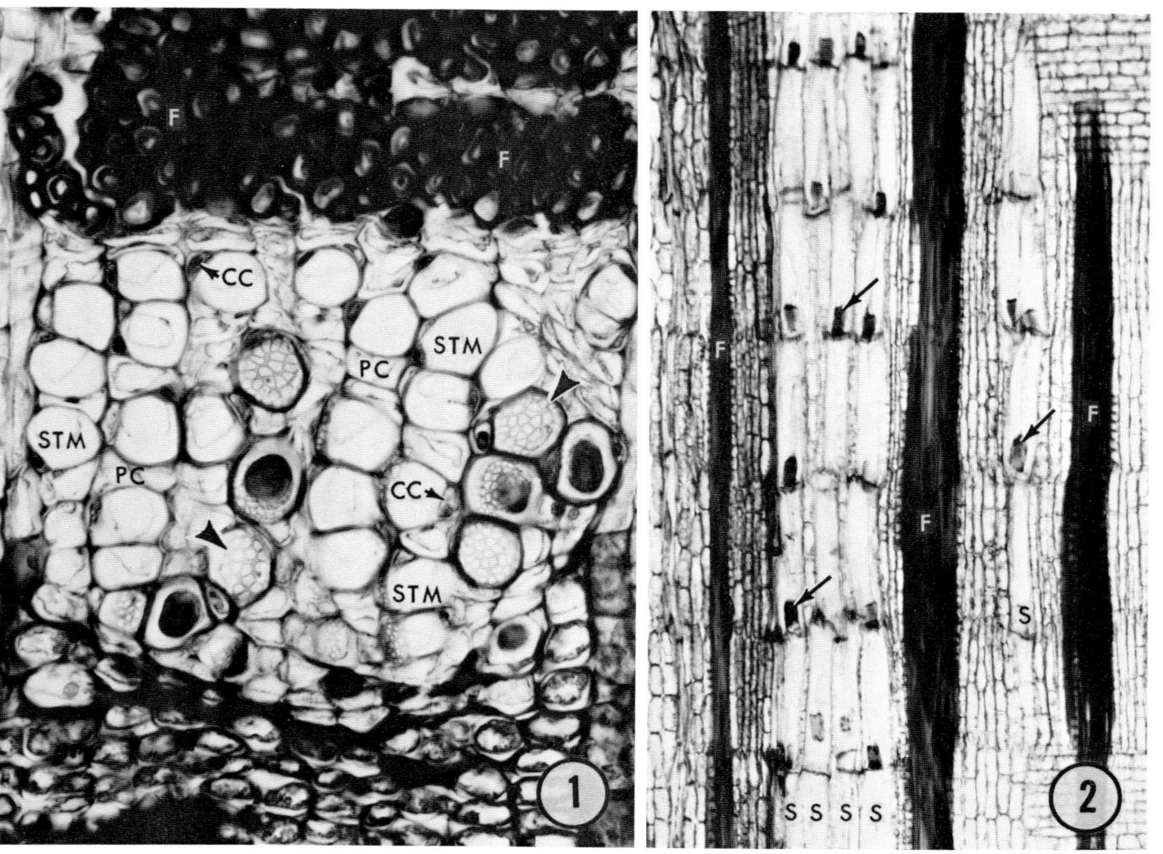

F
F
CC
STM
PC
STM
PC
CC
STM
1
F
F
F
S
S S S S
2

1981; Thorsch and Esau, 1981b), or remnants of nuclear envelope and/or chromatin may persist indefinitely (Eleftheriou and Tsekos, 1982a,c), depending upon the species (Evert, 1977).

The behavior of the nucleus during degeneration is variable and depends in part upon the species and partly upon the location of the sieve-tube member in the plant (Esau, 1969). In some sieve-tube members nuclear degeneration involves a gradual loss of stainable contents (chromatin and nucleoli) and eventual rupture of the nuclear envelope (Esau, 1978b; Hoefert, 1980). This kind of nuclear degeneration, called chromatolysis, contrasts with the pycnotic type found in other sieve-tube members and in which the chromatin forms a very dense mass prior to rupture of the nuclear envelope (Danilova and Telepova, 1981; Thorsch and Esau, 1981c).

For several years it was believed that the nucleoli in a number of dicotyledonous sieve-tube members were released from

FIGS. 1 and 2. Photomicrographs of secondary phloem of black locust (Robinia pseudo-acacia). *Fig. 1. Transverse section showing sieve-tube members (STM), companion cells (CC), and phloem parenchyma cells (PC). Sieve plates of some sieve-tube members can be seen in face view (arrowheads). The dark, thick-walled cells, above, are fibers (F). X 370. Fig. 2. Longitudinal (radial) section showing several sieve tubes (S) which consist of sieve-tube members arranged end-on-end. The elongate bodies (unlabeled arrows) found in each of the sieve tubes (one per sieve-tube member) are regarded as a form of P-protein and appear as crystalline structures with the electron microscope. In the secondary phloem of black locust, groups of thick-walled fibers (F) alternate with groups of relatively thin-walled cells, including sieve-tube members, companion cells, and phloem parenchyma cells. X 125.*

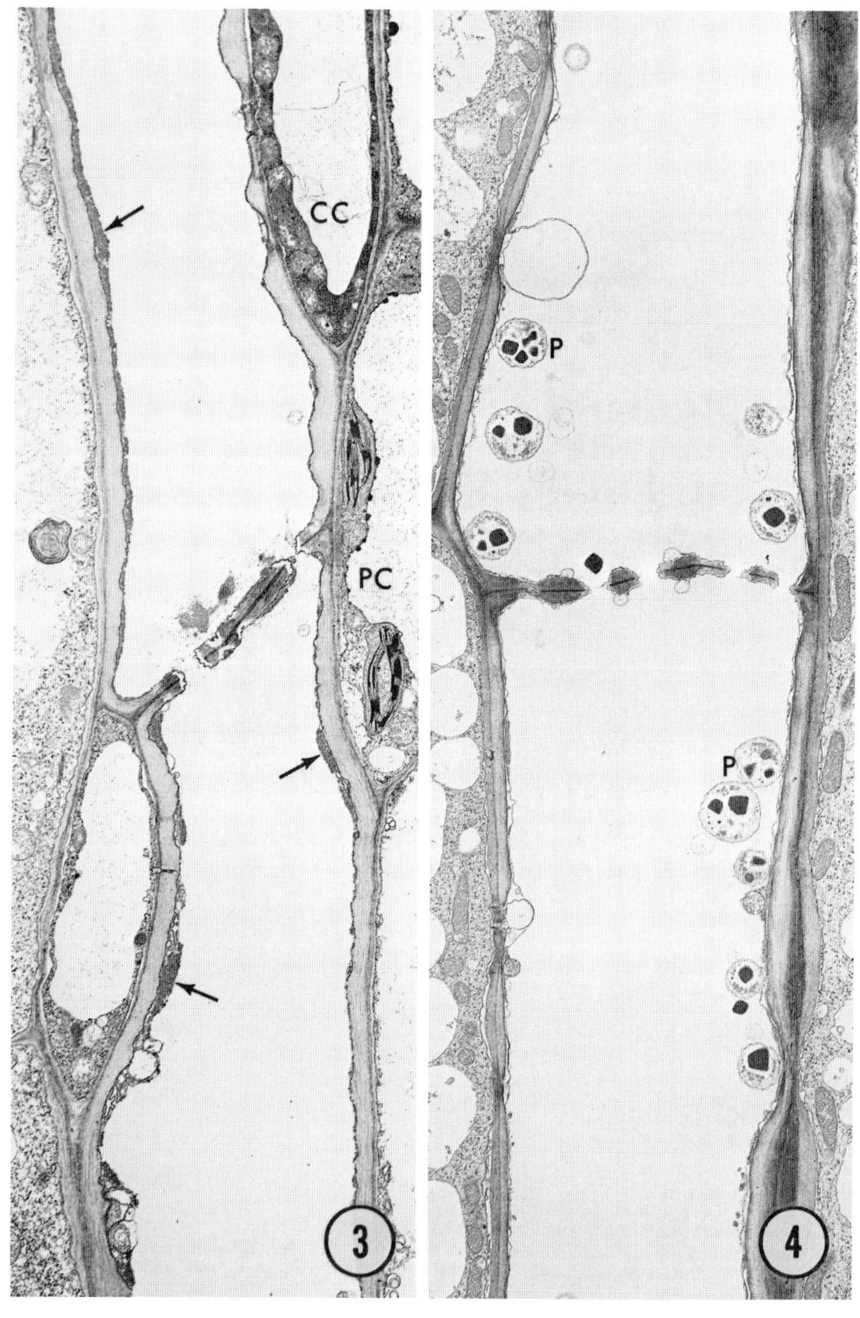
CC
PC
3
P
P
4

the degenerating nuclei and remained intact within the mature cell (Esau, 1969). It is now known, however, that the structures once regarded as "extruded nucleoli" are not nuclear in origin but are proteinaceous cytoplasmic inclusions similar to P-protein (FIGS. 14-16; Deshpande and Evert, 1970; Oberhäuser and Kollmann, 1977; Esau, 1978a; Nehls et al., 1978).

The nuclei of some sieve-tube members (Boraginaceae) contain crystalline, proteinaceous inclusions which resemble the crystalline form of P-protein that arises in the cytoplasm of fabaceous sieve-tube members (Esau and Thorsch, 1982; Esau and Magyarosy, 1979a,b). During nuclear degeneration, these crystalloids are released into the cell lumen where they may remain intact or become fragmented.

b. *Endoplasmic reticulum.* In young sieve-tube members the ER is rough-surfaced, cisternal, and fairly evenly dispersed throughout the cytoplasm. During nuclear degeneration, but while the nucleus is still intact, the ER cisternae increase in number and begin to form stacks, some of which may become applied to the nuclear envelope (Melaragno and Walsh, 1976). [In cotton, complexes of microtubules and ER become associated with the nuclear envelope (Thorsch and Esau, 1981b)]. During

FIGS. 3 and 4. Longitudinal sections of portions of mature sieve-tube members, showing parietal distribution of cytoplasmic components and sieve plates with unoccluded pores. FIG. 3. Cucurbita maxima. *Unlabeled arrows point to P-protein. CC, companion cell; PC, parenchyma cell. X 4,180. FIG. 4.* Zea mays. *Typical of monocotyledonous sieve-tube members, those of maize contain P-type plastids (P). X 8,500. (Fig. 3 from Evert, R.F., Eschrich, W., Eichhorn, S.E. 1973. Planta 109, 193-210. Fig. 4 courtesy of M.A. Walsh.)*

the stacking process the ER cisternae lose their ribosomes and become smooth, although some electron-dense material remains between contiguous cisternae.

The stacks of cisternae are most abundant at about the time the nucleus disappears and then occur near the cell wall, with the cisternae either parallel or perpendicular to the wall. As the sieve-tube members become more mature the ER may undergo further modification into convoluted, lattice-like and tubular forms (FIGS. 5-7; Oparka and Johnson, 1978; Esau and Hoefert, 1980; Thorsch and Esau, 1981a). In most fully mature sieve-tube members the ER is represented largely by a complex network — a parietal, anastomosing system — which lies next to the plasmalemma (FIG. 13; Esau, 1978b; Hoefert, 1980; Parthasarathy, 1980; Thorsch and Esau, 1981a). This is the most common form of ER found in mature sieve-tube members.

Wooding (1967) has suggested that the stacking of ER may represent a sequestering of the membrane system in an inactive form. Since then the localization of acid phosphatase and ATPase activity (Evert, 1977; Browning et al., 1980; Cronshaw, 1980) on both the anastomosing system and stacked ER has led to the suggestion that the ER may be an important source of enzymes involved in autophagic phenomena (Oparka et al., 1981) and possibly a specific cytoplasmic differentiation related to the conducting function of the cell (Esau and Charvat, 1975). With regard to the latter suggestion, it is pertinent to point out that the parietal ER of the mature sieve-tube member is in direct continuity with the ER of the companion cell via the numerous pore-plasmodesma connections in the sieve element-companion cell walls. Substances entering the sieve-tube member via such connections may be rapidly distributed along all wall surfaces within the anastomosing ER, and hence be provided with a greatly increased surface area for entry into the lumen of the sieve-tube member.

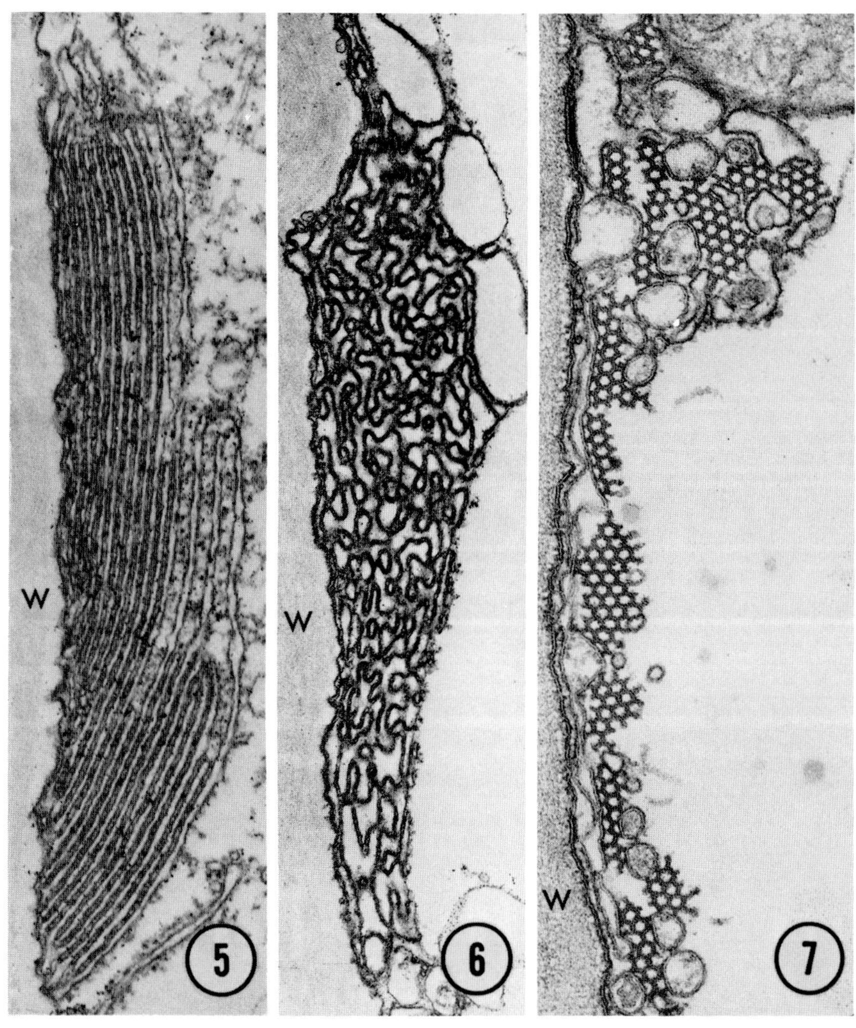

FIGS. 5-7. Sections of mature sieve-tube members showing different arrangements or forms of endoplasmic reticulum along walls. W, wall. FIG. 5. **Ulmus americana.** *Stacked cisternae. X 37,600. FIG. 6.* **Ulmus americana.** *Convoluted form. X 23,500. FIG. 7.* **Rhus glabra.** *Lattice-like form. X 78,300. (Figs. 5 and 6 from Evert, R.F., Deshpande, B.P. 1969. Protoplasma 68, 403-432. Fig. 7 courtesy of S.E. Eichhorn.)*

c. *Plastids and mitochondria.* The plastids of very young sieve-tube members are often difficult to distinguish from mitochondria, for in such cells both organelles contain a dense matrix and are often similar in size. In addition, in certain planes of section the internal membranes of the plastids appear numerous and often resemble those of mitochondria. During development of the sieve-tube member, inclusions characteristic of the plastid type appear and the plastid matrix becomes less electron opaque. In mature members, the matrix commonly is quite electron transparent and the internal membranes sparse (FIGS. 8-10).

The comparative ultrastructural morphology of the sieve element plastid is providing botanists with useful characters for seed plant systematics and phylogeny (Behnke, 1981). Two major kinds of plastid can be recognized: S-type plastids (that store starch) and P-type plastids (that elaborate protein inclusions and often store starch grains). Both S- and P-type plastids occur in dicotyledons (FIG. 10). Only P-type plastids have been recorded in monocotyledons (FIG. 4).

Unlike ordinary starch, sieve-tube starch stains brownish red rather than blue-black with iodine (I_2KI). Using sequential enzymatic digestion and electron microscopy, Palevitz and Newcomb (1970) concluded that the sieve-tube starch in *Phaseolus vulgaris* is a highly branched molecule of the amylopectin type with numerous $\alpha(1\rightarrow6)$ linkages at the branch points. The proteinaceous nature of the crystalline inclusions and of the filamentous material found in P-type plastids has been confirmed by use of proteolytic enzymes (Evert, 1977). The release of starch grains or other inclusions from sieve-element plastids has long been considered as a response to injury, and is a valuable indicator of disturbance in mature sieve-tube members (Esau, 1969).

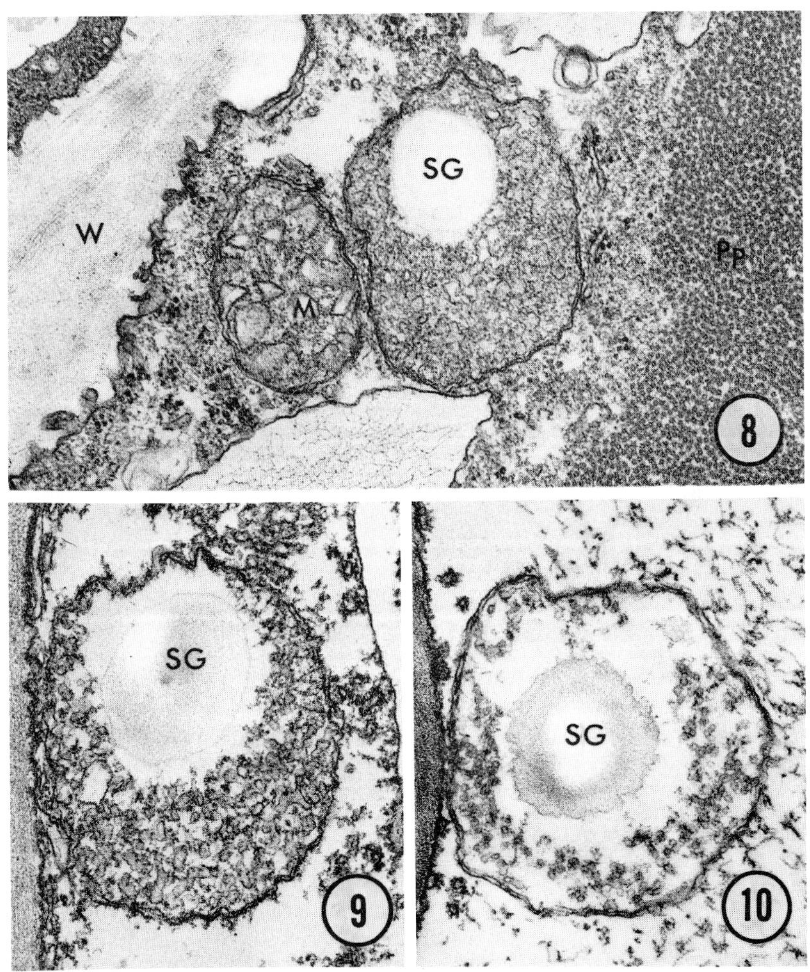

FIGS. 8-10. **Tilia americana.** *Portions of sieve elements illustrating plastids at different stages of development. FIG. 8. A plastid and mitochondrion (M) in an immature sieve element at time of formation of P-protein (Pp) bodies. X 43,500. FIG. 9. Plastid in sieve element nearing maturity. X 46,000. FIG. 10. Plastid in mature sieve element. X 39,000. SG, starch grain in plastid; W, wall. (From Evert, R.F., Deshpande, B.P. 1971. Planta 96, 97-100.)*

Among the surviving components of the sieve-tube member, the mitochondria undergo the least amount of structural change during differentiation. In well-fixed sieve-tube members the mitochondria are normal in appearance, although sometimes they have somewhat dilated cisternae. In long-lived sieve-tube members the mitochondria may degenerate with age.

d. Dictyosomes. Dictyosomes, or Golgi bodies, are abundant in young sieve-tube members, and they have clearly been implicated in cell wall formation (Evert, 1977). For example, in studies utilizing tritiated glucose, derivatives of the tritiated glucose are first found concentrated in the dictyosomes. The label then appears in smooth dictyosome vesicles, which subsequently fuse with the plasmalemma and discharge their contents into the wall.

Numerous smooth vesicles are produced throughout the period of cell wall formation (Danilova and Telepova, 1981; Eleftheriou and Tsekos, 1982c). Coated vesicles bud off the smooth ones, and both kinds of vesicle apparently transport substances utilized in the synthesis of cell wall constituents (Hoefert, 1979; Eleftheriou and Tsekos, 1982b). With the cessation of cell wall formation, the dictyosomes become reduced in number and gradually disappear altogether.

Dictyosomes and their vesicles have also been implicated in the formation of autophagic vacuoles in protophloem sieve-tube members of the barley root (Buvat and Robert, 1979).

e. Microtubules and microfibrils. In the young sieve-tube member cortical microtubules are relatively scarce and randomly oriented in relation to the long axis of the cell. With the initiation of cell wall thickening their numbers increase sharply, reach a plateau, and remain relatively constant in number throughout the period of wall thickening (Hardham and Gunning, 1979; Thorsch and Esau, 1982). During this period, the microtubules are transversely oriented to the long axis of the cell and parallel to the cellulose microfibrils of the cell wall. With the cessation of cell wall thickening, the number of microtubules gradually declines and, during the final stages of maturation, typically the microtubules disappear entirely from the cell.

Microfilaments comprising cytoplasmic fibers are common constituents of differentiating sieve-tube members (Parthasarathy and Pesacreta, 1980). They usually are found in the peripheral regions of the cell and are almost always oriented parallel to the long axis of the cell. With few exceptions they do not persist beyond the stage of nuclear degeneration.

f. Plasmalemma and tonoplast. In well-fixed sieve-tube members the plasmalemma has a distinct three-ply appearance in thin sections. In differentiating sieve-tube members, it commonly has an undulating outline, in part a reflection of dictyosome vesicle activity associated with the delivery of substances for cell wall synthesis. The plasmalemma maintains its integrity and differentially permeable properties throughout the life of the sieve-tube member.

The tonoplast, by contrast, breaks down during the final stages of maturation so that the delimitation between vacuole and cytoplasm ceases to exist. The absence of a tonoplast in mature sieve-tube members has been recorded for a wide variety of species examined with both light and electron microscopes (Esau, 1969). Occasional reports of the presence of intact tonoplasts and, hence, of vacuoles in mature sieve-tube members are difficult to evaluate (Walsh, 1980). Some such vacuoles may have been anomalies or instances in which the tonoplast was late in disappearing from the cell. In some woody dicotyledons, the tonoplast persists until very late in maturation of the sieve-tube member (Evert and Deshpande, 1969).

g. P-protein. Formerly known as slime, P-protein (phloem protein) is a characteristic component of the protoplasts of dicotyledonous sieve-tube members. With few exceptions [e.g., protophloem elements in the root of *N. tabacum* (Esau and Gill, 1972); metaphloem sieve elements of *Epifagus virginiana* (Walsh and Popovich, 1977)], P-protein has been encountered in the sieve-tube members of all dicotyledons examined thus far with the electron microscope, including those of such primitive Ranalean representatives as *Liriodendron tulipifera, Magnolia soulangeana, Degeneria vitiensis* (Friis and Dute, 1983), *Drimys granadensis* (Spanner and Moattari, 1978), and *Trochodendron aralioides* (Jørgensen et al., 1975). P-protein is not as widely distributed in the monocotyledons (Parthasarathy, 1980). It has been reported as lacking in sieve-tube members of many palms, *Lemna minor,* the native Greek grass *Aegilops comosa* (Eleftheriou and Tsekos, 1982a), and such familiar species as *Hordeum vulgare, Triticum aestivum, Oryza sativa, Saccharum officinarum,* and *Zea mays* (Evert, 1982).

As seen with the electron microscope, P-protein occurs in several morphological forms (amorphous, filamentous, tubular, and crystalline) that vary from species to species (Kollmann, 1980). In addition, conformational changes from one type of P-protein to another may take place within a differentiating sieve-tube member. Several workers have suggested that the various forms of P-protein are constructed of basically similar or possibly identical subunits which undergo a self-assembly process. The validity of this concept for P-protein interconvertibility has been questioned by Sabnis and co-workers (Sabnis and Hart, 1979; Sloan et al., 1976) who have reported that the protein subunits in sieve-tube exudate clearly differ from species to species. In view of these findings, Sabnis and Hart (1979) have suggested that the ultrastructural studies may need to be re-evaluated and extended in an attempt to correlate protein complements with specific ultrastructural arrangements.

With the light microscope, P-protein is first discernible in the cytoplasm of young sieve-tube members as discrete bodies, of which there may be one or more per cell (Esau, 1969). At the electron microscope level, these bodies often appear as aggregates of tubular components (FIG. 11). In some species, the tubular aggregates are preceded by aggregates of fibrous material, from which it has been suggested the tubules may be assembled. The origin of P-protein, however, is not known. Among the cytoplasmic components present at the time of P-protein synthesis, those most often implicated with that process are the ER, ribosomes, and spiny vesicles (Cronshaw, 1975a; Arsanto, 1982). It is pertinent to note that the implication of spiny vesicles in P-protein formation rests largely on their spatial association with P-protein in parenchyma cells associated with sieve-tube members (Esau, 1971b; Deshpande, 1974b). In sieve-tube members, such association has been demonstrated only once (in *Cucurbita maxima*, Cronshaw and Esau, 1968).

As differentiation of the sieve-tube member progresses, the P-protein bodies increase in size. Then, at about the time degenerative changes have begun and the ER is being reorganized, the P-protein bodies begin to spread out in the cytoplasm, the tubules often assuming a striated appearance in the process (FIG. 12). [It has been suggested that the striated appearance results from stretching of the "super-double helix" proposed by some investigators as comprising the

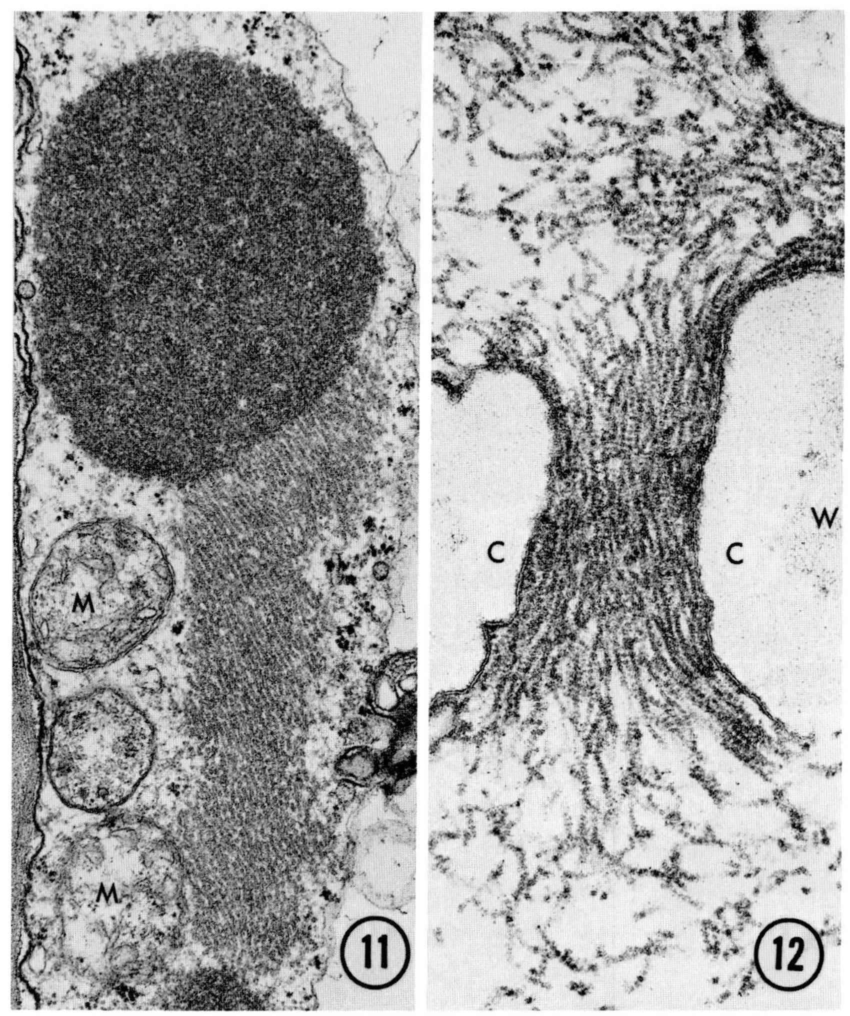

FIGS. 11 and 12. P-protein. FIG. 11. **Cucurbita maxima.** *P-protein bodies in young sieve element. The upper body is composed of fibrillar components and the lower, elongate body of tubular components. M, mitochondrion. X 38,000. FIG. 12.* **Rhus glabra.** *P-protein accumulated in sieve-plate pore shows striated appearance. The pore is lined with callose (C). W, wall. X 61,500.*

structure of P-protein tubules (Lawton and Johnson, 1976; Arsanto, 1982).] According to some accounts, the individual P-protein filaments move apart and become dispersed in the surrounding cytoplasm, after which they become evenly distributed in the lumen of the sieve-tube member following disappearance of the tonoplast. Other accounts indicate that the dispersal of the P-protein bodies results in a parietal network of very fine strands that remains in the parietal position after the tonoplast disappears (Evert, 1977, 1982).

The question as to what constitutes the normal distribution of P-protein within the lumen and sieve-plate pores of the mature sieve-tube member has been the subject of numerous investigations and long has represented the most controversial aspect of sieve-element structure (Evert, 1982). The main problem is the extreme sensitivity of the sieve-element protoplast, with its high hydrostatic pressure, to manipulation and chemical fixation, and the surging phenomenon associated with its injury and resultant pressure release at the time of sampling.

In well-preserved sieve-tube members showing little or no effects of surging (see Evert, 1977, or 1982, for criteria used in judging quality of preservation), the P-protein either occurs as a loose network of filaments occupying the entire lumen of the cell, or it is parietal in position (FIGS. 3 and 13). It is quite clear that P-protein does not normally exist in the form of transcellular strands or as a component of such strands, as hypothesized by some workers (Evert, 1977, 1982).

A parietal distribution for P-protein in mature sieve-tube members has now been reported for a number of species (Evert, 1977, 1982; Lawton and Newman, 1979). In addition, P-protein often is found in a parietal position in senescing sieve-tube members whose sieve-plate pores have been occluded with definitive callose (Evert, 1977, 1982; Lawton and Newman, 1979).

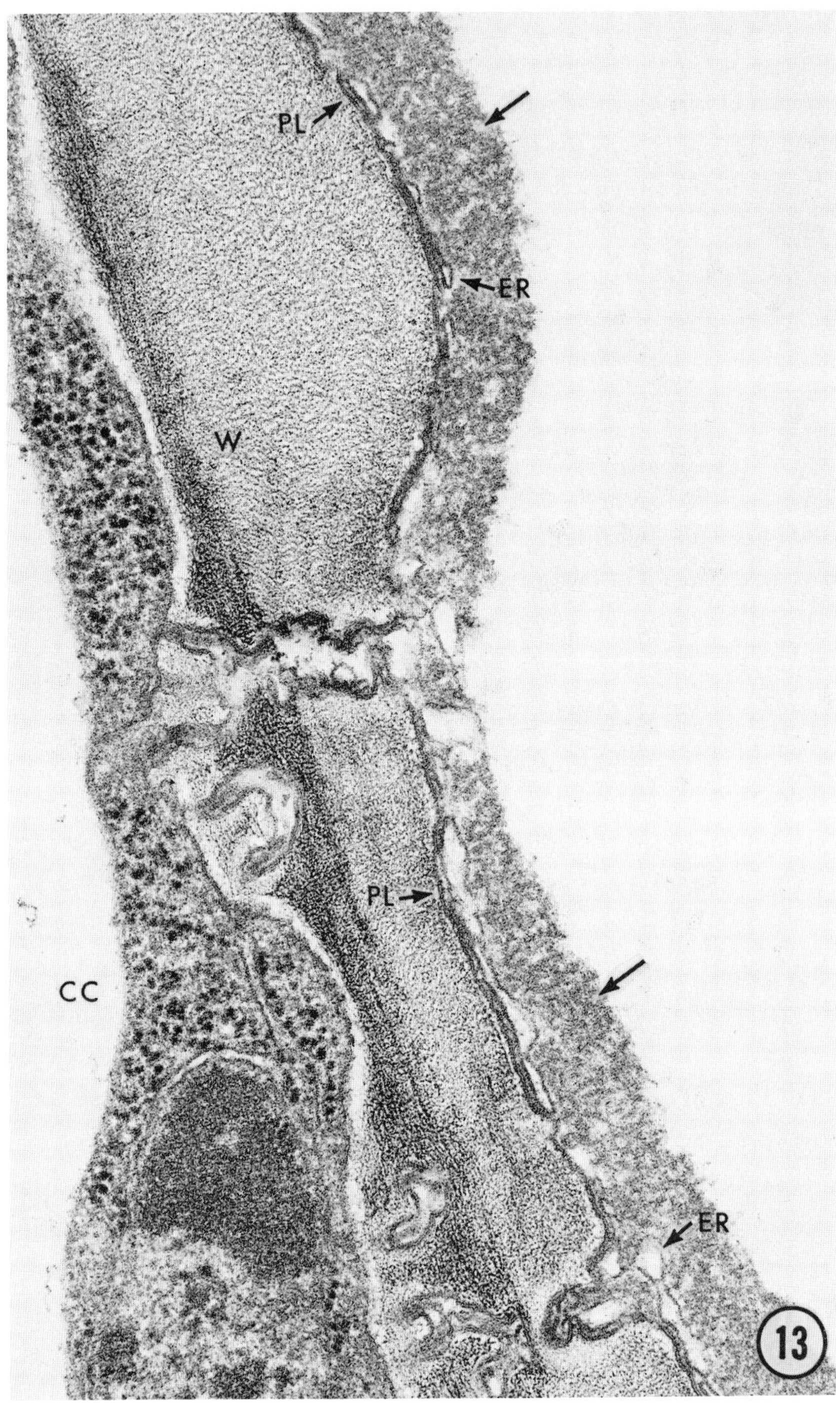
PL
ER
W
PL
CC
ER
13

Such elements might be expected to be least affected by the surging phenomenon that occurs when sieve tubes are severed, reflecting the normal distribution of P-protein within the lumen.

In some taxa, the P-protein bodies disperse only partially or not at all (Behnke, 1981). The cytoplasmic inclusions once regarded as extruded nucleoli are examples of such bodies. Most notable examples are the crystalline P-protein bodies of the Fabaceae. Long regarded as "persistent slime bodies" by light microscopists (FIG. 2; Esau, 1969), the results of early electron microscope investigations on fabaceous sieve-tube members indicated that these inclusions dispersed into a mass of fine fibrils or filaments in the mature cell. The results of more recent studies indicate, however, that these inclusions do in fact remain intact (Fisher, 1975; Esau, 1978b; Lawton, 1978a,b).

Esau and Thorsch (1982), noting the similarity of the fabaceous P-protein bodies to the crystalloids of nuclear origin in sieve-tube members of the Boraginaceae, raised an important question regarding the classification of proteinaceous inclusions in sieve elements. Inasmuch as common P-protein may arise in both cytoplasm and nucleus of the same species (*Tilia americana*, Evert and Deshpande, 1970), should

FIG. 13. Cucurbita maxima. *Transverse section showing portion of wall (W) and parietal layer of cytoplasm in mature sieve element. The parietal layer consists of plasmalemma (PL), discontinuous profiles of parietal network of endoplasmic reticulum (ER), and a meshwork of fibrillar P-protein (unlabeled arrows). Note pore-branched plasmodesma connections in the sieve element-companion cell (CC) wall. X 78,000. (From Evert, R.F., Eschrich, W., Eichhorn, S.E. 1973. Planta 109, 193-210.)*

the different origin — cytoplasmic vs nuclear — of the crystalloid inclusions be a deciding factor in whether they are classified as P-protein? Nuclear crystalloids reminiscent of fabaceous P-protein bodies also occur in the sieve elements of ferns, a group of plants considered as lacking P-protein (Evert and Eichhorn, 1974a). This raises still another question: If the nuclear crystalloids of the Boraginaceae were classified as P-protein would those of the ferns also be so classified? Crystalline inclusions also occur in parenchyma cell nuclei in a wide variety of vascular plants, including the mesophyll cell nuclei of fern leaves (Wergin et al., 1970). It is highly unlikely that anyone would agree to their being classified as P-protein. And what about the spherical cytoplasmic inclusions once considered to be extruded nucleoli but classified today as forms of crystalline P-protein (FIGS. 14-16; Behnke, 1981)? Except for their different shape and smaller size, they are similar to both the fabaceous P-protein bodies and the nuclear crystalloids of sieve elements. In view of these questions and those raised by Sabnis and co-workers, it should be obvious that use of the protein category designated as P-protein is in need of further scrutiny (Esau and Magyarosy, 1979b).

For information on the chemistry of P-protein, the reader is referred to Kollmann (1980) and McEuen et al. (1981) and literature cited therein; suffice to note here that the bulk of biochemical data so far amassed clearly indicates that P-protein is neither actin- nor tubulin-like, indicating that P-protein does not play a force-generating role in assimilate transport. Suggestions that P-protein may play such a role were prompted largely by the apparent morphological similarity of P-protein filaments to microtubules and filamentous structures associated with cytoplasmic streaming and motion in both plant and animal cells.

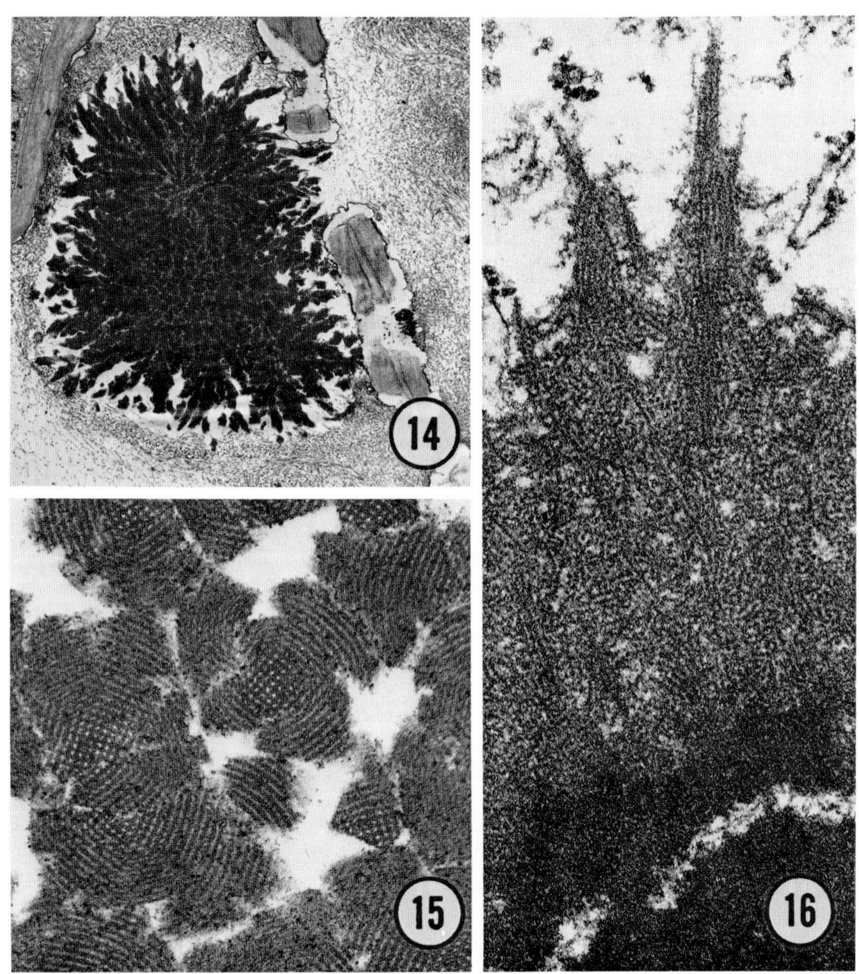

FIGS. 14-16. Crystalline P-protein. FIG. 14. **Quercus alba.** *Compound spherical body near sieve plate in mature sieve element. X 5,900. FIG. 15.* **Quercus alba.** *Detail of spherical body. X75,000. FIG. 16.* **Tilia americana.** *Portion of spherical inclusion. The peripheral region (above) is composed of rodlike components, while the more dense central region (below) shows little or no substructure. X 49,700. (Fig. 14 courtesy of S.E. Eichhorn. Figs. 15 and 16 from Deshpande, B.P., Evert, R.F. 1970. J. Ultrastruct. Res. 33, 483-494.)*

2. *The Wall.* The walls of sieve-tube members commonly are described as primary and, although occasional reports of lignified walls appear in the literature (Esau, 1969; Kuo and O'Brien, 1974), standard microchemical tests of the walls generally give positive reactions for only cellulose and pectin. Although variable in thickness, the walls of sieve-tube members commonly are thicker than those of contiguous parenchyma cells. In many species, the sieve-tube member wall consists of two morphologically distinct layers, a relatively thin outer layer and a more or less thick inner layer. In fresh sections, the distinct inner layer exhibits a glistening or pearly appearance and, hence, is described as nacreous or nacré (Esau, 1969). In some sieve-tube members the nacreous layers become so thick as almost to occlude the lumen of the cells. Both wall layers give positive reactions in tests for cellulose and pectin but, by comparison with the other wall layer, the nacreous layer is less cellulosic and pectin-poor (Esau and Cheadle, 1958; Botha and Evert, 1981).

The behavior of the nacreous layer is quite variable. In some sieve-tube members it is transitory in nature and becomes reduced in thickness as the cell ages. This is especially true of primary sieve-tube members, in which "shrinkage" of the wall has usually been attributed to dehydration (Esau, 1969). In secondary phloem, the nacreous layer may or may not be reduced in thickness as the cell ages (Esau and Cheadle, 1958).

Several workers (Danilova and Telepova, 1978, 1981; Eleftheriou and Tsekos, 1982b,c) have suggested that the "wave-like" wall thickenings of root protophloem sieve-tube members may represent stored wall material which is utilized in later stages of development when wall material is needed but the synthesizing mechanism (Golgi apparatus) is no longer present. Redistribution of the stored material, with a resultant decrease in wall thickness, would allow the protophloem sieve-tube members to keep pace with the active elongation of surrounding cells. Eleftheriou and Tsekos (1982b) noted that the pertinent wall thickenings in their *Aegilops* roots neither glistened nor were distinct from the outer wall layer and, hence, could not be termed nacreous. Danilova and Telepova (1981), on the other hand, characterized as nacreous the wall thickenings in the protophloem sieve-tube members of their *Hordeum* roots.

Utilizing mild extraction procedures for the removal of noncellulosic wall components, Deshpande (1976b) was able to demonstrate that the nacreous thickenings in sieve-tube members of *Cucurbita* petioles are polylamellate, while the outermost part of the wall is similar to that of ordinary parenchyma cells in having lamellae composed of thinly distributed microfibrils (Deshpande, 1976a). Whereas the outer wall microfibrils were readily separated by certain treatments such as pectinase extraction, the lamellae of the inner (nacreous) layer, with its densely packed microfibrils, did not separate readily. By means of extraction procedures, Catesson (1982) also found the inner parts of secondary sieve-tube walls in *Populus* and *Acer* to have a polylamellated structure. Behnke (1971) suggested that the predominant parallel arrangement of the fibrils in nacreous walls of *Annona* and *Myristica* may be one condition for their pearly luster. He also noted that the fibrils appeared to be especially tightly packed together at the boundary of the sieve-tube lumen.

Use of the term nacreous has been extended to describe a distinctive inner wall layer revealed in sieve-tube members by the electron microscope (Esau, 1969; Cronshaw, 1975b). This layer, which may even be found in sieve-tube members with relatively thin walls, shows a distinct striate pattern when sectioned obliquely. In addition, after fixation with glutaraldehyde-osmium tetroxide, it may be considerably more electron opaque than the rest of the wall. In a recent electron microscope and histochemical study of the sieve-element walls in minor veins of *Beta vulgaris* and 13 other species, Lucas and Franceschi (1982) demonstrated the presence of a thin, dense, apparently pectin-rich layer of non-microfibrillar material adjacent to the plasmalemma. Although this pectin-rich layer appears to correspond to the layer exhibiting striations — that is, the nacreous layer in the extended use of the term — Lucas and Franceschi did not compare the two. They noted, however, that the pectin-rich layer does not represent nacreous wall as defined by Esau and Cheadle (1958) because it extends into the region of the sieve plate.

3. The Sieve Plate. As mentioned previously, sieve-tube members are characterized by the presence of sieve plates which typically occur on the end walls. Some sieve plates, called simple sieve plates, bear only a single sieve area while others, called compound sieve plates, bear two or more.

In most preparations of functional phloem, the sieve-plate pores are lined with variable amounts of the wall substance callose, a carbohydrate (β-1,3 glucan) that stains blue with resorcin blue or aniline blue and yields mainly glucose when hydrolyzed (Eschrich, 1975). Once considered to be a normal constitutent of the sieve plates and lateral sieve areas of functional sieve elements, it is now assumed that most if not all of the callose associated with the pores of such elements is deposited there in response to mechanical injury or some other kind of stimulation (Esau, 1969; Eschrich, 1975). That is not to imply, however, that all callose associated with sieve plates and lateral sieve areas is wound callose. It has been thoroughly documented that callose normally accumulates at the sieve plates and lateral sieve areas of senescing sieve elements, often forming large pads over them. This definitive callose disappears sometime after the sieve element dies. Callose may also accumulate at the sieve plates and lateral sieve areas of sieve elements that function for more than one season. Typically, this dormancy callose is deposited in the fall and then removed during reactivation in early spring. Callose also appears to play a role in the development of the sieve plate and lateral sieve areas.

Sieve-plate development has been thoroughly documented in ultrastructural studies, and a fairly uniform picture of its development has emerged (Deshpande, 1974a, 1975). In the young sieve-tube member, the sieve area (or sieve areas) is penetrated by variable numbers of plasmodesmata, each of which is associated with ER cisternae. The pore sites first become distinguishable from the rest of the wall by the appearance of callose around each plasmodesma on both sides of the wall. The paired callose deposits assume the form of platelets or cones, which undergo rapid enlargement, initially exceeding the rest of the wall in their rate of thickening (FIGS. 17 and 18). Soon, however, thickening of the wall between pore sites overtakes the callose platelets and the pore sites then appear as depressions in the plate. Perforation begins at about the time of nuclear degeneration, with the removal of wall material in the region of the middle lamella surrounding the plasmodesmata. Further simultaneous removal of the callose platelets and of the wall substance sandwiched between them results in formation of the pore. The ER cisternae remain closely appressed to the plasmalemma bordering the platelets throughout pore development but are removed from the pore sites as the pores attain their full size. It is pertinent to

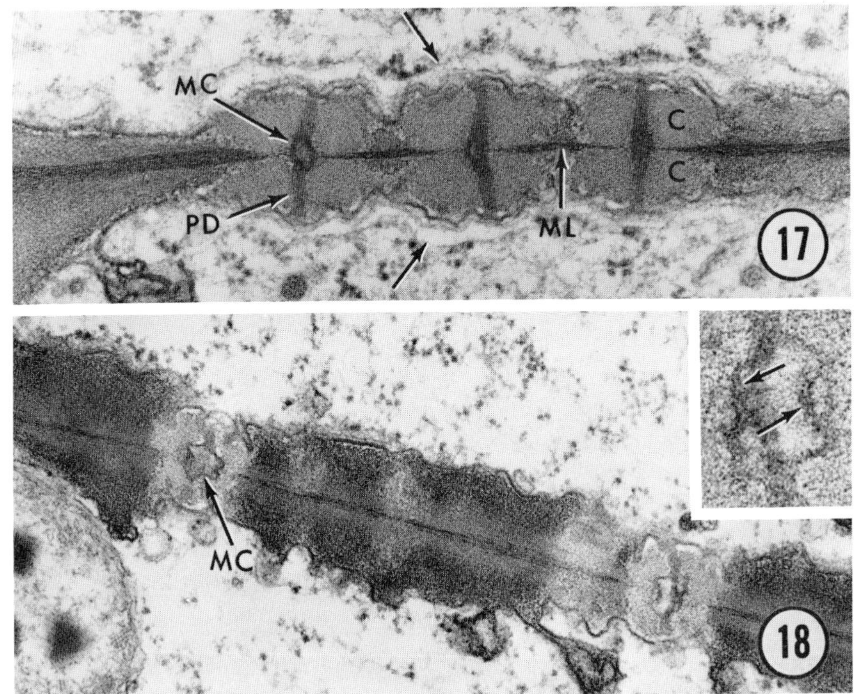

FIGS. 17 and 18. Two stages in development of the sieve plate in Zea mays. *FIG. 17. Sieve plate at early stage of pore formation. Perforation begins with formation of median cavity (MC) around plasmodesma (PD) in region of the middle lamella (ML) between callose platelets (C). Unlabeled arrows point to endoplasmic reticulum. X 50,500. FIG. 18. Sieve plate at later stage of pore formation. Median cavity has enlarged and is lined by plasmalemma (arrows in inset). Cellulosic portion of wall has undergone substantial increase in thickness. X 43,800. (From Walsh, M.A., Evert, R.F. 1975. Protoplasma 83, 365-388.)*

note that in at least one species (*Lemna minor*) callose platelets are not associated with sieve-plate development (Walsh and Melaragno, 1976).

At maturity, the plasmalemma-line sieve-plate pores in most well-preserved sieve-tube members contain little or no callose. If present at all, the callose forms a narrow cylinder outside the plasmalemma. Some elements of ER may traverse the pores closely appressed to the plasmalemma. Although fairly large quantities of ER have been recorded in the sieve-plate pores of protophloem sieve elements in palms and of some sieve-tube members in *Hordeum vulgare*, these apparently are exceptions (Evert, 1977). The distribution of P-protein within the pores of well-preserved sieve-tube members showing no evidence of surging generally reflects the distribution of the P-protein within the lumina of those cells, i.e., either in a loose filamentous network or in a parietal position within the pore (FIG. 19). Nevertheless, some workers still contend that the pores of at least a fair proportion of sieve plates are normally occluded with P-protein (Spanner, 1978a,b; see also discussion in Evert, 1982).

As mentioned previously, not all angiospermous sieve-tube members contain P-protein. Those of a great many monocotyledons lack P-protein and of some dicotyledons as well. Moreover, in some sieve-tube members the P-protein bodies fail to disperse (e.g., the extrafascicular sieve elements in *Cucurbita*, Evert et al., 1973a; and, as mentioned previously, the crystalloid P-protein bodies of the Fabaceae). With few exceptions, the sieve-plate pores in such cells are unoccluded by any cytoplasmic substances.

4. *The Lateral Sieve Areas*. The lateral sieve areas may occur on all walls between contiguous sieve-tube members. In

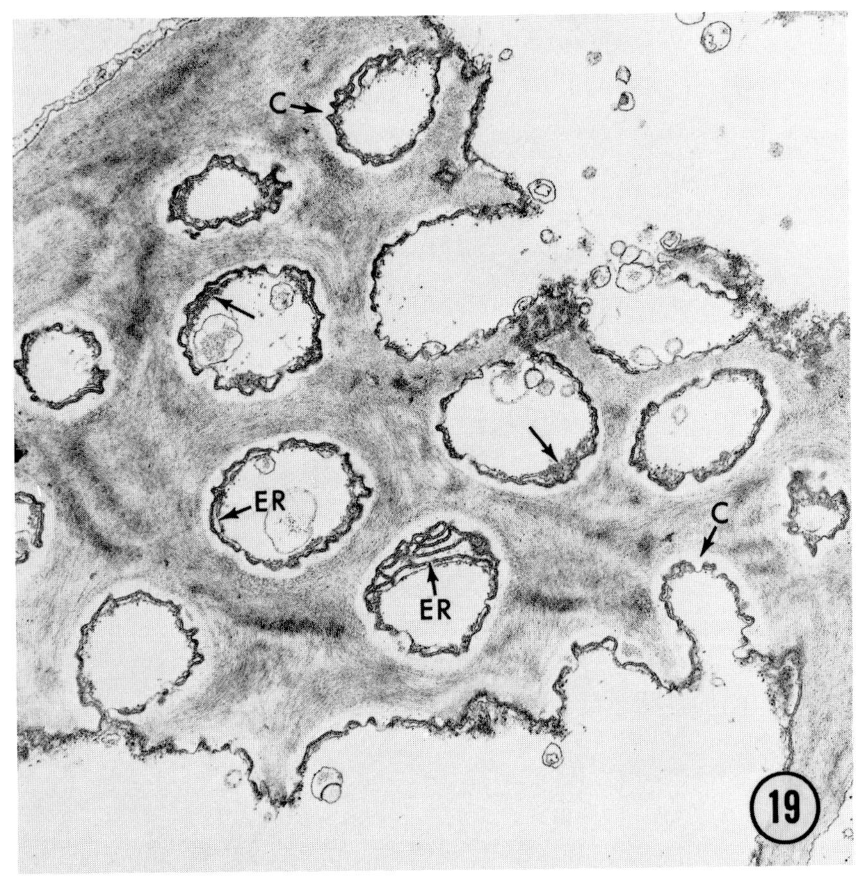

FIG. 19. Face view of portion of Cucurbita maxima *sieve plate. The unoccluded pores are lined by narrow callose cylinders (C) and plasmalemma (not labeled). Segments of endoplasmic reticulum (ER) and P-protein (unlabeled arrows) are also along margins of pores. The vesicle-like structures within some pores are blebs from the plasmalemma. X 21,100.*

the walls between two such cells a sieve area on one side of the wall is opposed by a sieve area on the opposite side of the wall. Sieve areas may also occur on the walls between sieve-tube members and parenchyma cells, including companion cells (Esau, 1969). Under these circumstances, the sieve areas occur only on the side of the sieve-tube member, the connections between the two cell types consisting of pores on the sieve-tube member side and plasmodesmata on the side of the parenchyma cell (FIG. 13).

Relatively little attention has been given to the structure and development of lateral sieve areas. Apparently the development of lateral sieve areas between contiguous sieve-tube members is essentially similar to that of the sieve plate. In the secondary phloem of several woody dicotyledons, the pores of the lateral sieve areas were found to be basically similar to those of the sieve plates, differing from them primarily in size (Evert et al., 1971). Much more information is needed on both the structure and development of the lateral sieve areas of angiospermous sieve-tube members.

B. *Parenchyma Cells Associated with Sieve-Tube Members*

The phloem of angiosperms contains variable numbers of parenchyma cells that differ from one another both structurally and functionally and in their degree of specialization in relation to the sieve-tube members (Esau, 1969). Of the different kinds of parenchyma cells associated with sieve-tube members, the companion cell is the one most intimately related to the sieve-tube member.

Commonly derived from the same mother cell, the companion cell and sieve-tube member typically have numerous connections consisting of pores on the sieve-tube member side of the wall and much-branched plasmodesmata on the companion cell side (FIG. 13; Esau, 1969). As the companion cell develops, its protoplast increases in density, due in part to an increase in the ribosome population and partly to an increase in density of the ground substance itself (Behnke, 1975; Esau, 1978a). In addition to the high ribosome population, the nucleate companion cell contains numerous mitochondria, rough ER, and plastids, which typically lack starch. Some companion cells and phloem parenchyma cells also contain P-protein (Esau, 1969; Evert, 1977).

Because of their numerous cytoplasmic connections with the sieve-tube members and their general ultrastructural resemblance to secretory cells, it is believed that the companion cells play a role in the delivery of sugar to the sieve tube (loading of the sieve tube). In addition, it has been suggested that the companion cells may maintain the enucleate sieve-tube members through the transfer of informational molecules (Gunning, 1976) or other substances such as ATP (Lehmann, 1979) from the companion cells to their associated sieve-tube members via the many connections in their common walls. The interdependence of these two cells is further supported by the fact that both cease to function and die at the same time. Companion cells frequently are absent from the protophloem in some roots and shoots (Esau, 1969), a factor which may be associated with the short life span of many sieve-tube members in this earliest part of the primary phloem (Eleftheriou and Tsekos, 1982b).

Companion cells and other parenchyma cells of the phloem intergrade with one another, and the two are not always unequivocally distinguishable from one another even at the electron microscope level (Esau, 1969). Some parenchyma cells may arise from the same mother cells as the sieve-tube members and companion cells. The more closely related (ontogenetically) they are to the sieve-tube members, the more closely they resemble companion cells, both in appearance and in frequency of plasmodesmatal connections with the sieve-tube members. Moreover, such parenchyma cells may die at the same time as their associated sieve-tube members. Although the ontogenetic relation is usually stressed in the definition of a companion cell, it is becoming increasingly evident that this is but a secondary feature in the establishment of a specialized functional relation between a sieve-tube member and a contiguous parenchyma cell.

Both parenchyma cells and companion cells are interconnected with one another and among themselves by plasmodesmata, as are the ray parenchyma cells and axial parenchyma cells in the secondary phloem of dicotyledons (Esau, 1969). Typically, the companion cells of the secondary phloem abut the rays (Evert, 1963a), presumably acting as intermediaries in the transfer of assimilates from the sieve-tube member to the ray parenchyma cells and vice versa.

In the functional phloem, the parenchyma cells and companion cells apparently have unlignified primary walls. After the tissue ceases to be involved with long-distance transport, the parenchyma cell may remain relatively unchanged or become sclerified. Two instances have been recorded of sclerification of companion cells in old phloem, one in the secondary phloem of *Tilia americana* (Evert, 1963b), and the other in the metaphloem of the perennial monocotyledon *Smilax rotundifolia* (Ervin and Evert, 1967).

C. *Minor Vein Phloem*

In most parts of the plant the sieve elements commonly are wider than the contiguous parenchyma cells. By contrast, the minor vein phloem of dicotyledonous leaves is characterized by the presence of diminutive sieve-tube members and large parenchyma cells, many of which have dense, organelle-rich protoplasts (FIG. 20; Evert, 1980; Fisher and Evert, 1982).

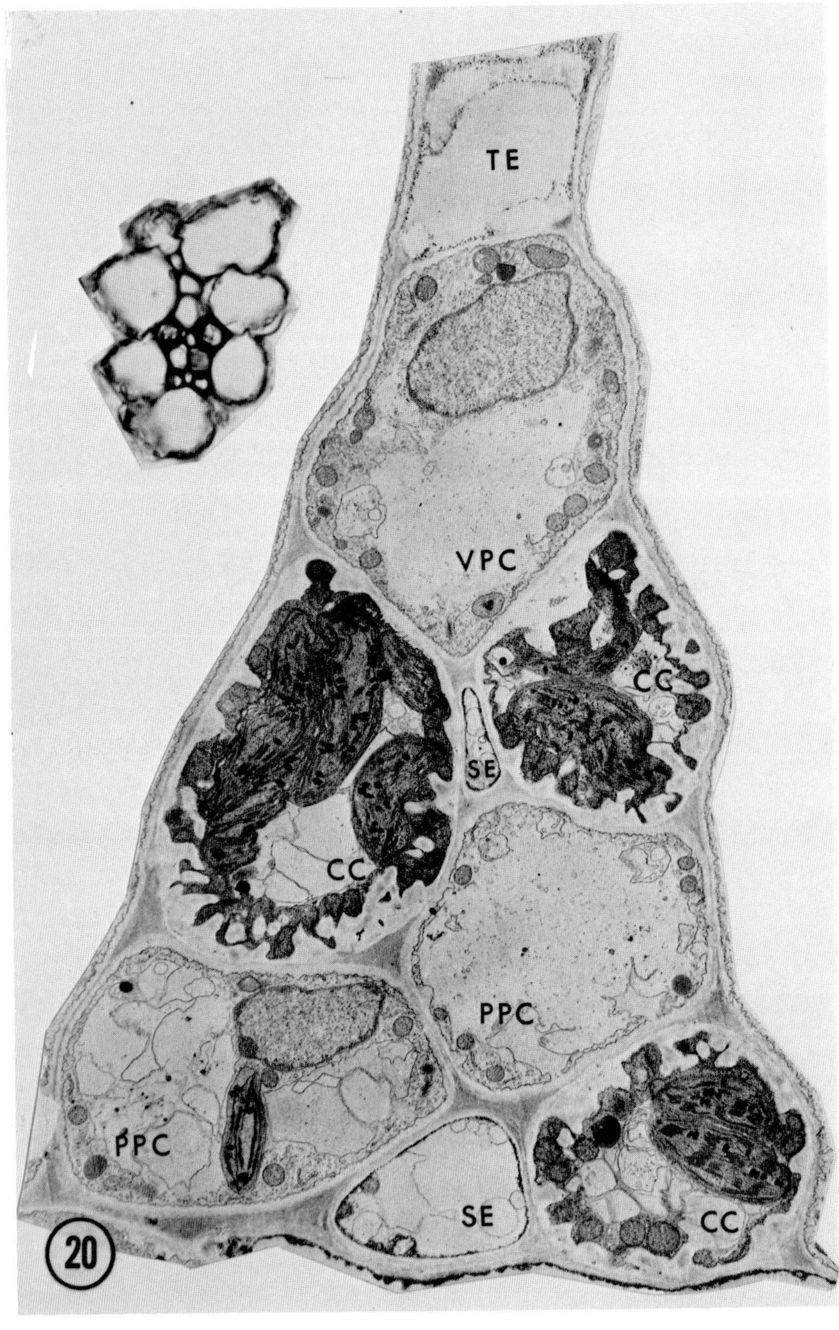
TE
VPC
CC
SE
CC
PPC
PPC
SE
CC
20

The dense parenchyma cells have numerous plasmodesmatal connections with their associated sieve-tube members, and commonly are considered to be companion cells. Fischer (1884) called these dense cells "Übergangszellen" or "intermediary cells" (Esau, 1969) because he thought they served as intermediaries in the transfer of photosynthates between mesophyll cells and sieve elements of the leaf. Less dense parenchyma cells of the minor vein phloem may also be involved in the transfer or loading of photosynthates into the sieve tubes. Although these cells may also be termed intermediary (Esau, 1977), they commonly are called phloem parenchyma cells to distinguish them from companion cells.

In the minor veins of some herbaceous dicotyledons the parenchymatous cells of the phloem develop wall ingrowths that greatly increase the surface of the plasmalemma (FIG. 20). Such cells are called transfer cells and are of two types, A and B (Pate and Gunning, 1969, 1972). A-type transfer cells are companion cells with typically dense protoplasts and wall ingrowths on all walls, but often less abundant on the wall where plasmodesmata connect the transfer cell with the sieve

FIG. 20. Transverse section of minor vein from Tagetes patula *leaf. The sieve elements (SE) are narrower than the parenchymatous cells bordering them. The companion cells (CC), which have electron-dense protoplasts, also have wall ingrowths and, therefore, are A-type transfer cells. The phloem parenchyma cells (PPC) lack wall ingrowths in the* T. patula *leaf. TE, tracheary element; VPC, vascular parenchyma cell. X 4,800. Inset, photomicrograph of minor vein of* Tagetes sp. *showing bundle sheath enclosing the vein. X 400. (From Evert, R.F. 1980. Ber. Dtsch. Bot. Ges. 93, 43-55.)*

tube. B-type transfer cells are phloem parenchyma cells with wall ingrowths best developed opposite the sieve tubes or the companion cells.

Two primary roles have been proposed for the transfer cells of minor veins: to collect and pass on photosynthates and to retrieve and recycle solutes that enter the leaf apoplast in the transpiration stream (Gunning et al., 1974). Pate and Gunning (1972) regard the wall-membrane apparatus of the transfer cell as an adaptation facilitating apoplast-symplast exchanges of solutes across the plasmalemma. The results of two recent histochemical studies on minor vein phloem indicate that the wall ingrowths of the transfer cells are rich in pectin (Botha and Evert, 1981; Lucas and Franceschi, 1982), suggesting that they are highly hydrated in vivo, a condition that would facilitate solute transport in the apoplast.

Two types of sieve tube — a thick-walled type and a thin-walled type — occur in the longitudinal vascular bundles of at least some Gramineae (FIG. 21; Evert, 1980; Botha et al., 1982; Colbert and Evert, 1982). In Z. *mays* only the thin-walled sieve tubes have companion cells (Evert, 1980). As one might expect, these two cell types have numerous pore-plasmodesma connections between them. The sieve tube-companion cell complexes either lack or have infrequent plasmodesmatal connections with other cell types. The thick-walled sieve tubes, which occur nearest the xylem and commonly abut the vessels, lack companion cells but have abundant connections with contiguous vascular parenchyma cells. Whether the two types of sieve tube in the other grasses have similar relationships with companion cells and vascular parenchyma cells remains to be determined.

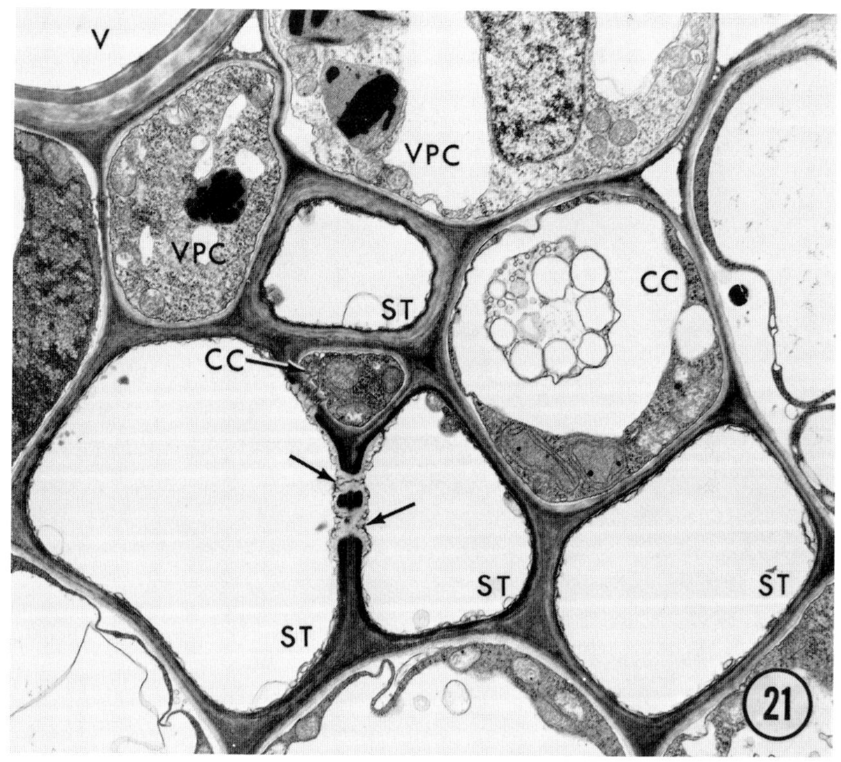

FIG. 21. Transverse section of portion of small (longitudinal) vascular bundle from Zea mays *leaf. This bundle contains one thick-walled sieve tube (separated from a vessel by vascular parenchyma cells) and three thin-walled sieve tubes. Unlabeled arrows point to callose-plugged sieve-plate pores in lateral wall between two thin-walled sieve tubes. CC, companion cell; ST, sieve tube; VPC, vascular parenchyma cell. X 9,500. (From Evert, R.F. 1980. Ber. Dtsch. Bot. Ges. 93, 43-55.)*

II. GYMNOSPERM PHLOEM

Most of the information available on the structure of gymnosperm phloem comes from studies on the secondary phloem of conifers (FIGS. 22 and 23). The sieve elements of gymnosperm phloem are sieve cells, which typically are associated with highly specialized parenchyma cells called albuminous cells.

A. *The Sieve Cell*

Sieve cells are sieve elements in which all of the sieve areas are of the same degree of specialization. In gymnosperms, the sieve areas commonly are more numerous on the overlapping ends of the cells.

1. *The Protoplast.* As with young sieve-tube members, young sieve cells contain all of the cellular components characteristic of young, nucleate plant cells. Similarly, the sieve cell undergoes a selective autophagy during maturation resulting in disorganization and/or disappearance of many cellular components, including nucleus, ribosomes, dictyosomes, microtubules, microfilaments, and tonoplast.

a. *Nucleus.* Nuclear degeneration in gymnospermous sieve cells is pycnotic, with the necrotic nucleus commonly persisting as an electron dense mass (FIG. 24), sometimes with portions of the nuclear envelope still intact (Evert et al., 1973b; Neuberger and Evert, 1976). Such degeneration has now been recorded in sieve cells of conifers, Gnetales (Alosi and Alfieri, 1972; Evert et al., 1973b; Behnke and Paliwal, 1973) and cycads (Parthasarathy, 1975). During nuclear degeneration in *Welwitschia mirabilis* (Evert et al., 1973b), a close spatial relationship develops between the nucleus and mitochondria. Many mitochondria become closely appressed to the nuclear envelope, while others are nearly or entirely surrounded by the nucleus.

b. *Endoplasmic reticulum.* At about the time the chromatin of the nucleus has begun to aggregate into dense clumps, numerous elements of smooth, tubular ER begin to appear in the cytoplasm of the differentiating sieve cell (Evert et al., 1973b; Neuberger and Evert, 1974). At this time rough ER is still present and the ribosome population is still high. As the sieve cell approaches maturity, proliferation of the tubular ER continues and numerous three-dimensional aggregates of tubular ER appear throughout the cytoplasm. By this time the ribosome population has diminished greatly and the rough ER of the young sieve cell apparently has been incorporated into the newly-formed system of tubular ER.

At maturity, the cell contains a vast and complex network of smooth, tubular ER. Between sieve areas the ER is mostly parietal in distribution and varies from one to several layers in depth (FIG. 25). Most of the ER in mature sieve cells, however, occurs opposite the sieve areas. Within a given sieve cell the ER forms a continuous system, with the massive aggregates at the sieve areas interconnected longitudinally by the parietal ER (Neuberger and Evert, 1974).

c. *Plastids.* Both S- and P-type plastids occur in gymnospermous sieve cells. The S-type plastids are the most common and occur throughout the Pinopsida (with the exception of the Pinaceae), Ginkgopsida, Cycadopsida, and Gnetopsida. All Pinaceae examined thus far have P-type plastids (Behnke, 1981). In well-preserved sieve cells the plastids occur intact (FIG. 25).

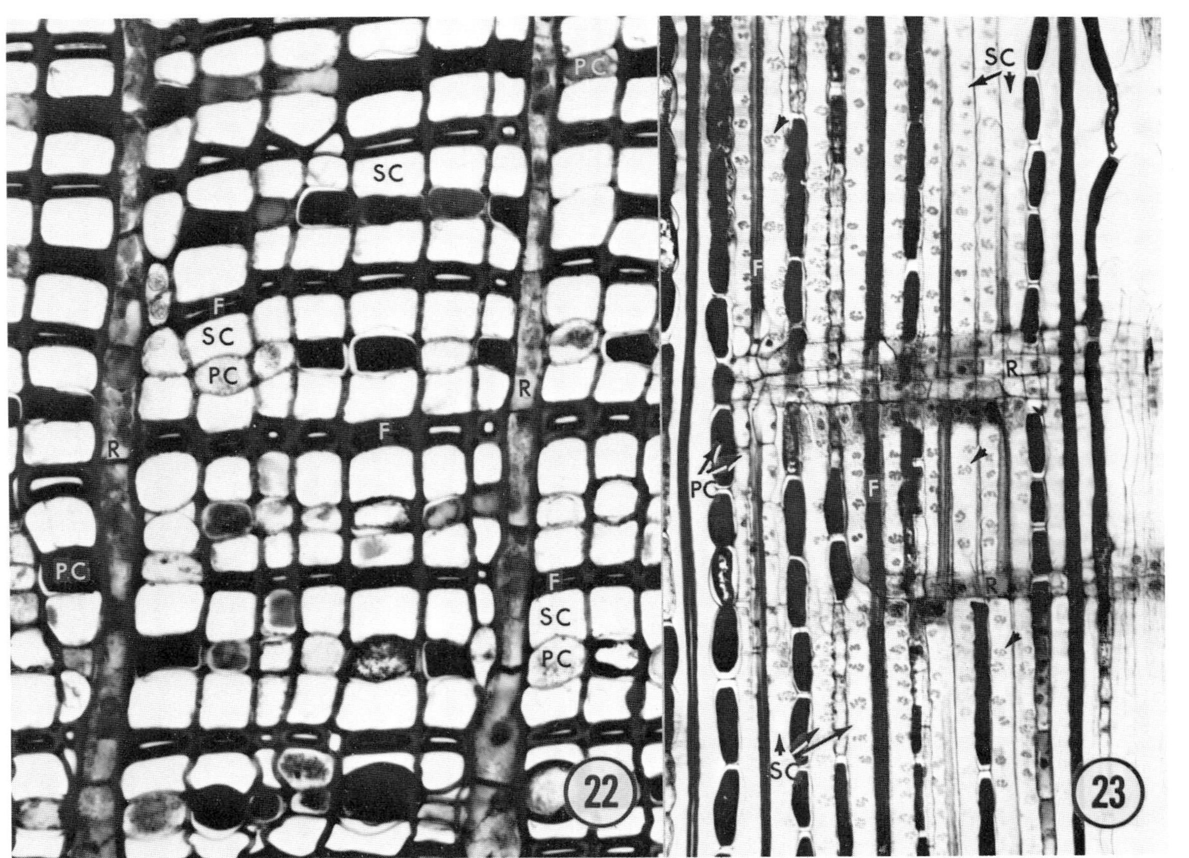

PC
SC
F
SC
PC
R
F
R
PC
F
SC
PC
22
SC
F
R
PC
F
R
SC
23

d. *Other protoplasmic constituents.* The mitochondria of gymnospermous sieve cells undergo no structural modifications during cellular differentiation and in mature cells are normal in appearance (Behnke and Paliwal, 1973; Neuberger and Evert, 1974, 1976).

Despite some reports to the contrary, the sieve cells of gymnosperms do not contain P-protein. Several investigators have erroneously interpreted the filamentous material from ruptured Pinaceae plastids as P-protein. In addition, necrotic nuclei and poorly-preserved ER aggregates have been misidentified as P-protein (see discussions in Neuberger and Evert, 1974; Parthasarathy, 1975).

2. *The Wall.* With the exception of the Pinaceae, the walls of gymnospermous sieve cells are characterized as primary (Esau, 1969). The sieve cells of the Pinaceae have thickened walls (FIGS. 24 and 25), which were designated as true secondary walls by Abbe and Crafts (1939). In *Pinus strobus* the thickened sieve-cell wall has a polylamellate structure (Chafe and Doohan, 1972).

FIGS. 22 and 23. Photomicrographs of secondary phloem of yew (Taxus canadensis). *FIG. 22. Transverse section showing sieve cells (SC), phloem parenchyma cells (PC), and fibers (F), which alternate in tangential bands. Two rays (R) can be seen in this view. X 285. FIG. 23. Longitudinal (radial) section showing the vertically oriented sieve cells (SC), strands of phloem parenchyma cells (PC), and fibers (F). Parts of two rays (R), which are horizontally oriented, can be seen traversing the vertically oriented cells. Arrowheads point to sieve areas on the walls of sieve cells. X 115.*

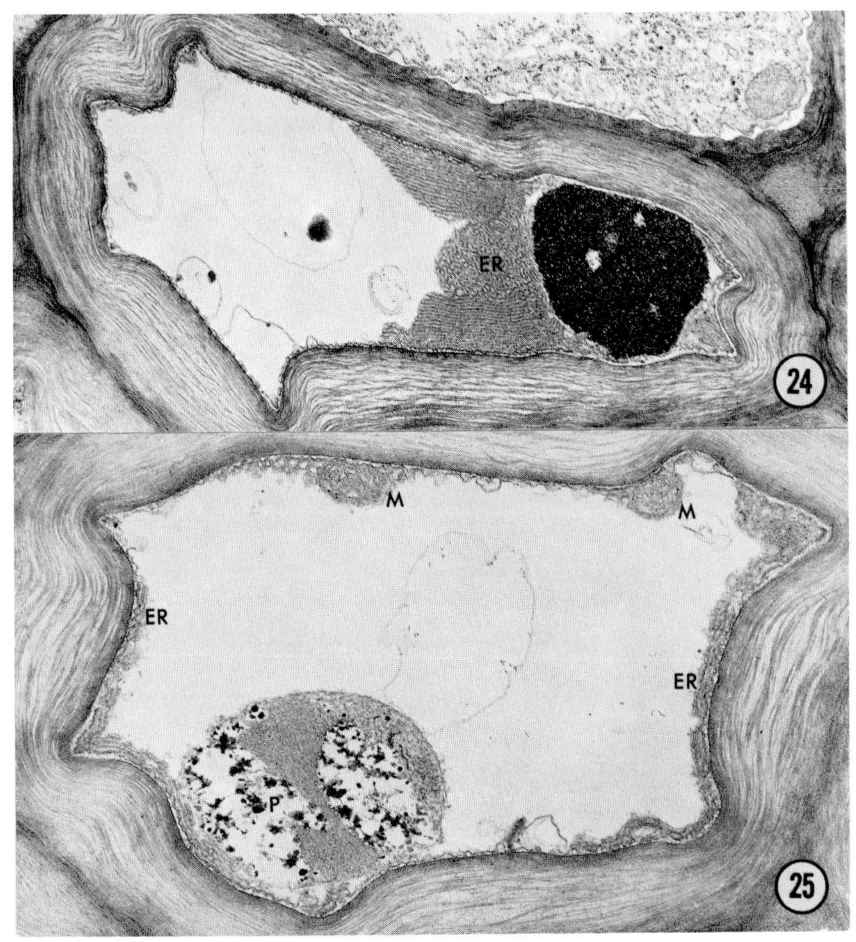

FIGS. 24 and 25. Transverse sections of mature sieve cells in the hypocotyl of **Pinus resinosa**. *In Fig. 24 the necrotic nucleus, represented by an aggregation of electron-dense chromatin material, is bordered by large aggregates of smooth endoplasmic reticulum. X 9,900. Fig. 25 illustrates the typical parietal distribution of cellular components of mature sieve cells. X 15,950. ER, endoplasmic reticulum; M, mitochondrion; P, plastid. (From Neuberger, D.S., Evert, R.F. 1974. Am. J. Bot. 61, 360-374.)*

3. *The Sieve Areas.* Compared with pore development in angiosperm sieve plates, few details have been recorded on the development of sieve-area pores in gymnosperms (Evert et al., 1973c; Neuberger and Evert, 1975). It is apparent, however, that neither ER cisternae nor callose platelets are involved with development of the sieve-area pores in gymnosperms, as they are with the sieve-plate pores in angiosperms. The sieve areas of gymnosperms develop from portions of the wall that are traversed by numerous plasmodesmata. Prior to widening of the plasmodesmatal canals into pores, the portion of the wall bearing the sieve area increases in thickness. Pore formation is accompanied by formation of a median cavity in the middle of the wall. The first-formed aggregates of tubular ER arise in the cytoplasm opposite the developing sieve areas (Neuberger and Evert, 1976).

At maturity the sieve areas consist of plasmalemma-lined pores joined in the middle of the wall by large median cavities (Behnke and Paliwal, 1973; Evert et al., 1973c; Neuberger and Evert, 1975). Numerous membranes extend into the pores from the aggregates of smooth, tubular ER opposite the sieve areas and merge in the median cavities (FIG. 26). Hence, the complex ER networks of contiguous sieve cells are interconnected via the sieve areas. Callose may or may not line the pores (Evert et al., 1973c; Neuberger and Evert, 1975).

B. *Parenchyma Cells Associated with Sieve Cells*

The counterpart of the companion cell in gymnosperm phloem is the albuminous cell. Unlike the companion cell, the albuminous cell commonly is not ontogenetically related to its associated sieve element, the sieve cell (Esau, 1969). The principal feature distinguishing the albuminous cell from other parenchyma cells of the phloem is its connections with

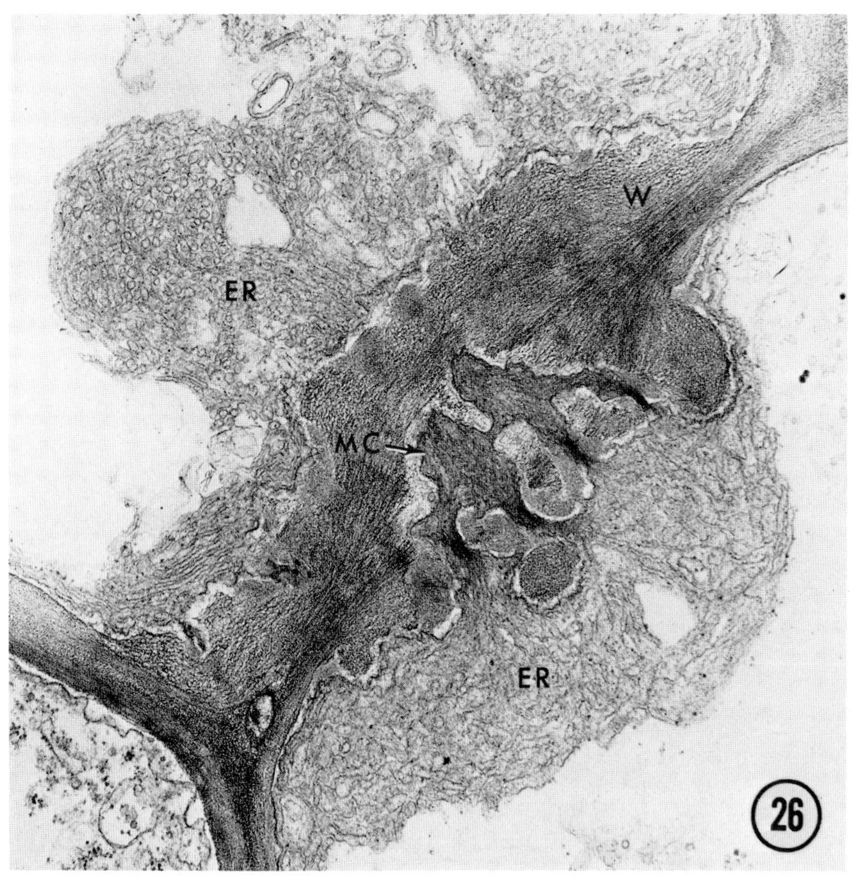

FIG. 26. Oblique section of sieve area in wall (W) between mature sieve cells in hypocotyl of **Pinus resinosa**. *Massive aggregates of tubular endoplasmic reticulum (ER) border on both sides of the wall. ER can be seen traversing the pores (below) and entering the median cavity (MC), which contains much ER. X 27,300. (From Neuberger, D.S., Evert, R.F. 1975. Protoplasma 84, 109-125.)*

the sieve cells (FIG. 27), which are reminiscent of the connections between companion cells and sieve-tube members: pores on the sieve-cell side and branched plasmodesmata on the side of the albuminous cell (Neuberger and Evert, 1975; Sauter et al., 1976; Sauter, 1980). The sieve cell-albuminous cell connections have fairly large median cavities, containing numerous elements of smooth ER, in the region of the middle lamella. Like companion cells, at maturity albuminous cells are free of starch and contain numerous mitochondria and a large ribosome (polysome) population, in addition to other cellular components characteristic of nucleate plant cells (Neuberger and Evert, 1975; Sauter et al., 1976). In addition, like companion cells, the albuminous cells cease to function and die at the same time as their associated sieve elements. Impressive histochemical data have been amassed by Sauter and co-workers that strongly implicate the albuminous cell with a role in loading and unloading of sieve cells and/or in the sustaining of their metabolism (Sauter, 1980).

The location of albuminous cells may differ considerably among taxa (den Outer, 1967). In some gymnosperms, the albuminous cells occur exclusively in the axial system, while in others they are found almost exclusively in the rays. In still other taxa, they occur in both systems.

Several workers (Sauter and Braun, 1968; Parameswaran and Liese, 1970; Carde, 1974; Sauter et al., 1976) have adopted the term "Strasburger cell" in preference to the term "albuminous cell." The move was begun by Sauter and Braun (1968) who pointed out that the term albuminous cell is a misnomer because it implies the cell is rich in protein. During their study of the phloem in *Larix decidua*, they found the albuminous cells associated with mature sieve cells to contain a somewhat higher protein and DNA content than neigh-

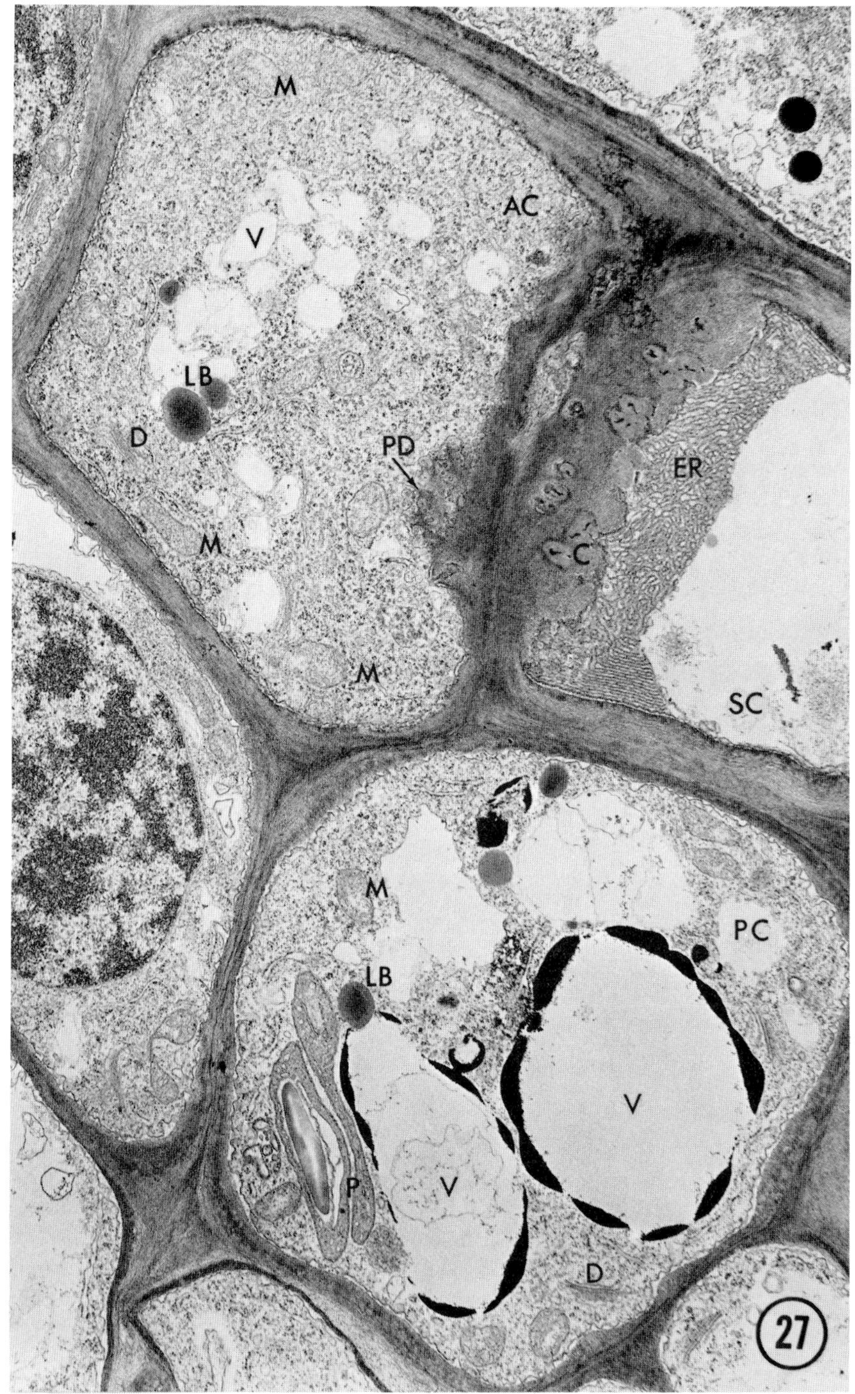
M
AC
V
LB
D
PD
ER
M
C
M
SC
M
PC
LB
V
P
V
D
27

boring ray parenchyma cells, but a much lower one than that of very young albuminous cells. Actually, Strasburger's original definition of the "Eiweisszell" (albuminous cell) referred not to the visible contents of the cell but to the presumed function of the cell at that time: to transport proteins between the sieve cell and other cells (Esau, 1971a). Srivastava (1970) has suggested using the term "companion cell" for both companion cells and albuminous cells, but only in a physiological sense. Alosi and Alfieri (1972) found many of the albuminous cells in the secondary phloem of *Ephedra* to be ontogenetically related to the sieve cells, further emphasizing the intergradation between companion cells, phloem parenchyma cells, and albuminous cells.

III. THE PHLOEM OF LOWER VASCULAR PLANTS

With few exceptions, the sieve elements of lower vascular plants, or vascular cryptogams, may be regarded as sieve cells; that is, the sieve-area pores on the end walls are not measurably different from those on the lateral walls. Unlike their counterparts among the seed plants, the sieve elements

FIG. 27. Transverse section of secondary phloem from hypocotyl of Pinus resinosa *showing connections between albuminous cell (AC) and mature sieve cell (SC). Plasmodesmata (PD) occur on albuminous-cell side of wall and pores on sieve-cell side. The pores are occluded with callose (C) and bordered by massive aggregates of endoplasmic reticulum (ER). Vacuoles (V) of axial parenchyma cell (PC) are lined with electron-dense material. D, dictyosome; LB, lipid body; M, mitochondrion; P, plastids. X 12,000. (From Neuberger, D.S., Evert, R.F. 1975. Protoplasma 84, 109-125.)*

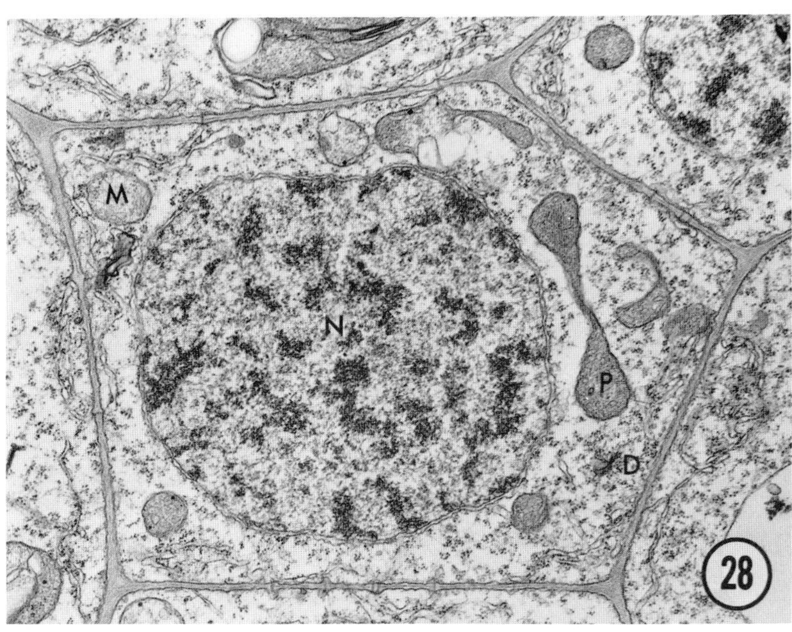
M
N
P
D
28

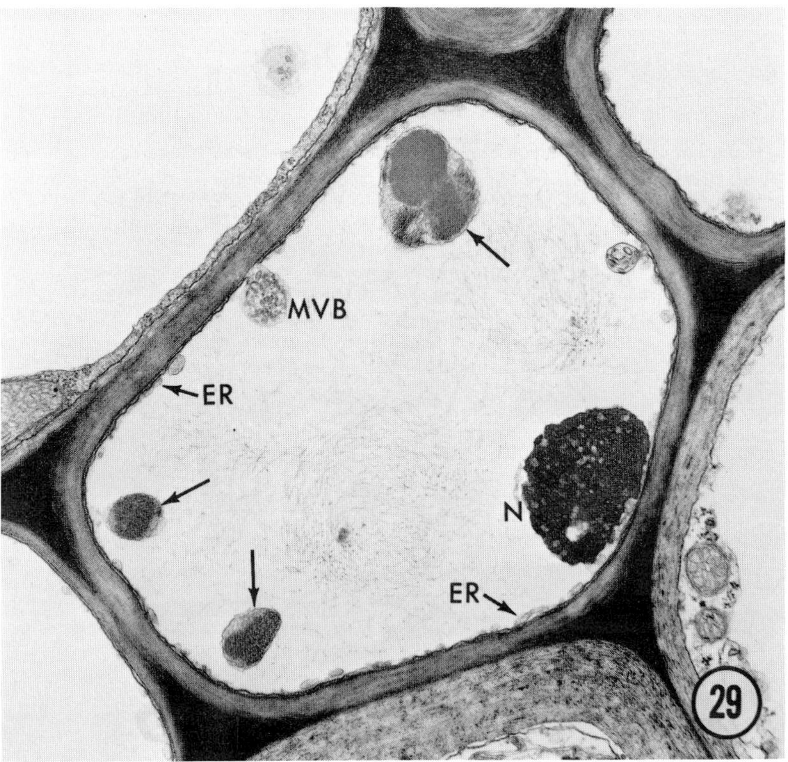
MVB
ER
N
ER
29

of lower vascular plants are not associated with specialized parenchyma cells analogous to companion cells or albuminous cells.

A. *The Sieve Element*

Typically, the sieve elements of lower vascular plants are quite long, in some species reportedly reaching lengths of 30 to 40 mm (Lamoureux, 1961). The degree of inclination of their end walls ranges from transverse to very oblique, and apparently may vary from plant part to plant part within the same individual. In addition, the degree of inclination of the end walls may be greatly influenced by the rate of growth of the plant part (Hébant, 1969). The more rapid the growth, the longer the sieve cells and the more greatly inclined the end walls.

1. *The Protoplast.* Like their counterparts among seed plants, the sieve elements of vascular cryptogams undergo a selective autophagy during maturation, resulting in degeneration of many cellular components and reorganization of others (FIGS. 28 and 29; 35-37). At maturity their protoplasts con-

FIG. 28 and 29. Transverse sections of sieve elements in leaf of Isoetes muricata. *FIG. 28. Young sieve element prior to initiation of morphological changes in nucleus. X 28,400. Fig. 29. Mature sieve element. Most of the endoplasmic reticulum (ER) occurs along the wall as an anastomosing network. Three dilated ER cisternae (unlabeled arrows) can be seen with crystalline and/or fibrillar contents, while some fibrillar material from ruptured cisternae occurs free in the lumen. X 18,000. D, dictyosome; P, plastid; MVB, multivesicular body; N, nucleus. (From Kruatrachue, M., Evert, R.F. 1974. Am. J. Bot. 61, 253-266.)*

sist of a plasmalemma, a parietal network of smooth ER, and variable numbers of plastids and mitochondria. The sieve elements of some species may have occasional aggregates or stacks of ER, in addition to the parietal ER network. Remnants of nuclei commonly persist in some.

a. *Nucleus.* Nuclear degeneration in the sieve elements of lower vascular plants may be either pycnotic or chromatolytic, or by a process intermediate between the two. During a light-microscope study of some tropical ferns, Hébant (1969) noted that nuclear degeneration in protophloem sieve elements was generally of the pycnotic type. In the metaphloem, however, the process of chromatolysis apparently occurred as often as pycnotic degeneration. Both types of nuclear degeneration were observed in the sieve elements of aerial shoots of *Equisetum hyemale* (Dute and Evert, 1978), but only an intermediate process was encountered in those of the *E. hyemale* roots (Dute and Evert, 1977a).

The details of nuclear degeneration may differ markedly from one species to the next. For example, in *Selaginella kraussiana* (Burr and Evert, 1973), the nucleus undergoes unique structural changes during maturation, ultimately being converted into a mass of tubules which persists in the mature sieve element. In leaves and roots of *Isoetes* (Kruatrachue and Evert, 1974, 1978), the very long ("filiform"), persistent nucleus consists of dense chromatin material partly delimited from the rest of the protoplast by remnants of the nuclear envelope (FIG. 29). In the fern *Platycerium bifurcatum* (Evert and Eichhorn, 1974a), crystalloids, which may eventually extend almost the entire length of the nucleus, arise in the matrix of the sieve-element nucleus (FIGS. 30-32). At the time of nuclear degeneration, which is intermediate between pycnotic and chromatolytic, the crystalloids are liberated into the lumen of the cell and eventually degenerate. [Nuclear crystalloids have long been known to occur in the sieve elements of certain species of ferns (e.g., Poirault, 1893).] According to Shah and Nair (1978), nucleoli are sometimes extruded from degenerating nuclei in some fern sieve elements. The nucleoli do not persist, however, but degenerate during differentiation of the sieve element.

b. *Endoplasmic reticulum.* As mentioned previously, at maturity the sieve elements of lower vascular plants contain a parietal network of smooth, tubular ER. In young sieve elements the ER is rough-surfaced and cisternal and fairly evenly distributed throughout the cytoplasm. With increasing age, the ER commonly increases in quantity. In some species (e.g., *Selaginella kraussiana*, Burr and Evert, 1973; *Platycerium bifurcatum, Phlebodium aureum*, Evert and Eichhorn, 1974a; *Psilotum nudum,* Perry and Evert, 1975; *Equisetum hyemale*, Dute and Evert, 1977a,b, 1978), the ER eventually accumulates in stacks and most of the stacks become parietal in distribution before being reduced in amount or disappearing from the cell altogether (FIGS. 33 and 34). Some stacking of ER may occur against the nuclear envelope. In root and aerial shoot sieve elements of *Equisetum hyemale* (Dute and Evert, 1977a, 1978) massive aggregates of ER are formed in the cytoplasm prior to stacking.

Disappearance of ER from the sieve elements of polypodiaceous ferns may occur in different ways (Evert and Eichhorn, 1976). In *Platycerium bifurcatum*, many ER membranes apparently are deposited outside the plasmalemma, where they degenerate and gradually intergrade in appearance with the fibrillar material comprising the nacreous thickening. In *Phlebodium aureum, Polypodium schraderi*, and *Microgramma lycopodioides* the ER forms multivesicular bodies (MVBs). As the

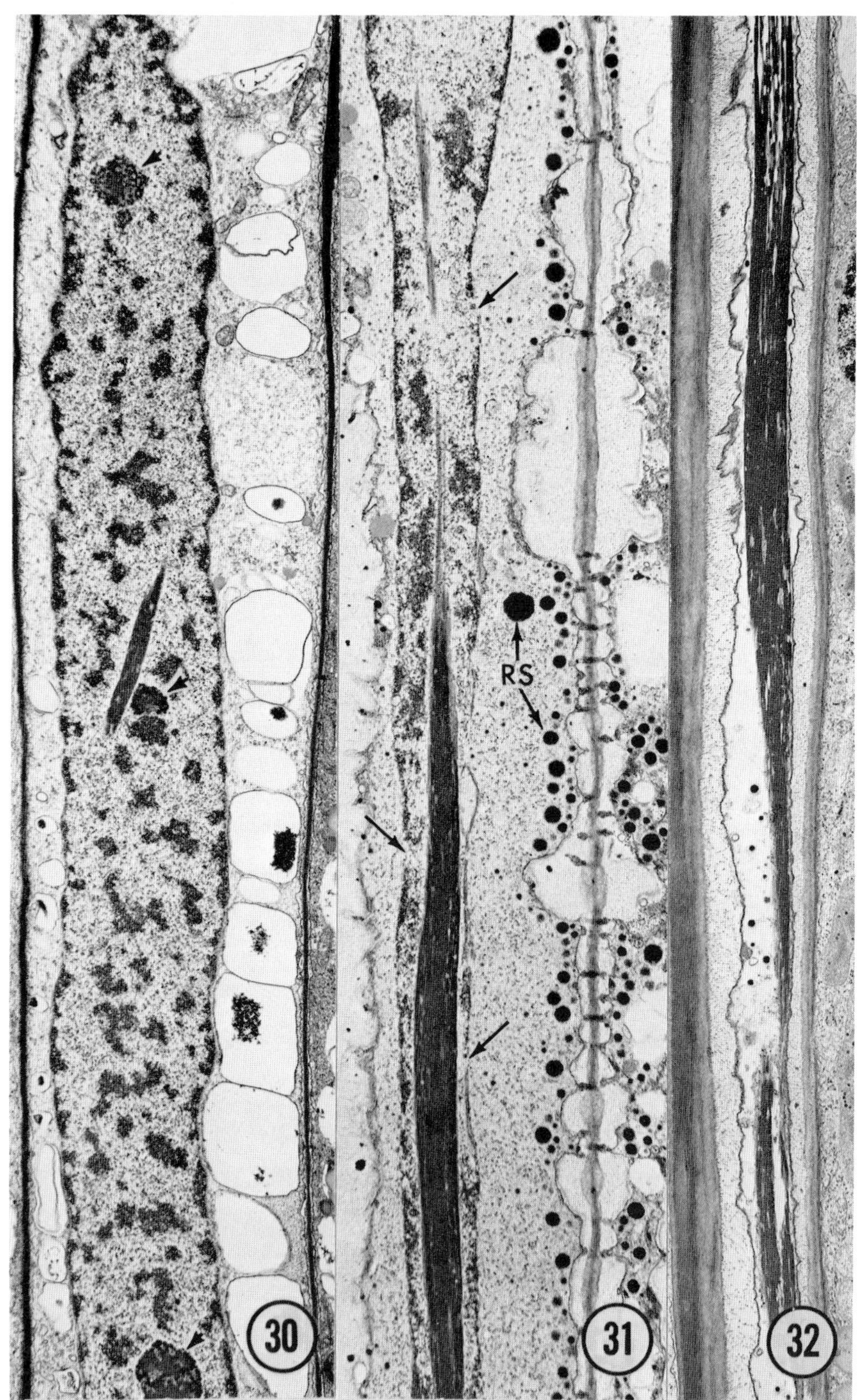
RS
30
31
32

cell approaches maturity, the membranes delimiting the MVBs fuse with the plasmalemma and their vesicular contents gradually disappear. These events indicate that autolytic phenomena associated with degeneration of the ER occur in the region of the nacreous thickening.

c. Refractive spherules. With the exception of the sieve elements of lycopods (*Selaginella*, Burr and Evert, 1973; Hébant et al., 1980; *Isoetes*, Kruatrachue and Evert, 1974, 1977, 1978; *Lycopodium*, Warmbrodt and Evert, 1974a; Hébant et al., 1978), the most distinctive feature of the sieve-element protoplast of vascular cryptogams is the presence of refractive spherules. The spherules, which appear highly refractive when viewed in unstained sections with the light microscope, appear as electron-dense bodies, each surrounded by a single membrane, with the electron microscope (FIGS. 31, 33-38). Being proteinaceous in nature, refractive spherules often have been compared with the P-type plastids of monocotyledons

FIGS. 30-32. Longitudinal sections showing different stages in nuclear development in sieve elements of Platycerium bifurcatum. *FIG. 30. Nucleus in young sieve element with a crystalline inclusion. Arrowheads point to nucleoli. X 6,000. FIG. 31. Crystalloid-containing nucleus is beginning to degenerate. Unlabeled arrows point to discontinuities in nuclear envelope. Note numerous refractive spherules (RS) bordering developing sieve areas on either side of the wall, on right. X 6,300. FIG. 32. View of crystalloid free in the lumen of nearly mature sieve element. X 5,200. (From Evert, R.F., Eichhorn, S.E. 1974. Planta 119, 301-318.)*

(Behnke, 1967). That the plastids and refractive spherules are distinct entities was first demonstrated by Maxe (1964) for the sieve elements in *Polypodium vulgare*.

It is pertinent to note that refractive spherules have been reported as lacking in sieve elements of *Pteris longifolia* (Shah and Fotedar, 1974b). During that light-microscope study, "minute granules" were observed in the parietal cytoplasm but were judged not to be refractive spherules because they did not show a positive reaction with mercuric bromphenol blue. Since then, however, refractive spherules have been found in the sieve elements of other Pteridaceae examined with the electron microscope (Warmbrodt and Evert, 1979b). The sieve elements in *P. longifolia* need to be reexamined for the presence of refractive spherules.

Both ER and Golgi apparatus have been implicated with the formation of refractive spherules. In *Psilotum nudum* (Perry and Evert, 1975), *Botrychium virginianum* (Evert, 1976), and *Equisetum hyemale* (Dute and Evert, 1977a, 1978), for example, formation of the refractive spherules is preceded by the appearance of granular or flocculent material in dilated portions of predominately tubular ER (FIG. 38). Aggregation or condensation of the material apparently results in formation of the refractive spherules.

Only dictyosomes could be implicated with refractive spherule formation in *Platycerium bifurcatum* and *Phlebodium aureum* (Evert and Eichhorn, 1974b). Numerous contacts were encountered, however, between the ER and dictyosomes, and it was often difficult to distinguish between membranes of the Golgi apparatus and ER in the vicinity of the dictyosomes. Since then both the Golgi apparatus and ER have been implicated with the formation of refractive spherules in sieve elements of *Davallia fijiensis* (Fisher and Evert, 1979).

Numerous connections were found between smooth tubular ER and peripheral tubules of the dictyosomes, indicating that these two cytoplasmic components are parts of a single endomembrane system. It seems likely that reports of either ER or dictyosome involvement in refractive spherule formation are not contradictory in fact but, as Hébant (1976) suggested, "contradictory in appearance only."

Many refractive spherules persist in the parietal layer of cytoplasm of mature sieve elements, while others are deposited into the region of the wall (Evert and Eichhorn, 1974b; Perry and Evert, 1975; Evert, 1976; Dute and Evert, 1977a, 1978; Kruatrachue, 1978). The latter phenomenon is accomplished by fusion of the delimiting membrane of the refractive spherule with the plasmalemma of the mature sieve element. Liberman-Maxe (1978) observed a similar phenomenon in protophloem sieve elements only of *Polypodium vulgare*, but there the pertinent bodies were composed of polysaccharides and not of protein. Further cytochemical study will be necessary to determine whether two types of membrane-bound body — one consisting of polysaccharides and the other of protein — exist side-by-side in sieve elements of other vascular cryptogams. The exact chemical composition and function of refractive spherules remain to be determined.

Substantial quantities of crystalline and fibrillar proteinaceous material have been encountered in numerous, relatively small vacuoles of *Selaginella* (Burr and Evert, 1973) and in cisternae of rough ER of *Isoetes* (Kruatrachue and Evert, 1974, 1978). In *Isoetes*, the dilated portions of ER enclosing the proteinaceous substance become smooth-surfaced and migrate to the cell wall (FIG. 29). Along the way they may apparently form multivesicular bodies which then fuse with the plasmalemma, discharging their contents to the outside.

It is quite clear that the crystalline or fibrillar material in *Selaginella* and *Isoetes* is not homologous to the P-protein of angiospermous sieve elements. P-protein is lacking in sieve elements of lower vascular plants.

d. Plastids. Relatively little variation exists in the appearance of plastids in mature sieve elements among the lower vascular plants and, therefore, plastid structure in this group of plants does not provide useful taxonomic characters as in the case of seed plants (Behnke, 1981). The plastids are delimited by one or two membranes and contain variable numbers of internal membranes, plastoglobuli, and spherical lamellar inclusions (FIG. 34). Although chloroplasts have been identified in young sieve elements of *Lycopodium lucidulum* (Warmbrodt and Evert, 1974a), only the mature sieve-element plastids of *Polypodium schraderi* and *Microgramma lycopodioides* have been found with internal membranes organized into grana and intergrana lamellae (Evert and Eichhorn, 1976).

e. Other protoplasmic constituents. As with their counterparts in seed plants, the sieve-element mitochondria of lower vascular plants undergo little, if any, structural change during sieve-element differentiation. At maturity they are generally round in outline and contain variable numbers of cristae.

Dictyosomes and dictyosome vesicles, both smooth and coated, are most abundant during the period of cell-wall formation. Both types of vesicle apparently are involved with the transport of substances to the developing wall. When cell-wall formation is completed, the dictyosomes and cortical microtubules gradually decrease in number and finally disappear (Evert, 1976; Evert and Eichhorn, 1976; Dute and Evert,

1977a). Other protoplasmic components such as ribosomes and microfilaments also disappear as the sieve element approaches maturity. At maturity, the sieve elements also lack vacuoles.

2. *The Wall*. Typically, the sieve elements in lower vascular plants develop walls that are thicker than those of the parenchyma cells bordering them. Although the walls frequently become quite thick, they do not always consist of morphologically distinct inner and outer layers. In a restricted sense, such walls cannot be designated nacreous (Esau and Cheadle, 1958). This is true, for example, of the sieve-element walls in *Isoetes muricata* (Kruatrachue and Evert, 1974, 1978), *Lycopodium lucidulum* (Warmbrodt and Evert, 1974a), and *Equisetum hyemale* (Dute and Evert, 1977a).

During his comparative studies of the phloem in vascular cryptogams, Lamoureux (1961) found true nacreous wall layers in only some species of Hymenophyllaceae and Polypodiaceae. In some of those species, nacreous walls were found in some plant parts and not in others. By contrast, in a recent ultrastructural study, nacreous thickenings were found in sieve elements of all nine families of homosporous leptosporangiate ferns examined (Warmbrodt and Evert, 1979b).

Hébant (1969) reported that nacreous wall development in tropical ferns was quite variable and dependent upon the type of organ, the age of the sieve elements, and the habitat. He found that ferns which go through periods of drought have well-developed nacreous walls, while aquatic types and those growing in regions of constant humidity lack such wall layers. Nevertheless, Evert and Eichhorn (1976) found nacreous walls in protophloem and metaphloem sieve elements of field-grown (*Polypodium schraderi* and *Microgramma lycopodioides*) and of greenhouse-grown (*Platycerium bifurcatum* and *Phlebodium*

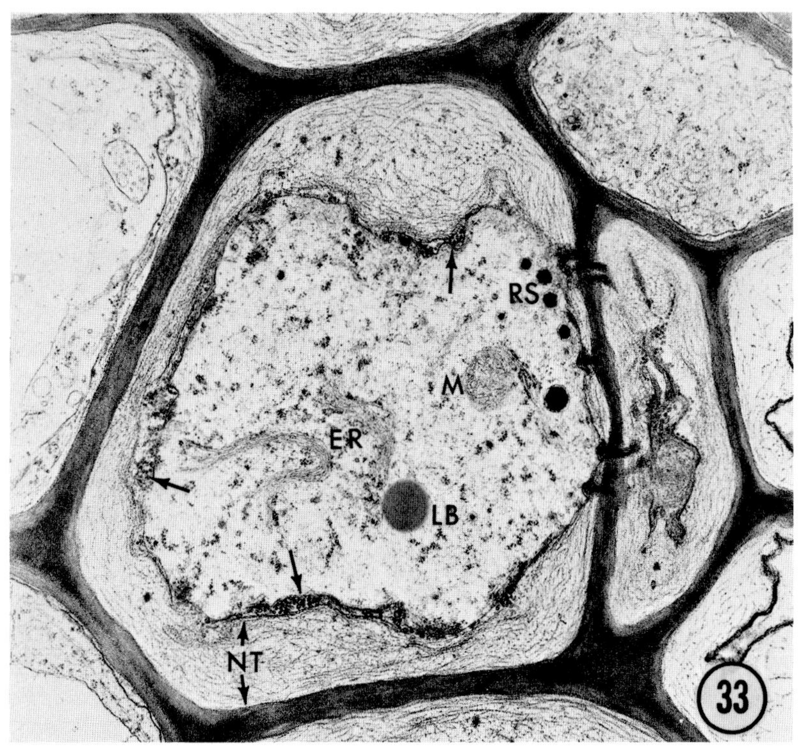
RS
M
ER
LB
NT
33

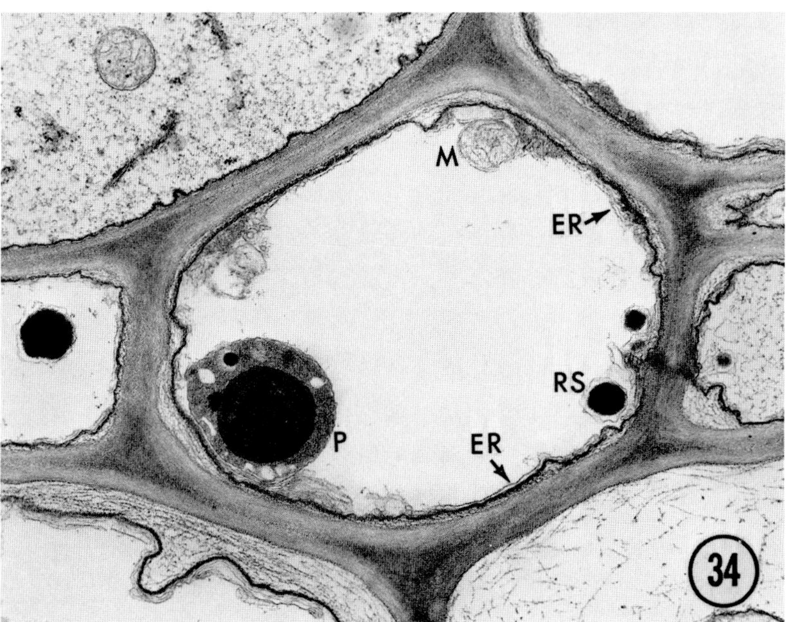
M
ER
RS
P
ER
34

aureum) plants. Liberman-Maxe (1978) found nacreous thickenings in only the metaphloem sieve elements of *Polypodium vulgare*.

Considerable variation exists in the structure and development of the so-called nacreous thickenings in vascular cryptogams. The results of two studies, one on the sieve elements of *Botrychium virginianum* (Evert, 1976) and the other on four species of polypodiaceous ferns (Evert and Eichhorn, 1976), serve to illustrate this point. Whereas the nacreous thickenings in *B. virginianum* gave positive tests for cellulose and pectin, those in the polypodiaceous ferns were negative for these substances. When viewed with polarized light, the nacreous thickenings in *B. virginianum* were strongly anisotropic, while those of the polypodiaceous ferns were isotropic. In addition, the nacreous thickenings in *B. virginiana* had a polylamellate structure, with the cellulose microfibrils arranged parallel with one another and more or

FIGS. 33 and 34. Transverse sections of sieve elements in leaf of Platycerium bifurcatum. *FIG. 33. Immature sieve element with numerous extracytoplasmic membranes (arrowheads) in region of nacreous thickening (NT). Most of the endoplasmic reticulum (ER) in this cell is stacked. X 15,400. FIG. 34. Mature or nearly mature sieve element in which most of the nacreous thickening has disappeared. X 18,800. LB, lipid body; M, mitochondrion; P, plastid; RS, refractive spherule. (From Evert, R.F., Eichhorn, S.E. 1976. Am. J. Bot. 63, 30-48.)*

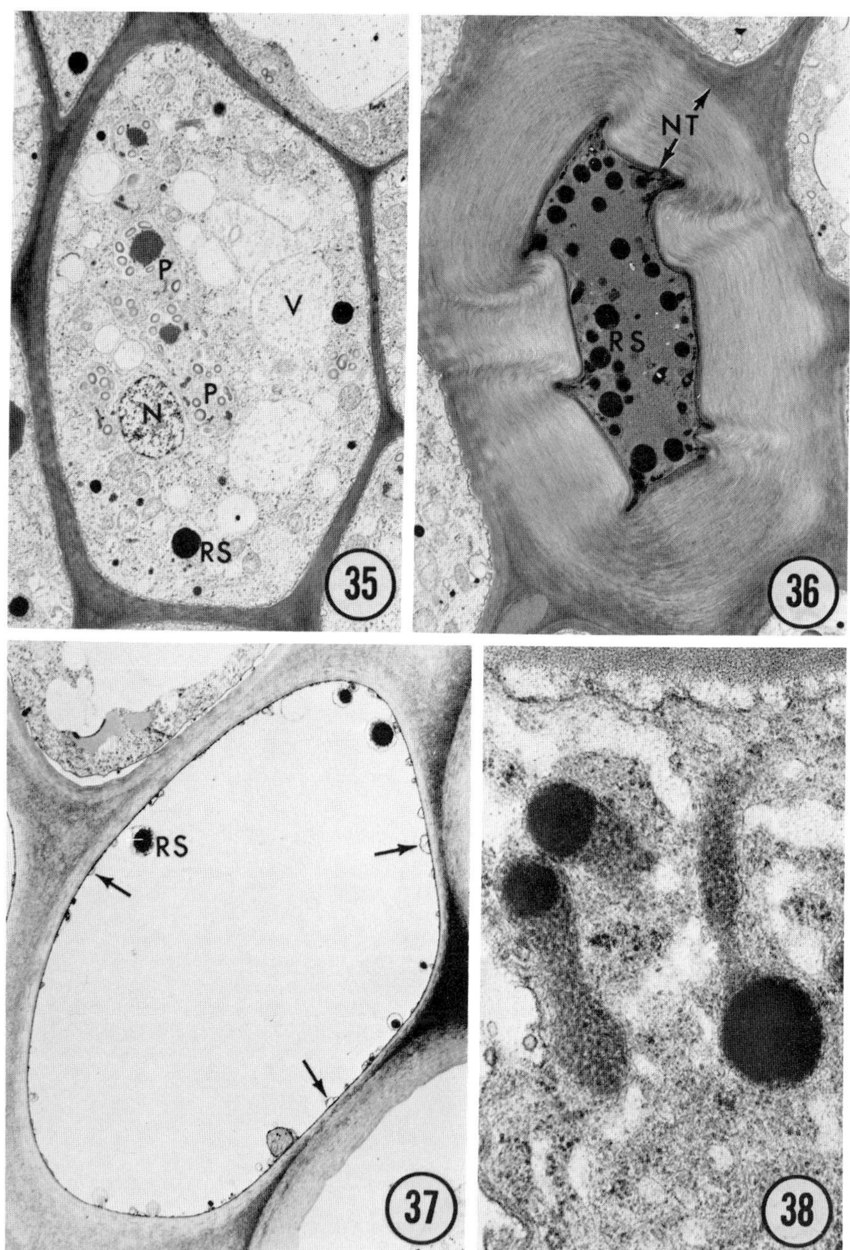
P
V
P
N
RS
35
NT
RS
36
RS
37
38

less at right angles to the long axis of the cell (FIG. 36). By contrast, the nacreous thickenings in *Phlebodium aureum, Platycerium bifurcatum, Polypodium schraderi,* and *Microgramma lycopodioides* consisted of a coarse fibrillar material arranged in a loosely woven meshwork (FIG. 33; Evert and Eichhorn, 1976).

Nacreous walls similar to those in *B. virginianum* are found in *Psilotum nudum* (Perry and Evert, 1975), and walls similar to those in *Polypodium schraderi* and *M. lycopodioides* are found in *Polypodium vulgare* (Liberman-Maxe, 1978). (The sieve element walls in *Psilotum* are unique, however, in that they often become lignified.) In *P. schraderi, M. lycopodioides,* and *P. vulgare* the nacreous thickenings consist of two distinct regions: an outer, relatively wide region in which the fibrils are loosely arranged and an inner, electron-dense region consisting of a more compact fibrillar meshwork.

FIGS. 35-38. Transverse sections of sieve elements in leaf of Botrychium virginianum. *FIG. 35. Young sieve element prior to initiation of nacreous wall formation. X 5,100. FIG. 36. Immature sieve element with fully-formed nacreous thickening (NT). X 5,000. FIG. 37. Mature sieve element. All that remains of the endoplasmic reticulum (unlabeled arrows) is a sparse tubular network — thus the discontinuous profiles — along the wall. X 6,000. FIG. 38. Portion of a differentiating sieve element. Three refractive spherules can be seen in dilated portions of smooth ER filled with small granules which apparently aggregate to form refractive spherules. X 39,000. N, nucleus; P, plastid; RS, refractive spherule; V, vacuole. (From Evert, R.F. 1976. Isr. J. Bot. 25, 101-126.)*

Using Thiéry's test (Thiéry, 1967), Liberman-Maxe (1978) was able to show the polysaccharide nature of the fibrillar meshwork in *P. vulgare*.

And now, considering some differences in the development of the nacreous thickenings in *B. virginianum* and the polypodiaceous ferns *Platycerium* and *Phlebodium*, in the latter two ferns nacreous wall formation is closely correlated with the appearance of numerous membranes or vesicles in the region of the wall. These extracytoplasmic membranes, which apparently are derived from the plasmalemma, eventually degenerate and gradually intergrade in appearance with the fibrillar material comprising the nacreous thickening. Similar extracytoplasmic membranes are not associated with the nacreous thickening in *B. virginianum*. Other than microtubules, only dictyosome-derived vesicles and possibly ER could be implicated with nacreous wall formation in this eusporangiate fern.

In both *B. virginianum* and the pertinent polypodiaceous ferns (Evert and Eichhorn, 1976), the nacreous thickenings gradually decrease in thickness and disappear (FIGS. 33 and 34; 36 and 37). The inner, electron-dense portion of the nacreous thickening in the polypodiaceous ferns persists in the mature sieve element. Entire nacreous thickenings may persist, however, in some sieve elements of homosporous leptosporangiate ferns (Liberman-Maxe, 1978; Warmbrodt and Evert, 1979b).

It should be obvious from this brief discussion of nacreous thickenings in vascular cryptogams and the previous discussion on the nacreous thickenings in angiosperms that the term nacreous is not being uniformily applied to equivalent wall layers in the sieve elements of vascular plants. In

1969, Esau pointed to the need for systematic ultrastructural research on the nacreous wall. Fifteen years later that need is even greater.

3. The Sieve Areas. Considerable variation exists in the distribution, size, and contents of the cytoplasmic connections or pores between sieve elements of the lower vascular plants (Warmbrodt and Evert, 1978, 1979a,b). In some species all of the pores are clustered into sieve areas, while in others they occur only singly on both end and lateral walls (e.g., in *Selaginella*, Burr and Evert, 1973); in still other species, the pores occur in clusters on the end walls and only singly (e.g., in *Psilotum*, Perry and Evert, 1975; in *Isoetes* root, Kruatrachue and Evert, 1978) or not at all (e.g., in *Isoetes* leaf, Kruatrachue and Evert, 1974) on the lateral walls.

In most lower vascular plants the range in size of the pores on end and lateral walls is generally similar in individual sieve elements, so that these elements may be regarded as sieve cells. Sieve-tube members have been reported in the mid-internodal regions of the aerial stem in several species of *Equisetum* (Dute and Evert, 1978) and in the fern *Cyathea gigantea* (Shah and Fotedar, 1974a). In *E. hyemale* the average diameter of the pores in the end wall was 0.4 μm and that of lateral wall pores 0.27 μm. In *C. gigantea*, the pores of the end walls ranged from 2-5 μm in diameter, while those of the lateral walls measured less than 0.8 μm in diameter. Several authors have reported the presence, in *Marsilea*, of cells resembling sieve-tube members (Loyal and Maheshwari, 1979).

In mature sieve elements of the *Isoetes muricata* root (Kruatrachue and Evert, 1978), the occasional solitary pores (0.2 μm in diam.) in the lateral walls are considerably smaller than the pores in the end walls (0.4-0.7 μm in diam.). In sieve elements of the leaf (Kruatrachue and Evert, 1974), the end walls contain either plasmodesmata or pores (0.2-0.57 μm in diam.), or both, but only plasmodesmata occur in the lateral walls. In the corm of *I. muricata* (Kruatrachue and Evert, 1977), pores of variable size (0.2-1.0 μm in diam.) occur in both radial and transverse walls of the secondary sieve elements, but more or less uniformly small pores (0.15-0.2 μm in diam.) occur in the tangential walls. Because of the variability of pore size in both radial and transverse walls of the corm sieve elements, these elements might best be considered as sieve cells. On the other hand, the leaf sieve

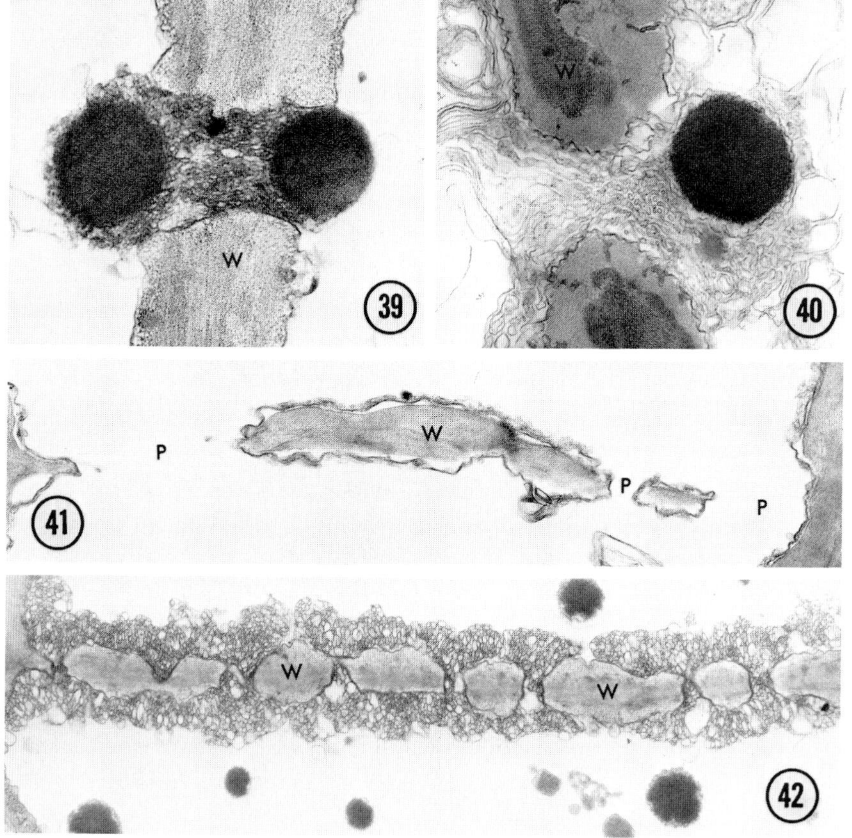

elements with end-wall pores and the sieve elements of the root could be called "sieve-tube members." The gradient in the nature of the end-wall connections in the *I. muricata* leaf can be interpreted as a reflection of the phylogenetic relationship between sieve elements and parenchyma cells.

The situation in the *I. muricata* leaf creates a problem in terminology, for it might be argued whether the cells containing only plasmodesmata in their end and lateral walls should be called sieve elements at all. A similar problem exists with regard to certain cells in the aerial stem of *Equisetum hyemale* (Dute and Evert, 1977b). There, the sieve

FIGS. 39-42. Electron micrographs of sieve-area pores in some lower vascular plants. FIG. 39. In the eusporangiate fern Botrychium virginianum *the pores are filled with numerous membranes, apparently tubular endoplasmic reticulum. The electron-dense bodies on either side of the pore are refractive spherules. X 34,500. FIG. 40. Endoplasmic reticulum filled pore in wall between mature sieve elements in aerial stem of the horsetail* Equisetum hyemale. *The electron dense body (right) is a refractive spherule. X 22,500. FIG. 41. Unoccluded pores (P) in wall between mature sieve elements in corm of the quillwort* Isoetes muricata. *X 15,400. FIG. 42. Endoplasmic reticulum filled pores in wall between mature sieve elements in aerial shoot of the psilopsid* Psilotum nudum. *Electron-dense bodies on both sides of the wall are refractive spherules. X 13,900. (Fig. 39 from Evert, R.F. 1976. Isr. J. Bot. 25, 101-126. Fig. 40 from Dute, R.R., Evert, R.F. 1978. Ann. Bot. 42, 23-32. Fig. 41 from Kruatrachue, M., Evert, R.F. 1977. Am. J. Bot. 64, 310-325. Fig. 42 from Perry, J.W., Evert, R.F. 1975. Am. J. Bot. 62, 1038-1052.) W, wall.*

elements located near the nodes contain very oblique end walls. During maturation, the plasmodesmata in those walls undergo little or no structural modification. In fact, the average diameter of the plasmodesmata before they are "widened" is 0.04 μm, whereas at maturity the cytoplasmic connections average only 0.058 μm in diameter. The sequence of events in the ontogeny of the protoplast of these primitive sieve elements is essentially similar to that of metaphloem sieve elements elsewhere in the aerial stem.

Callose is not associated with development of the sieve-area pores in *Selaginella* (Burr and Evert, 1973), *Lycopodium* (Warmbrodt and Evert, 1974a), or *Isoetes* (Kruatrachue and Evert, 1974, 1977, 1978); nor in *Psilotum* (Perry and Evert, 1975), *Equisetum* (Dute and Evert, 1977a, 1978), or *Botrychium* (Evert, 1976). Widening of the plasmodesmatal canals in these sieve elements occurs more or less uniformly along the entire length of the plasmodesmata. In *Psilotum*, the sites of the future pores are clearly indicated by the presence of ER aggregates that arise opposite them early in relation to pore development. In *Polypodium vulgare* (Liberman-Maxe, 1978) the pore sites become distinguishable from the rest of the wall by the appearance of paired callose collars or cone-shaped structures, one on either side of the wall, around each plasmodesma. Perforation begins in the region of the middle lamella and then the plasmodesmatal canal widens on both sides of the wall until the pore is fully-formed.

At maturity, the plasmalemma-lined sieve-area pores of *Psilotum* (Perry and Evert, 1975) and of the eusporangiate and protoleptosporangiate ferns (Warmbrodt and Evert, 1979a) contain numerous electron-dense membranes, apparently tubular elements of ER, which are structurally distinct from the parietal ER in the lumen of the cell (FIGS. 39 and 42).

Callose has not been identified in association with the sieve-area pores of either *Psilotum* or the eusporangiate ferns examined with the electron microscope.

The sieve-area pores in the root (Dute and Evert, 1977a) and mid-internodal regions of the aerial stems (Dute and Evert, 1978) of *Equisetum* are traversed by large numbers of membranes (FIG. 40). The nature of the membranes could not be determined in the root largely because of the presence of large quantities of callose (presumably wound callose). In the mid-internodal region the many membranes apparently are tubules of ER, which are continuous with those in the cell lumen.

The sieve-area pores in heterosporus (Warmbrodt and Evert, 1978) and homosporous leptosporangiate ferns (Warmbrodt and Evert, 1979b) are traversed by variable numbers of ER membranes, which generally do not occlude the pores; nor do they differ structurally from the parietal ER of the lumen. Continuity commonly is found between the pore membranes and those of the parietal network. Considerable variation exists in the quantity of callose associated with the pores in tissues prepared for electron microscopy, ranging from none to massive deposits.

By contrast to those of the other vascular cryptogams, the plasmalemma-lined sieve-area pores of *Selaginella* (Burr and Evert, 1973), *Isoetes* (Kruatrachue and Evert, 1974, 1977, 1978), and *Lycopodium* (Warmbrodt and Evert, 1974a) are virtually wide open, being traversed by only an occasional tubule of ER (FIG. 34). Although callose has been identified at the mature sieve areas of *Selaginella* and *Isoetes*, none could be detected in sieve elements of *Lycopodium* with the electron microscope. Occasional callose deposits have been found, however, in *Lycopodium* with the light microscope (Lamoureux, 1961).

B. *Parenchyma Cells Associated with Sieve Elements*

As mentioned previously, specialized parenchyma cells analogous to companion cells or albuminous cells are lacking in lower vascular plants. Nevertheless, cytoplasmic connections, consisting of a small pore on the sieve-element side of the wall and a single, unbranched plasmodesma on the parenchyma-cell side, are numerous between sieve elements and parenchyma cells of vascular cryptogams. Despite early reports of the presence of companion cells in the roots of *Equisetum* (Esau, 1969), the cells designated there as "companion cells" are not equivalent to the companion cells in angiosperms (Dute and Evert, 1977a).

Sieve-element associated parenchyma cells in vascular crytogams generally do not differ structurally from other parenchyma cells of the vascular cylinder or bundle. The most notable exception is found in the vascular cylinder of *Lycopodium lucidulum*, which contains two distinct types of parenchyma cells, one of which is always associated with sieve elements and the other with tracheids (Warmbrodt and Evert, 1974b). The sieve-element associated parenchyma cells have dense protoplasts and numerous connections with their associated sieve elements. By contrast, the tracheid-associated parenchyma cells are light in appearance and lack plasmodesmata in their walls. The remaining parenchyma cells have characteristics intermediate between the two extremes. Hence, in *L. lucidulum* distinct structural and probably physiological specialization exists among the parenchyma cells of the vascular cylinder.

Hébant and co-workers have demonstrated by histochemical methods the presence of intensive phosphatase and respiratory enzyme activities in parenchyma cells associated with sieve elements and tracheary elements in *Lycopodium clavatum* (Hébant et al., 1978), *Selaginella wildenowii* (Hébant et al., 1980), and several species of ferns (DeFaÿ and Hébant, 1981). By comparison to the stelar parenchyma cells, the parenchyma cells of the surrounding tissues showed much less intense staining. Hébant and co-workers interpreted these results as an indication of a functional association between the stelar parenchyma cells and the conducting cells.

Transfer cells occur in the vascular bundles of some fern leaves, but they are not the equivalents of the A- and B-type transfer cells in the phloem of some herbaceous dicotyledons. In the ferns, the wall ingrowths occur only opposite the tracheids or opposite sieve elements, tracheids, and other parenchyma cells (Warmbrodt and Evert, 1978, 1979b).

IV. LONGEVITY OF SIEVE ELEMENTS

Commonly, sieve elements are characterized as short-lived cells. The concept of short sieve-element longevity comes mainly from studies on the secondary phloem of woody dicotyledons and conifers, in which the sieve elements function for only one growing season or parts of two (Esau, 1969). *Tilia* has been cited as an exception, the sieve elements of *T. americana* living for as many as 5 years (Evert, 1962), and those of *T. cordata* for as many as 10 (Holdheide, 1951). Recently it has been shown that the sieve cells in needles of *Pinus longaeva* remain alive, on the average, for from 3.82 to 6.5 years (Ewers, 1982).

The longevity of sieve elements in palms and other perennial monocotyledons is often much greater than that in dicotyledons and conifers. For example, in the basal part of long-lived palms such as the Royal (*Roystonea*) and Palmetto (*Sabal*), the sieve elements may remain alive for 100 or more years (Parthasarathy, 1980). Sieve elements estimated to be

8-10 years old have been found in stems of the much smaller perennial monocotyledons *Smilax hispida* and *Polygonatum canaliculatum* (Ervin and Evert, 1970).

As noted by Parthasarathy (1975), the sieve elements in long-lived lower vascular plants, which lack secondary vascular tissue, may be expected to function for the life of the plant part in which they occur. Parthasarathy reported finding living sieve elements in the basal part of a 12-year-old tree fern (*Alsophila stipularis*). Living sieve elements 5 or more years old have been reported in the bracken fern, *Pteridium aquilinum* (Hume, 1912).

V. CONCLUSION

Research on the structure of phloem has been and continues to be motivated largely by the desire of investigators to reveal structure-function relationships essential to an understanding of transport phenomena in this complex tissue. No less important to other investigators has been the desire to reveal the structural modifications this tissue has undergone through the ages.

Comparison of the structure of mature, presumably functional sieve elements in the different groups of vascular plants should be useful to an understanding of both structure-function relationships and the possible trends of specializations of this conduit. When this is done, we find that the sieve elements in the different taxa of vascular plants show basic similarities: during maturation their protoplasts undergo a selective autophagy and develop a greater degree of continuity with one another. In all taxa, the mature sieve-element protoplast contains a plasmalemma, a parietal network of smooth ER, and variable numbers of plastids and mitochondria. The greater degree of continuity of protoplasts, which is brought about through the development of sieve-area pores from plasmodesmata, and the modifications to the protoplast presumably are features that facilitate the transport of assimilates from cell to cell.

The highly specialized sieve-tube members of angiosperms, with their relatively large, unoccluded sieve-plate pores and lumina lined by a thin layer of cytoplasm, are especially suited for the transport of assimilates at fairly high velocities — typically 50-100 cm per hour in angiosperms. Moreover, they appear ideally suited for the operation of an osmotically generated presure-flow mechanism, which requires a conduit relatively free of obstructions that would otherwise provide resistance to solution flow. Presently, the osmotically generated pressure-flow mechanism is generally regarded as the only viable explanation for the long-distance transport of assimilates through the sieve tube (Evert, 1982). With regard to that mechanism, the most important protoplasmic component of the sieve-tube member is its differentially permeable plasmalemma.

Sieve-tube members, which are typical of angiosperms, are lacking entirely in gymnosperms and occur as exceptions among the lower vascular plants. Only the sieve-tube members of angiosperms contain P-protein. Although lacking in many monocotyledonous sieve-tube members, the presence of P-protein differentiates most angiospermous sieve elements from those of lower taxa. If P-protein serves to seal the sieve-plate pores of injured sieve tubes, as is widely believed, its presence in angiosperms is probably related to the large size and greater penetrability of those pores compared with the small, often membrane-filled sieve-area pores, of gymnosperms and lower vascular plants. Plugging of sieve-plate pores with P-protein is an almost instantaneous reaction to pressure release in active sieve tubes and may serve to prevent the loss of assimilates until sufficient wound callose has been deposited to occlude the pores.

By comparison with the open condition of the sieve-plate pores of angiospermous sieve-tube members, the sieve-area pores of gymnospermous sieve cells are largely occupied with tubular elements of ER. Although velocities of 48-60 cm per hour have been recorded for the translocation of assimilates in the phloem of *Metasequoia glyptostroboides* (Willenbrink and Kollmann, 1966), much slower velocities have been recorded for other conifers. Kollmann and Dörr (1966) found the exudation of phloem sap from aphid stylets terminating in sieve cells of *Juniperus communis* to occur at rates only about one-tenth those reported for the exudation of sap from aphid stylets terminating in sieve tubes of *Salix viminalis* (Weatherley et al., 1959). They concluded that the differences in the rate of sap flow between *Salix* and *Juniperus* must be related to the differences in structure of the phloem in angiosperms and gymnosperms. Similarly, Thompson et al. (1979) found the speed of translocation movement in the phloem of *Picea mariana* and *Pinus banksiana* apparently to be much slower than that in the phloem of the broad-leaved angiosperms *Fraxinus americana* and *Ulmus americana*. In addition, during a study on the effect of cooling on the rate of translocation in the phloem of *Picea sitchensis* and *Abies procera*, the rates of progress of the profile of radioactivity (^{14}C-translocate) was only 2.7 and 5.7 cm per hour, respectively, in control stems (Watson, 1980). Moreover, the results of that study indicated that a free flow of translocate can occur in the sieve cells and that the pathway may therefore offer a low resistance.

What mechanism provides the driving force for long distance assimilate transport in gymnosperm phloem? As mentioned previously, the parietal ER of mature gymnospermous sieve cells is continuous with the ER of the albuminous cells via the numerous pore-plasmodesma connections in their common walls. Substances entering the sieve cells via these connections may remain largely in the ER system and move from sieve cell to sieve cell within the interconnected tubules, which form a continuum in the interrelated albuminous cell-sieve cell complexes of the phloem. It is not difficult to envisage an osmotically generated pressure-flow mechanism operating in gymnosperm phloem. Whereas translocation in angiosperm phloem involves virtually the entire cross-sectional area of the sieve-tube member, with the plasmalemma serving as the principal membrane across which osmosis takes place, in gymnosperm phloem it may be the many tubules of ER that are primarily involved in assimilate transport. If the fluid contents of the ER tubules have a higher osmotic potential than those of the cell lumen, osmosis across the ER membranes would provide the driving force for solution flow along the tubules. Such a system would not preclude a relatively slow solution flow within the cell lumina and across the intermembrane spaces of the sieve-area pores of contiguous sieve cells. Thompson et al. (1979) have also been tempted to speculate that some assimilate may move inside the channels of ER in gymnospermous sieve cells, although by some type of peristaltic movement.

Among the lower vascular plants, only the lycopods lack refractive spherules, a condition noted by Hébant et al. (1978) as being in agreement with the hypothesis that the lycopods constitute a relatively isolated group of primitive vascular plants. At maturity, the sieve elements of the lycopods are further characterized by the presence of plasmalemma-lined sieve-area pores that are unoccluded by any cytoplasmic material, being traversed by only an occasional tubule of ER.

With their relatively unoccluded sieve-area pores, it is not difficult to envisage operation of an osmotically generated pressure-flow mechanism in the phloem of the lycopods or of the leptosporangiate ferns. In addition, if an osmotically generated pressure-flow mechanism is possible in association with the tubular ER of gymnosperms, one might expect it to be possible in association with the tubular ER of the *Equisetum* sieve elements. It is difficult, however, to envisage such a mechanism providing the driving force for assimilate flow in *Psilotum* and the eu- and protoleptosporangiate ferns, in which the sieve-area pores are occluded with membranes not connected with the rest of the ER of the cell.

At this stage in our understanding of phloem structure in lower vascular plants it would be very helpful to know the translocation velocities of assimilates in the phloem of representative members of this diverse group of plants. If the velocities proved to be low, as they have been reported to be in the bracken fern (*Pteridium aquilinum*) at 5 cm per hour (Whittle, 1964), the resistance to solution flow provided by the membrane-filled sieve-area pores of *Psilotum* and the eu- and protoleptosporangiate ferns may not preclude operation of a pressure-flow mechanism. Moreover, if the velocities are very low, diffusion might be sufficient to explain assimilate movement in the phloem of these relatively primitive vascular plants.

The most primitive sieve elements encounted among the lower vascular plants are those in the leaf of *Isoetes muricata* that contain only plasmodesmata in their walls (Kruatrachue and Evert, 1974) and the primitive-like metaphloem sieve elements near the nodes of the *Equisetum hyemale* aerial stems (Dute and Evert, 1977b). Little wider than plasmodesmata, the connections between the *E. hyemale* elements are reminiscent of those interconnecting the leptoids in the leafy gametophytes of certain species of Polytrichales (Hébant, 1977). The similarity between the *E. hyemale* sieve elements and the leptoids does not end with the connections in their walls. Both protoplasts undergo a controlled autolysis resulting in degeneration of their nuclei and a general clearing of the cells, the surviving cytoplasmic components becoming largely parietal in distribution. The leptoids also contain inclusions that originate within the ER and are similar in appearance to refractive spherules.

Sieve elements are associated with parenchyma cells in all vascular plants — an association interpreted as a reflection of the close functional interdependence between these two cell types, and probably a result of the protoplasmic specialization of the sieve element. The highest level of specialization is found between the sieve-tube member and companion cell. Not only are these two cells derived from the same mother cell but they cease to function at the same time. Although not derived from the same mother cell, the sieve cells and albuminous cells of gymnosperms are also closely related physiologically. The albuminous cells have many cytologic similarities to companion cells; moreover, like the companion cells, they have numerous cytoplasmic connections with, and cease to function at the same times as, their associated sieve elements. Among the vascular cryptogams the sieve-element associated parenchyma cells are rarely distinctive cytologically but are very active metabotically and have numerous cytoplasmic connections with the sieve elements.

ACKNOWLEDGMENTS

Thanks are expressed to Susan E. Eichhorn for her assistance with preparation of the manuscript, including the figures, and to Jean T. Arnold, typist. The preparation of this article was supported in part by NSF grant PCM80-03855.

REFERENCES

Abbe, L.B. and Crafts, A.S. (1939). Phloem of white pine and other coniferous species. *Bot. Gaz.* 100, 695-722.

Alosi, M.C. and Alfieri, F.J. (1972). Ontogeny and structure of the secondary phloem in *Ephedra*. *Am. J. Bot.* 59, 818--827.

Aronoff, S., Dainty, J., Gorham, P.R., Srivastava, L.M. and Swanson, C.A., eds. (1975). "Phloem Transport." Plenum, New York.

Arsanto, J.-P. (1982). Observations on P-protein in dicotyledons. Substructural and developmental features. *Am. J. Bot.* 69, 1200-1212.

Behnke, H.-D. (1967). Über den Aufbau der Siebelement-Plastiden einiger Dioscoreaceen. *Z. Pflanzenphysiol.* 57, 243-254.

____ (1971). Über den Feinbau verdickter (nacré) Wände und der Plastiden in der Siebröhren von *Annona* und *Myristica*. *Protoplasma* 72, 69-78.

____ (1975). Companion cells and transfer cells. See Aronoff et al., pp. 153-175.

____ (1981). Sieve-element characters. *Nord. J. Bot.* 1, 381-400.

____ and Paliwal, G.S. (1973). Ultrastructure of phloem and its development in *Gnetum gnemon*, with some observations on *Ephedra campylopoda*. *Protoplasma* 78, 305-319.

Botha, C.E.J. and Evert, R.F. (1981). Studies on *Artemisia afra* Jacq.: The phloem in stem and leaf. *Protoplasma* 109, 217-231.

____, ____, Cross, R.H.M. and Marshall, D.J. (1982). Comparative anatomy of mature *Themeda triandra* Forsk. leaf blades: A correlated light and electron microscope study. *J.S. Afr. Bot.* 48, 311-328.

Browning, A.J., Hall, J.L. and Baker, D.A. (1980). Cytochemical localization of ATPase activity in phloem tissues of *Ricinus communis*. *Protoplasma* 104, 55-65.

Burr, F.A. and Evert, R.F. (1973). Some aspects of sieve-element structure and development in *Selaginella kraussiana*. *Protoplasma* 78, 81-97.

Buvat, R. and Robert, G. (1979). Activités golgiennes et origine des vacuoles dan les cellules criblées du protophloème de la racine de l'orge (*Hordeum sativum*). *Ann. Sci. Nat., Bot.* Ser. 13. 1, 51-66.

Carde, J.-P. (1974). Le tissue de transfert (=cellules de Strasburger) dans les aiguilles du pin maritime (*Pinus pinaster* Ait.). II. Caractères cytochimiques et infrastructuraux de la paroi et des plasmodesmes. *J. Microsc. (Paris)* 20, 51-72.

Catesson, A.M. (1982). Cell wall architecture in the secondary sieve tubes of *Acer* and *Populus*. *Ann. Bot.* 49, 131-134.

Chafe, S.C. and Doohan, M.E. (1972). Observations on the ultrastructure of the thickened sieve cell wall in *Pinus strobus* L. *Protoplasma* 75, 67-78.

Colbert, J.T. and Evert, R.F. (1982). Leaf vasculature in sugarcane (*Saccharum officinarum* L.). *Planta* 156, 136-151.

Cronshaw, J. (1975a). P-proteins. See Aronoff et al., pp. 79-115.

____ (1975b). Sieve element walls. See Aronoff et al., pp. 129-147.

____ (1980). ATPases in mature and differentiating phloem and xylem. *J. Histochem. Cytochem.* 28, 375-377.

____ (1981). Phloem structure and function. *Annu. Rev. Plant Physiol.* 32, 465-484.

____ and Esau, K. (1968). P-protein in the phloem of *Cucurbita*. I. The development of P-protein bodies. *J. Cell Biol.* 38, 25-39.

Danilova, M.F. and Telepova, M.N. (1978). Differentiation of protophloem sieve elements in seedling roots of *Hordeum vulgare*. *Phytomorphology* 28, 418-431.

____ and ____ (1981). Differentiation of proto- and metaphloem sieve elements in roots of *Hordeum vulgare* (Poaceae). *Bot. Zh. (Leningrad)* 66, 169-178.

DeFaÿ, E. and Hébant, C. (1981). Respiratory and phosphatase activities in parenchyma cells associated with conducting elements of ferns. *Ann. Bot.* 47, 703-707.

Deshpande, B.P. (1974a). Development of the sieve plate in *Saxifraga sarmentosa* L. *Ann. Bot.* 38, 151-158.

____ (1974b). On the occurrence of spiny vesicles in the phloem of *Salix*. *Ann. Bot.* 38, 865-868.

____ (1975). Differentiation of the sieve plate of *Cucurbita*: A further view. *Ann. Bot.* 39, 1015-1022.

____ (1976a). Observations on the fine structure of plant cell walls. II. The microfibrillar framework of the parenchymatous cell wall in *Cucurbita*. *Ann. Bot.* 40, 439-442.

____ (1976b). Observations on the fine structure of plant cell walls. III. The sieve tube wall of *Cucurbita*. *Ann. Bot.* 40, 443-446.

____ and Evert, R.F. (1970). A reevaluation of extruded nucleoli in sieve elements. *J. Ultrastruct. Res.* 33, 483-494.

Dute, R.R. and Evert, R.F. (1977a). Sieve-element ontogeny in the root of *Equisetum hyemale*. *Am. J. Bot.* 64, 421-438.

____ and ____ (1977b). Primitive-like metaphloem sieve elements in the aerial stem of *Equisetum hyemale*. *Protoplasma* 91, 257-266.

____ and ____ (1978). Sieve-element ontogeny in the aerial shoot of *Equisetum hyemale L*. *Ann. Bot.* 42, 23-32.

Eleftheriou, E.P. and Tsekos, I. (1982a). The ultrastructure of protophloem sieve elements in leaves of *Aegilopus comosa* var. *thessalica*. *Ann. Bot.* 49, 557-567.

____ and ____ (1982b). Developmental features of cell wall formation in sieve elements of the grass *Aegilopus comosa* var. *thessalica*. *Ann. Bot.* 50, 519-529.

____ and ____ (1982c). Development of protophloem in roots of *Aegilops comosa* var. *thessalica*. *II. Sieve-element differentiation*. *Protoplasma* 113, 221-233.

Ervin, E.L. and Evert, R.F. (1967). Aspects of sieve element ontogeny and structure in *Smilax rotundifolia*. *Bot Gaz.* 128, 138-144.

____ and ____ (1970). Observations on sieve elements in three perennial monocotyledons. *Am. J. Bot.* 57, 218-224.

Esau, K. (1969). The phloem. *In* "Encyclopedia of Plant Anatomy" (Eds. W. Zimmermann, P. Ozenda, H.D. Wulff), Vol. 5, Part 2. Gebrüder Borntraeger, Berlin-Stuttgart.

____ (1971a). The sieve element and its immediate environment: Thoughts on research of the past fifty years. *J. Indian Bot. Soc.* 50A, 115-129.

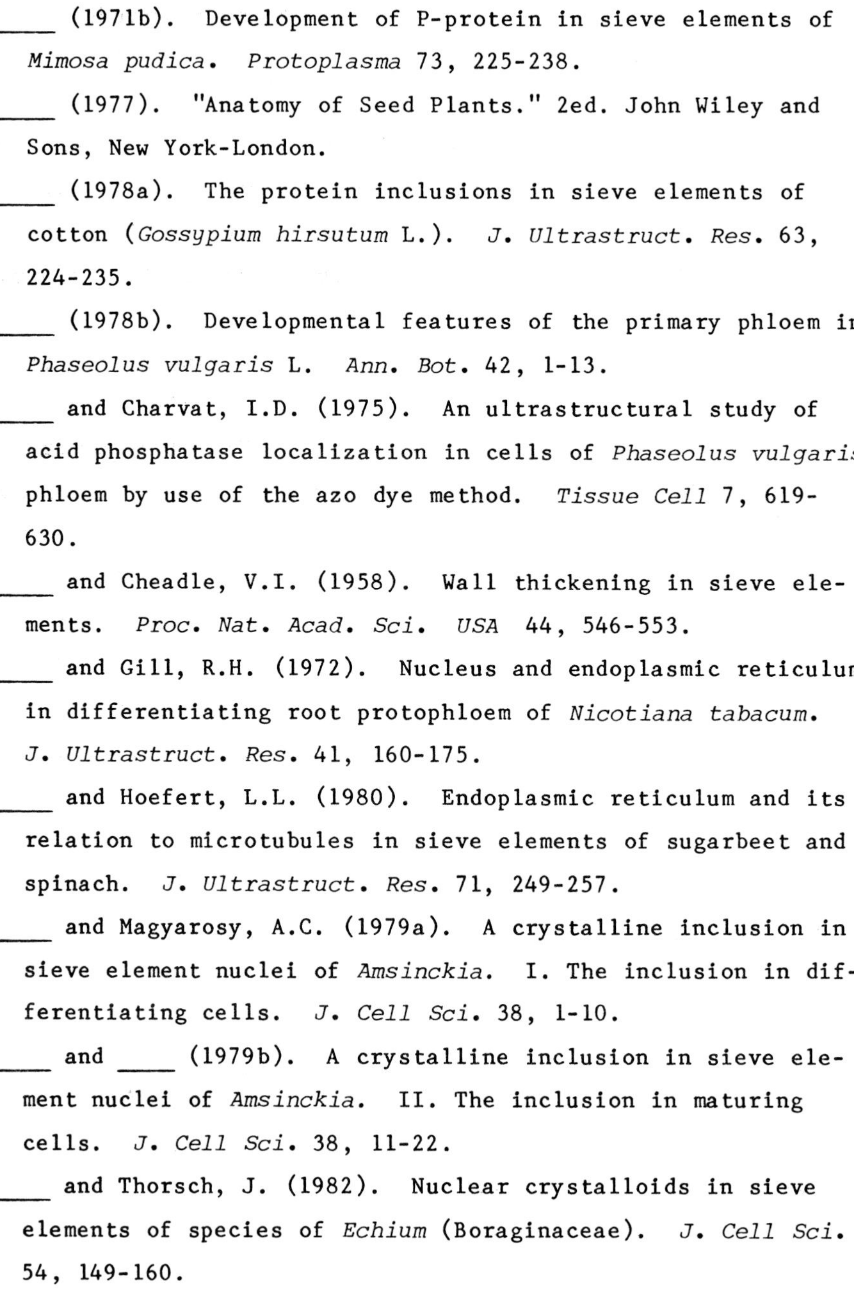

____ (1971b). Development of P-protein in sieve elements of *Mimosa pudica*. *Protoplasma* 73, 225-238.

____ (1977). "Anatomy of Seed Plants." 2ed. John Wiley and Sons, New York-London.

____ (1978a). The protein inclusions in sieve elements of cotton (*Gossypium hirsutum* L.). *J. Ultrastruct. Res.* 63, 224-235.

____ (1978b). Developmental features of the primary phloem in *Phaseolus vulgaris* L. *Ann. Bot.* 42, 1-13.

____ and Charvat, I.D. (1975). An ultrastructural study of acid phosphatase localization in cells of *Phaseolus vulgaris* phloem by use of the azo dye method. *Tissue Cell* 7, 619-630.

____ and Cheadle, V.I. (1958). Wall thickening in sieve elements. *Proc. Nat. Acad. Sci. USA* 44, 546-553.

____ and Gill, R.H. (1972). Nucleus and endoplasmic reticulum in differentiating root protophloem of *Nicotiana tabacum*. *J. Ultrastruct. Res.* 41, 160-175.

____ and Hoefert, L.L. (1980). Endoplasmic reticulum and its relation to microtubules in sieve elements of sugarbeet and spinach. *J. Ultrastruct. Res.* 71, 249-257.

____ and Magyarosy, A.C. (1979a). A crystalline inclusion in sieve element nuclei of *Amsinckia*. I. The inclusion in differentiating cells. *J. Cell Sci.* 38, 1-10.

____ and ____ (1979b). A crystalline inclusion in sieve element nuclei of *Amsinckia*. II. The inclusion in maturing cells. *J. Cell Sci.* 38, 11-22.

____ and Thorsch, J. (1982). Nuclear crystalloids in sieve elements of species of *Echium* (Boraginaceae). *J. Cell Sci.* 54, 149-160.

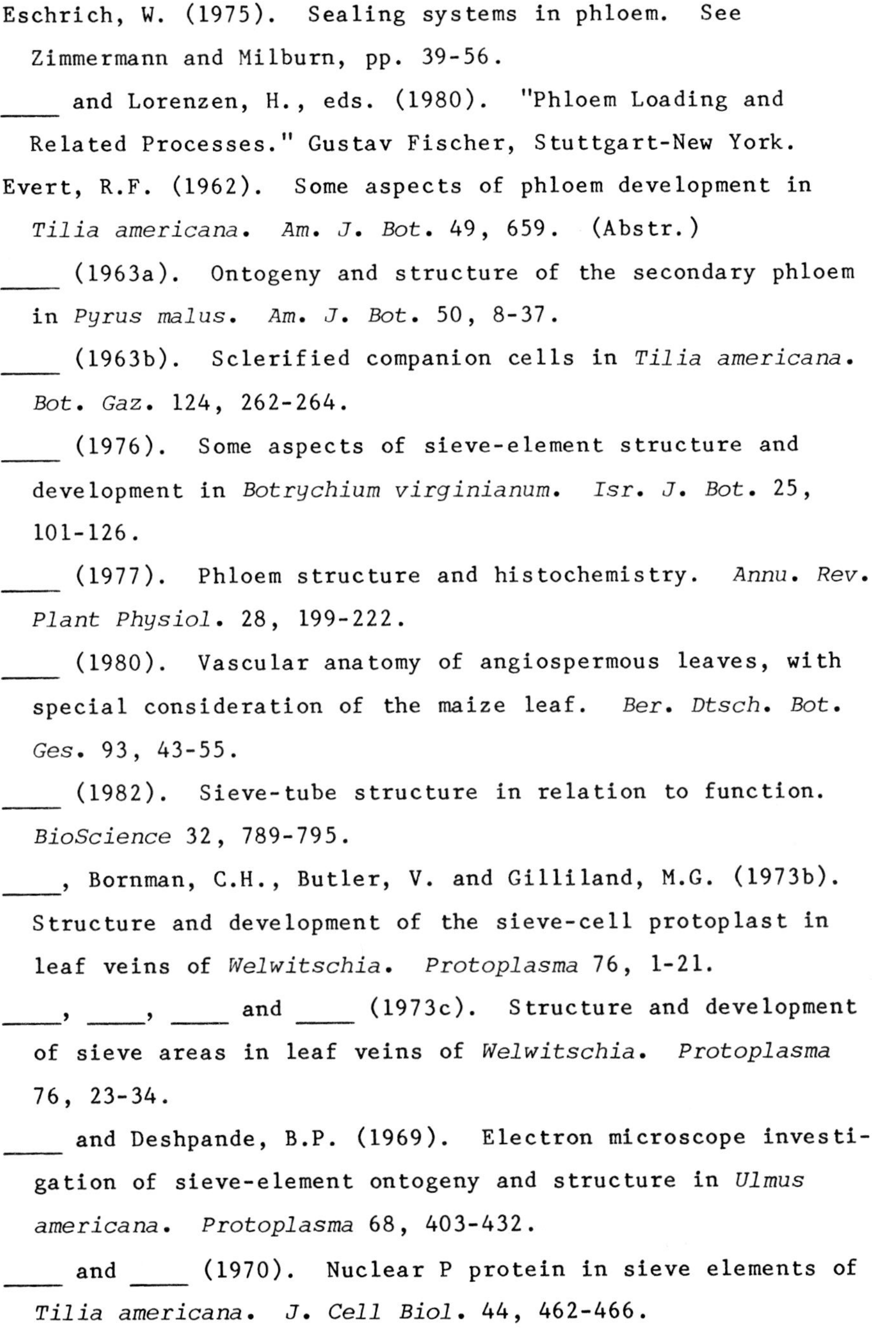

Eschrich, W. (1975). Sealing systems in phloem. See Zimmermann and Milburn, pp. 39-56.

____ and Lorenzen, H., eds. (1980). "Phloem Loading and Related Processes." Gustav Fischer, Stuttgart-New York.

Evert, R.F. (1962). Some aspects of phloem development in *Tilia americana*. *Am. J. Bot.* 49, 659. (Abstr.)

____ (1963a). Ontogeny and structure of the secondary phloem in *Pyrus malus*. *Am. J. Bot.* 50, 8-37.

____ (1963b). Sclerified companion cells in *Tilia americana*. *Bot. Gaz.* 124, 262-264.

____ (1976). Some aspects of sieve-element structure and development in *Botrychium virginianum*. *Isr. J. Bot.* 25, 101-126.

____ (1977). Phloem structure and histochemistry. *Annu. Rev. Plant Physiol.* 28, 199-222.

____ (1980). Vascular anatomy of angiospermous leaves, with special consideration of the maize leaf. *Ber. Dtsch. Bot. Ges.* 93, 43-55.

____ (1982). Sieve-tube structure in relation to function. *BioScience* 32, 789-795.

____, Bornman, C.H., Butler, V. and Gilliland, M.G. (1973b). Structure and development of the sieve-cell protoplast in leaf veins of *Welwitschia*. *Protoplasma* 76, 1-21.

____, ____, ____ and ____ (1973c). Structure and development of sieve areas in leaf veins of *Welwitschia*. *Protoplasma* 76, 23-34.

____ and Deshpande, B.P. (1969). Electron microscope investigation of sieve-element ontogeny and structure in *Ulmus americana*. *Protoplasma* 68, 403-432.

____ and ____ (1970). Nuclear P protein in sieve elements of *Tilia americana*. *J. Cell Biol.* 44, 462-466.

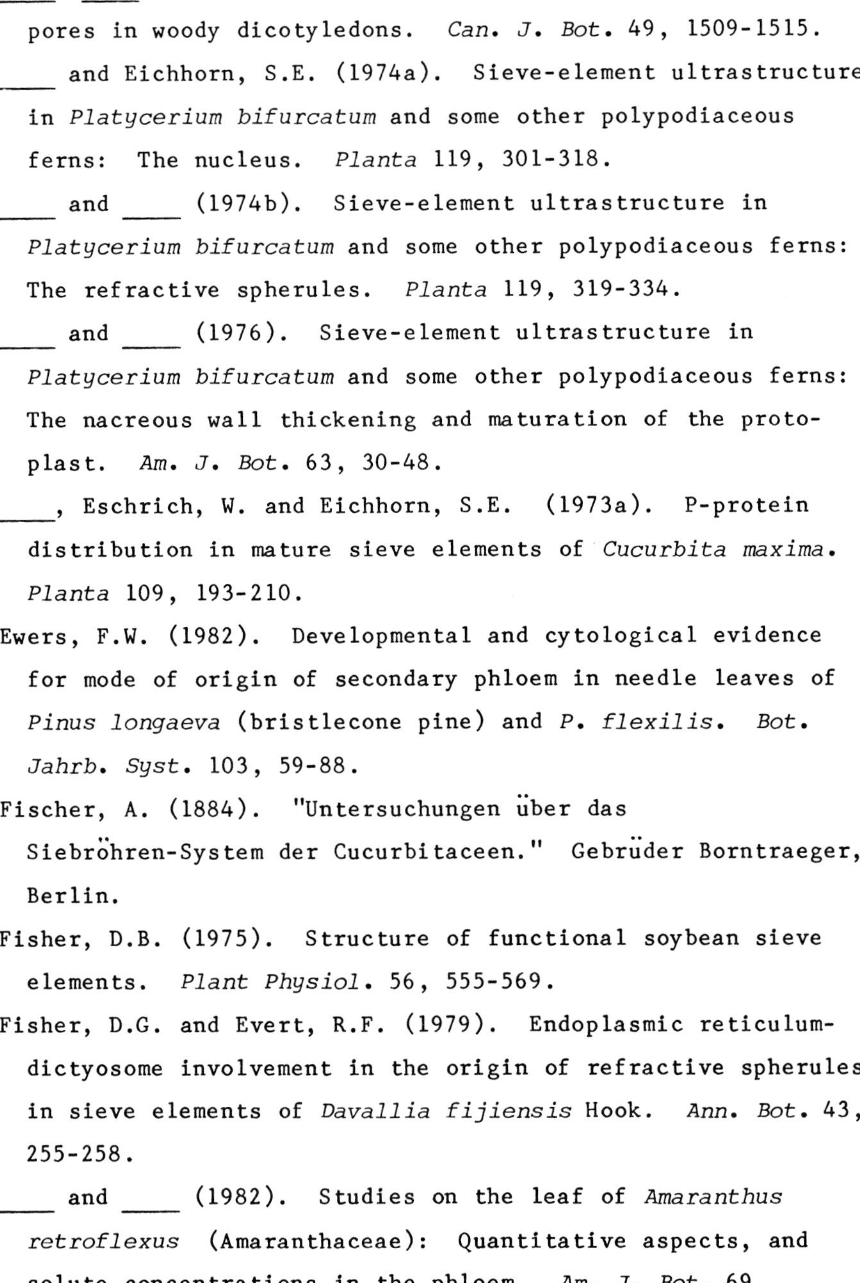

____, ____ and Eichhorn, S.E. (1971). Lateral sieve-area pores in woody dicotyledons. *Can. J. Bot.* 49, 1509-1515.

____ and Eichhorn, S.E. (1974a). Sieve-element ultrastructure in *Platycerium bifurcatum* and some other polypodiaceous ferns: The nucleus. *Planta* 119, 301-318.

____ and ____ (1974b). Sieve-element ultrastructure in *Platycerium bifurcatum* and some other polypodiaceous ferns: The refractive spherules. *Planta* 119, 319-334.

____ and ____ (1976). Sieve-element ultrastructure in *Platycerium bifurcatum* and some other polypodiaceous ferns: The nacreous wall thickening and maturation of the protoplast. *Am. J. Bot.* 63, 30-48.

____, Eschrich, W. and Eichhorn, S.E. (1973a). P-protein distribution in mature sieve elements of *Cucurbita maxima*. *Planta* 109, 193-210.

Ewers, F.W. (1982). Developmental and cytological evidence for mode of origin of secondary phloem in needle leaves of *Pinus longaeva* (bristlecone pine) and *P. flexilis*. *Bot. Jahrb. Syst.* 103, 59-88.

Fischer, A. (1884). "Untersuchungen über das Siebröhren-System der Cucurbitaceen." Gebrüder Borntraeger, Berlin.

Fisher, D.B. (1975). Structure of functional soybean sieve elements. *Plant Physiol.* 56, 555-569.

Fisher, D.G. and Evert, R.F. (1979). Endoplasmic reticulum-dictyosome involvement in the origin of refractive spherules in sieve elements of *Davallia fijiensis* Hook. *Ann. Bot.* 43, 255-258.

____ and ____ (1982). Studies on the leaf of *Amaranthus retroflexus* (Amaranthaceae): Quantitative aspects, and solute concentrations in the phloem. *Am. J. Bot.* 69, 1375-1388.

Friis, J.M. and Dute, R.R. (1983). Phloem of primitive angiosperms. II. P-protein in selected species of the Ranalean complex. *Proc. Iowa Acad. Sci.* (in press)

Gunning, B.E.S. (1976). The role of plasmodesmata in short distance transport to and from the phloem. *In* "Intercellular Communication in Plants: Studies on Plasmodesmata" (Eds. B.E.S. Gunning and A.W. Robards), pp. 203-227. Springer, Berlin-Heidelberg-New York.

____, Pate, J.S., Minchin, F.R. and Marks, I. (1974). Quantitative aspects of transfer cell structure in relation to vein loading in leaves and solute transport in legume nodules. *In* "Transport at the Cellular Level" (Eds. M.A. Sleigh and D.H. Jennings), pp. 87-126. Cambridge Univ. Press, Cambridge-London-New York.

Hardham, A.R. and Gunning, B.E.S. (1979). Interpolation of microtubules into cortical arrays during cell elongation and differentiation in roots of *Azolla pinnata*. *J. Cell Sci.* 37, 411-442.

Hébant, C. (1969). Observations sur le phloème de quelques Filicinées tropicales. *Nat. Monspel. Ser. Bot.* 20, 135-196.

____ (1976). Studies on the development of the conducting tissue-system in gametophytes of some Polytrichales. III. Further observations on leptoids, with particular reference to their endoplasmic reticulum. *Protoplasma* 87, 79-90.

____ (1977). "The Conducting Tissues of Bryophytes." J. Cramer, Vaduz, Liechtenstein.

____, Guiraud, R., Barthonnet, J. and Ba, A.T. (1978). Le phloème de *Lycopodium clavatum*: Organisation, ultrastructure et histochimie. *Can. J. Bot.* 23, 2973-2980.

____, ____ and Martin, M.G. (1980). Le phloème de *Selaginella willdenowii*: histophysiologie comparée. *Plant Syst. Evol.* 135, 159-169.

Hoefert, L.L. (1979). Ultrastructure of developing sieve elements in *Thlaspi arvense* L. I. The immature state. *Am. J. Bot.* 66, 925-932.

____ (1980). Ultrastructure of developing sieve elements in *Thlaspi arvense* L. II. Maturation. *Am. J. Bot.* 67, 194-201.

Holdheide, W. (1951). Anatomie mitteleuropäischer. *In* "Handbuch der Mikroskopie in der Tecknik" (Ed. H. Freund), Vol. 5, Part 1, pp. 193-367. Umschau-Verlag, Frankfurt am Main.

Hume, E.M.M. (1912). The histology of the sieve tubes of *Pteridium aquilinum*, with some notes on *Marsilea quadrifolia* and *Lygodium dichotomum*. *Ann. Bot.* 26, 573-587.

Jørgensen, L.B., Møller, J.D. and Wagner, P. (1975). Secondary phloem of *Trochodendron aralioides*. *Bot. Tidsskr.* 69, 217-238.

Kollmann, R. (1980). Fine structure and biochemical characterization of phloem proteins. *Can. J. Bot.* 58, 802-806.

____ and Dörr, I. (1966). Lokalisierung funktionstüchtiger Siebzellen bei *Juniperus communis* mit Hilfe von Aphiden. *Z. Pflanzenphysiol.* 55, 131-141.

Kruatrachue, M. (1978). Some aspects of phloem structure and development in *Marsilea quadrifolia* L. *J. Sci. Soc. Thailand* 4, 127-138.

____ and Evert, R.F. (1974). Structure and development of sieve elements in the leaf of *Isoetes muricata*. *Am. J. Bot.* 61, 253-266.

____ and ____ (1977). The lateral meristem and its derivatives in the corm of *Isoetes muricata*. *Am. J. Bot.* 64, 310-325.

____ and ____ (1978). Structure and development of sieve elements in the root of *Isoetes muricata* Dur. *Ann. Bot.* 42, 15-21.

Kuo, J. and O'Brien, T.P. (1974). Lignified sieve elements in the wheat leaf. *Planta* 117, 349-353.

Lamoureux, C.H. (1961). Comparative studies on phloem of vascular cryptogams. Ph.D. Diss. University of California, Davis.

Lawton, D.M. (1978a). P-protein crystals do not disperse in uninjured sieve elements in roots of runner bean (*Phaseolus multiflorus)* fixed with glutaraldehyde. *Ann. Bot.* 42, 353-361.

____ (1978b). Ultrastructural comparison of the tailed and tailless P-protein crystals respectively of runner bean *(Phaseolus multiflorus)* and garden peas (*Pisum sativum*) with tilting stage electron microscopy. *Protoplasma* 97, 1-11.

____ and Johnson, R.P.C. (1976). A superhelical model for the ultrastructure of "P-protein tubules" in sieve elements of *Nymphoides peltata*. *Cytobiologie* 14, 1-17.

____ and Newman, Y.M. (1979). Ultrastructure of phloem in young runner-bean stem: Discovery, in old sieve elements on the brink of collapse, of parietal bundles of P-protein tubules linked to the plasmalemma. *New Phytol.* 82, 213-222.

Lehmann, J. (1979). Nachweis von ATP und ATP-ase in den Siebröhren von *Cucurbita pepo*. *Z. Pflanzenphysiol.* 94, 331-338.

Liberman-Maxe, M. (1978). La paroi des cellules criblées dans le phloème d'une Fougère, le Polypode. *Biol. Cell.* 31, 201-210.

Loyal, D.S. and Maheshwari, A.K. (1979). Sieve cells vs sieve tube members in *Marsilea* — a reevaluation. *Phytomorphology* 29, 68-70.

Lucas, W.J. and Franceschi, V.R. (1982). Organization of the sieve-element walls of leaf minor veins. *J. Ultrastruct. Res.* 81, 209-221.

Maxe, M. (1964). Aspects infrastructuraux des cellules criblées de *Polypodium vulgare* (Polypodiacée). *C. R. Acad. Sci.* Ser. D 258, 5701-5704.

McEuen, A.R., Hart, J.W. and Sabnis, D.D. (1981). Calcium-binding protein in sieve tube exudate. *Planta* 151, 531-534.

Melaragno, J.E. and Walsh, M.A. (1976). Ultrastructural features of developing sieve elements in *Lemna minor* L. — the protoplast. *Am. J. Bot.* 63, 1145-1157.

Nehls, R., Schaffner, G. and Kollmann, R. (1978). Feinstruktur des Protein-Einschlusses in den Siebelementen von *Salix sachalinensis* Fr. Schmidt. *Z. Pflanzenphysiol.* 87, 113-127.

Neuberger, D.S. and Evert, R.F. (1974). Structure and development of the sieve-element protoplast in the hypocotyl of *Pinus resinosa*. *Am. J. Bot.* 61, 360-374.

____ and ____ (1975). Structure and development of sieve areas in the hypocotyl of *Pinus resinosa*. *Protoplasma* 84, 109-125.

____ and ____ (1976). Structure and development of sieve cells in the primary phloem of *Pinus resinosa*. *Protoplasma* 87, 27-37.

Oberhäuser, R. and Kollmann, R. (1977). Cytochemische Charakterisierung des sogenannten "Freien Nucleolus" als Proteinkörper in den Siebelementen von *Passiflora coerulea*. *Z. Pflanzenphysiol.* 84, 61-75.

Oparka, K.J. and Johnson, R.P.C. (1978). Endoplasmic reticulum and crystalline fibrils in the root protophloem of *Nymphoides peltata*. *Planta*, 143, 21-27.

____, ____ and Bowen, J.D. (1981). Sites of acid phosphatase in the differentiating root protophloem of *Nymphoides peltata* (S.G. Gmel.) O. Kuntze. *Plant Cell Environ.* 4, 27-35.

Outer, R.W. den (1967). Histological investigations of the secondary phloem of gymnosperms. *Meded. Landbouwhogesch. Wageningen* 67-7, 1-119.

Palevitz, B.A. and Newcomb, E.H. (1970). A study of sieve element starch using sequential enzymatic digestion and electron microscopy. *J. Cell Biol.* 45, 383-398.

Parameswaran, N. and Liese, W. (1970). Zur Cytologie der Strasburger Zellen in Coniferennadeln. *Naturwissenschaften* 1, 45-46.

Parthasarathy, M.V. (1975). Sieve-element structure. See Zimmermann and Milburn, pp. 1-38.

____ (1980). Mature phloem of perennial monocotyledons. *Ber. Dtsch. Bot. Ges.* 93, 57-70.

____ and Pesacreta, T.C. (1980). Microfilaments in plant vascular cells. *Can. J. Bot.* 58, 807-815.

Pate, J.S. and Gunning, B.E.S. (1969). Vascular transfer cells in angiosperm leaves. A taxonomic and morphological survey. *Protoplasma* 68, 135-156.

____ and ____ (1972). Transfer cells. *Annu. Rev. Plant Physiol.* 23, 173-196.

Perry, J.W. and Evert, R.F. (1975). Structure and development of the sieve elements in *Psilotum nudum*. *Am. J. Bot.* 62, 1038-1052.

Poirault, G. (1893). Recherches anatomiques sur les Cryptogames vasculaires. *Ann. Sci. Nat., Bot.* Ser. 7. 18, 113-256.

Sabnis, D.D. and Hart, J.W. (1979). Heterogeneity in phloem protein complements from different species. *Planta* 145, 459-466.

Sauter, J.J. (1980). The Strasburger cells — equivalents of companion cells. *Ber. Dtsch. Bot. Ges.* 93, 29-42.

____ and Braun, H.J. (1968). Histologische und cytochemische Untersuchungen zur Funktion der Baststrahlen von *Larix decidua* Mill., unter besonderer Berücksichtigung der Strasburger-Zellen. *Z. Pflanzenphysiol.* 59, 420-438.

____, Dörr, I. and Kollmann, R. (1976). The ultrastructure of Strasburger cells (= albuminous cells) in the secondary phloem of *Pinus nigra* var. *austriaca* (Hoess) Badoux. *Protoplasma* 88, 31-49.

Shah, J.J. and Fotedar, R.L. (1974a). Sieve-tube members in the stem of *Cyathea gigantea*. *Am. Fern J.* 64, 27-28.

____ and ____ (1974b). Structure and development of phloem in the rachis of *Pteris longifolia*. *Phytomorphology* 24, 107-113.

____ and Nair, M.N.B. (1978). Nuclear autolysis in the young sieve cells of some ferns. *Nucleus (Calcutta)* 21, 161-168.

Singh, A.P. (1980). On the ultrastructure and differentiation of the phloem in sugarcane leaves. *Cytologia* 45, 1-31.

Sloan, R.T., Sabnis, D.D. and Hart, J.W. (1976). The heterogeneity of phloem exudate proteins from different plants: A comparative survey of ten plants using polyacrylamide gel electrophoresis. *Planta* 132, 97-102.

Spanner, D.C. (1978a). Sieve-plate pores, open or occluded? A critical review. *Plant Cell Environ.* 1,7-20.

____ (1978b). The Münch hypothesis, freeze substitution and the structure of sieve-plate pores. *Ann. Bot.* 42, 485-488.

____ and Moattari, F. (1978). The significance of P-protein and endoplasmic reticulum in sieve elements in light of evolutionary origins. *Ann. Bot.* 42, 1469-1472.

Srivastava, L.M. (1970). The secondary phloem of *Austrobaileya scandens*. *Can. J. Bot.* 48, 341-359.

Thiéry, J.P. (1967). Mise en évidence des polysaccharides sur coupes fines en microscopie électronique. *J. Microsc. (Paris)* 6, 987-1018.

Thompson, R.G., Fensom, D.S., Anderson, R.R., Drouin, R. and Leiper, W. (1979). Translocation of ^{11}C from leaves of *Helianthus, Heracleum, Nymphoides, Ipomoea, Tropaeolum, Zea, Fraxinus, Ulmus, Picea,* and *Pinus*: Comparative shapes and some fine structural profiles. *Can. J. Bot.* 57, 845-863.

Thorsch, J. and Esau, K. (1981a). Changes in the endoplasmic reticulum during differentiation of a sieve element in *Gossypium hirsutum*. *J. Ultrastruct. Res.* 74, 183-194.

____ and ____ (1981b). Nuclear degeneration and the association of endoplasmic reticulum with the nuclear envelope and microtubules in maturing sieve elements of *Gossypium hirsutum*. *J. Ultrastruct. Res.* 74, 195-204.

____ and ____ (1981c). Ultrastructural studies of protophloem sieve elements in *Gossypium hirsutum*. *J. Ultrastruct. Res.* 75, 339-351.

____ and ____ (1982). Microtubules in differentiating sieve elements of *Gossypium hirsutum*. *J. Ultrastruct. Res.* 78, 73-83.

Walsh, M.A. (1980). Preservation of the tonoplast in metaphloem sieve elements of embryonic roots of *Zea mays* L. *Ann. Bot.* 46, 557-565.

____ and Melaragno, J.E. (1976). Ultrastructural features of developing sieve elements in *Lemna minor* L. — sieve plate and lateral sieve areas. *Am. J. Bot.* 63, 1174-1183.

____ and Popovich, T.M. (1977). Some ultrastructural aspects of metaphloem sieve elements in the aerial stem of the holoparasitic angiosperm *Epifagus virginiana* (Orobanchaceae). *Am. J. Bot.* 64, 326-336.

Warmbrodt, R.D. and Evert, R.F. (1974a). Structure and development of the sieve element in the stem of *Lycopodium lucidulum*. *Am. J. Bot.* 61, 267-277.

____ and ____ (1974b). Structure of the vascular parenchyma in the stem of *Lycopodium lucidulum*. *Am. J. Bot.* 61, 437-443.

____ and ____ (1978). Comparative leaf structure of six species of heterosporous ferns. *Bot. Gaz.* 139, 393-429.

____ and ____ (1979a). Comparative leaf structure of six species of eusporangiate and protoleptosporangiate ferns. *Bot. Gaz.* 140, 153-167.

____ and ____ (1979b). Comparative leaf structure of several species of homosporous leptosporangiate ferns. *Am. J. Bot.* 66, 412-440.

Watson, B.T. (1980). The effect of cooling on the rate of phloem translocation in the stems of two gymnosperms, *Picea sitchensis* and *Abies procera*. *Ann. Bot.* 45, 219-223.

Weatherley, P.E., Peel, A.J. and Hill, G.P. (1959). The physiology of the sieve tube. Preliminary experiments using aphid mouth parts. *J. Exp. Bot.* 10, 1-16.

Wergin, W.P., Gruber, P.J. and Newcomb, E.H. (1970). Fine structural investigation of nuclear inclusions in plants. *J. Ultrastruct. Res.* 30, 533-557.

Whittle, C.M. (1964). Translocation in *Pteridium*. *Ann. Bot.* 28, 331-338.

Willenbrink, J. and Kollmann, R. (1966). Über den Assimilattransport in Phloem von *Metasequoia*. *Z. Pflanzenphysiol.* 55, 42-53.

Wooding, F.B.P. (1967). Fine structure and development of phloem sieve tube content. *Protoplasma* 64, 315-324.

Zimmermann, M.H. and Milburn, J.A., eds. (1975). "Encyclopedia of Plant Physiology." Vol. I. Transport in Plants 1. Phloem Transport. Springer, Berlin-Heidelberg.

CELLULAR PARAMETERS OF LEAF MORPHOGENESIS IN MAIZE AND TOBACCO

Scott Poethig[1]

Department of Agronomy
University of Missouri
Columbia, Missouri

Clonal analysis of leaf morphogenesis in maize and tobacco presents a picture of this phenomenon that differs in several important respects from classical views of leaf morphogenesis. This approach demonstrates that 1) the leaf primordium and the leaf blade are derived from a relatively large group of cells in each of several tissue layers rather than from a small number of cells in 1 or 2 layers, 2) these founder cells do not function as "stem" cells; each founder cell lineage has a limited and characteristic destiny, 3) neither the orientation of cell division at the margin of the blade, nor the frequency of cell division in this region indicates the existence of a marginal meristem, and 4) both the rate and duration of cell division exhibit considerable spatial and temporal variation during the expansion of the leaf blade.

[1]Present address: University of Pennsylvania, Department of Biology, Philadelphia, PA 19104

ISBN 0-12-746620-7

I. INTRODUCTION

The shape of a plant organ is a result of spatial variation in the amount and orientation of growth. It is usually assumed that these growth processes can be explained at a cellular level in terms of the amount and orientation of cell division, but establishing a connection between specific cellular parameters and specific morphogenetic phenomena has not been easy. Because cell division and cell expansion coincide during much of development and resist experimental separation, their relative importance in morphogenesis remains unclear. The lack of accurate quantitative information about spatial and temporal patterns of cell division and cell expansion is another factor that has contributed to confusion concerning the cellular basis of plant morphogenesis. Most analyses of this problem are based on observations of sectioned specimens and have been influenced in several important ways by the limitations of this approach. In practice, it is only possible to examine a small amount of tissue using sectioned specimens. As a result, attention is often restricted to limited regions of a primordium, leading investigators to emphasize the role of localized meristems in plant morphogenesis. Compounding this regional bias is the fact that specimens are generally viewed in only one or two planes, and almost never paradermally. Thus, virtually nothing is known about the behavior of the epidermis during organogenesis, or about regional variation in cell behavior within planar structures such as the leaf blade. Finally there is the problem of tracing cell lineages in histological specimens. Although it is generally believed that cell lineages can be reliably reconstructed from cell patterns in histological specimens, recent studies make it clear that this is true only over short distances and only in organs blessed with a regular pattern of cell division, such as roots. Interpretations of cell lineage patterns during leaf morphogenesis, for example, have often produced more confusion than edification.

In recent years several new techniques have yielded detailed information about the spatial pattern of cell division during leaf and shoot morphogenesis, and have provided novel views of the cell lineage of these structures. These techniques involve the use of cleared rather than sectioned specimens (Fuchs, 1966, 1975, 1976; Thomasson, 1970; Jeune, 1972), vital dyes (Green and Lang, 1981; Green and Poethig, 1982) and induced somatic mutations (Stein and Steffensen, 1959; Steffensen, 1968; Dulieu, 1968, 1969; Coe and Neuffer, 1978). Although the functional significance of these parameters of cell behavior remains to be established, it is clear that many cherished ideas about the cellular basis of plant morphogenesis must be abandoned. Particularly uncertain are traditional concepts concerning the function of initial cells and the role of localized meristems in morphogenesis.

In this paper, I will describe some of the cellular parameters of leaf development in maize (*Zea mays* L.) and tobacco (*Nicotiana tabacum* L.) as determined from an analysis of x-ray induced somatic sectors (clonal analysis). Three parameters will be considered: 1) The number of founder cells that initiate a leaf primordium, 2) the fate of the lineages of these founder cells and 3) the frequency distribution of cell division within the leaf blade. It must be admitted that there is no conclusive evidence that these parameters directly determine leaf shape. To ascertain whether leaf shape is in any way dependent on cell lineage or on the pattern of cell division, changes in these factors must be correlated with changes in leaf morphology and the dynamics of leaf growth. Although feasible, this analysis has not yet been undertaken. The results of this study do, however, point to some factors that may be critical in leaf development and demonstrate that, despite their obvious difference in morphology, maize and tobacco leaves share remarkably similar patterns of cell division.

II. METHODOLOGY

The technique of clonal analysis depends on the ability of ionizing radiation to induce chromosomal aberrations in somatic cells. In a normal plant, such aberrations rarely produce a visible phenotype. When the plant is heterozygous for a recessive cell marker gene, however, loss of the dominant allele of that gene results in the expression of the recessive allele, yielding a visible somatic sector. The tobacco stock used in this study was heterozygous for two mutations (*a1* and *a2*) which interact in reducing the amount of chlorophyll. Leaf cells heterozygous for both mutations are yellow green. Loss of either the *a1* or *a2* mutation produces a green sector, while loss of the wild type allele of either gene produces a yellow or white sector (Dulieu, 1974). Only green sectors were scored in this study. The maize stock was heterozygous for *wd*, a recessive albino mutation, and hence yielded white sectors upon irradiation.

Heterozygous plants were exposed to 100-300R of x-rays and a representative sample of plants was then dissected to determine the developmental stage of the leaf of interest. The remaining plants were allowed to grow until that leaf reached maturity, and then the location, size and shape of clones were recorded photographically or on a photocopy of the leaf. Observations were taken on leaves at the same 3 successive nodes in each experiment, and these data were pooled in order to recreate the developmental history of a single "typical" leaf.

In the case of tobacco, an analysis of periclinal chimeras was also undertaken. The chimeras described here were obtained by pruning heteroplastidic plants carrying the plastid mutation Dpl. The structure of these chimeras has been described by Stewart and Burk (1970).

III. LEAF MORPHOGENESIS IN TOBACCO

A. Leaf Initiation

Avery (1933) originally attributed the initation of the tobacco leaf to a single subapical initial, but more recent investigations (Dulieu, 1968; Stewart and Dermen, 1975) have demonstrated that the leaf is derived instead from a group of cells encompassing at least 3 cell layers. Evidence for this conclusion is provided by periclinal chimeras--plants in which one of the three fundamental cell layers in the apical meristem (designated LI, LII and LIII) is genotypically different from the other layers. The fate of these lineages is most readily seen when the subepidermal layer carries an albino mutation, a condition symbolized GWG (green-white-green). In such chimeras, the LI contributes primarily (if not solely) to the epidermis of the leaf, the LII forms one or more subepidermal layers in the midrib and lamina and is the source of all the mesophyll tissue near the leaf margin, while the LIII forms the core of the midrib and the middle mesophyll layers in the central region of the leaf blade (Dulieu, 1968; Stewart and Burk, 1970). Since the leaf contains derivatives of all 3 fundamental cell layers, it must arise from at least 3 layers of cells. But recent data show that in some cases the leaf primordium actually extends deeper into the apical meristem. This conclusion is based on observations of GWG chimeras displaying an extensive GWWG sector. The GWWG sectors examined here arose from a periclinal division in the LII at the flank of the meristem and extended for several nodes. If a tobacco leaf was derived from only 3 cell layers, then the internal tissue of a leaf arising in such a sector would be completely white. Instead, such leaves often have a central core of green tissue extending about halfway up the midrib. This indicates that a tobacco leaf is derived from at least 4 layers of the apical meristem as illustrated in Figure 1.

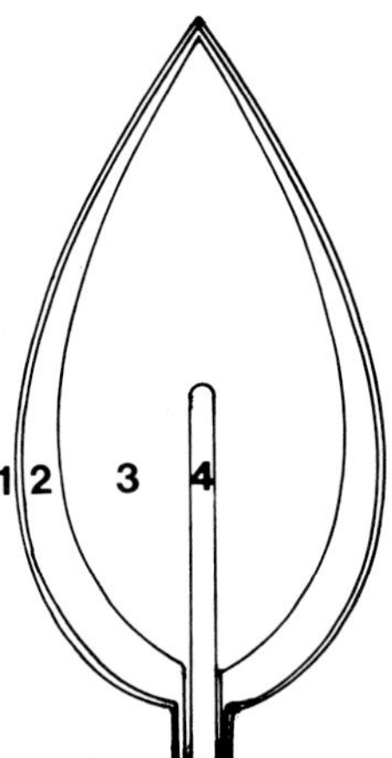

FIGURE 1. The distribution of tissue derived from 4 layers of the apical meristem in the median paradermal plane of a tobacco leaf.

Chimeras are useful for studying the fate of cell layers but say little about the behavior of individual cells in these layers. Information about the number and fate of the founder cells of a leaf primordium must be obtained by other techniques. Dulieu (1968) studied the cell lineage of the tobacco leaf, and the leaves of several other species, by treating seeds with EMS in order to induce albino sectors. In every species leaf sectors displayed the same pattern. Instead of emanating radially from the tip of the leaf--as would have been expected had the leaf been derived from a few apical cells--sectors ran from the base of the leaf to a point along the leaf margin. Only sectors located in the central plane of the midrib extended all the way to the tip of the leaf. The principal difference between the sectors in a broad leaf like that of tobacco, and a narrow leaf like that of flax, was the width of sectors in the lamina and their angle of inflection from the midrib.

Figure 2 is a schematic drawing of the type of sectors seen in tobacco leaves irradiated several plastochrons before initiation. These sectors share most of the features mentioned by Dulieu (1968) but are less deleterious than EMS-induced sectors. The sectors illustrated here are confined to the subepidermal layer of the leaf (except at the leaf margin) and form a continuous band extending from the internode on one side of the leaf to the internode on the other. Their shape and distribution confirms Dulieu's picture of the cell lineage of the tobacco leaf, as presented in Figure 3. It is apparent that the cells that initiate a leaf do not function in the manner of "stem cells" that reside at the apex of an indeterminate structure such as a shoot or root. Since the term initial cell generally connotes stem cell behavior, cells initiating a leaf or part of a leaf will be termed founder cells.

The number of founder cells in the horizontal and vertical dimensions of the leaf primordium can be estimated from the size and extent of sectors induced just prior to leaf initiation. These data indicate that the primordium encompasses 15 to 20 subepidermal cells in its horizontal dimension and at least 3 cells in its vertical dimension (Poethig, 1981). Assuming the number of founder cells in the other layers of the leaf primordium equals the number of founder cells in the subepidermal layer, this means that a tobacco leaf is derived from somewhere around 150 cells. On topological grounds, it is likely that the primordium encompasses fewer cells in the internal layers of the meristem than in the epidermal and subepidermal layers.

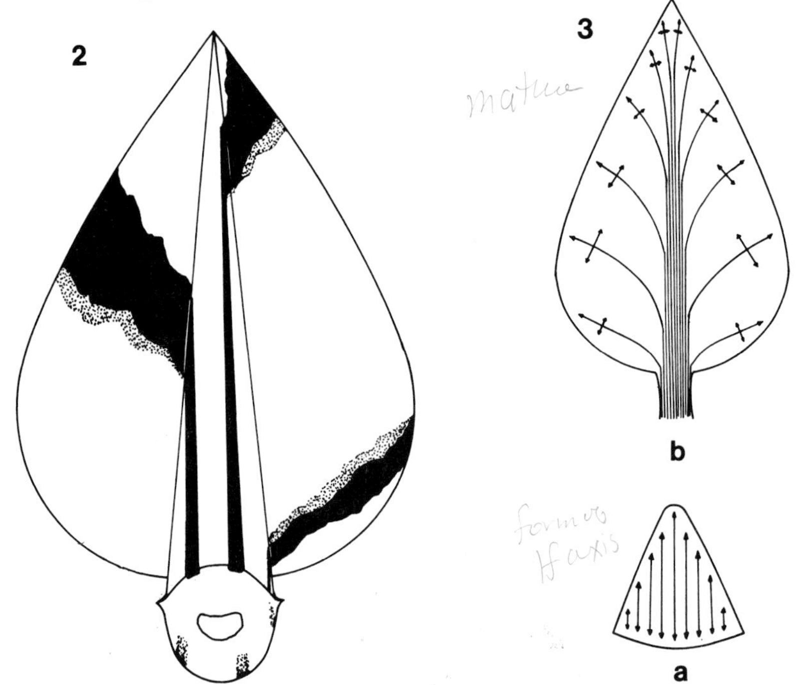

FIGURE 2. Subepidermal sectors induced 2-3 plastochrons before leaf initiation. Each sector extends in a continuous band from one side of the leaf to the other.

FIGURE 3. The fate of founder cell lineages during the formation of the leaf axis (a) and in a mature leaf (b). Redrawn from Dulieu and Bugnon (1966).

B. The Development of the Leaf Blade

The function of marginal cells in the initation and

expansion of the leaf blade is controversial. Some investigators have reported that the internal tissue of the leaf blade arises entirely from a single row of submarginal cells, and that the orientation of cell division at the leaf margin is both highly regulated and species specific (reviewed by Foster, 1936). Avery (1933) pictured submarginal cells as contributing in succession to the upper, middle and lower layers of the mesophyll. More recent investigators have challenged this interpretation, however, because quantitative analyses of the orientation and frequency of cell division at the leaf margin do not suggest that it plays a special role in leaf expansion (Maksymowych and Erickson, 1960; Fuchs, 1966, 1968, 1975, 1976; Dubuc-Lebreux and Sattler, 1980).

Clonal analysis supports the conclusion that cells at the margin of the leaf blade play a relatively minor role in its expansion. If a single row of submarginal cells initiated the blade, then submarginal sectors induced before the initiation of the blade would extend from the margin to the midrib, encompassing both the upper and lower layers of the blade. Instead they are usually confined to one surface of the leaf and tend to be elongated parallel to the margin (Figure 4). Sectors that do not abut the margin are usually not conspicuously elongated. Thus submarginal cells generally divide perpendicular to the margin rather than parallel to it, and hence contribute little to the transverse expansion of the blade. In addition, sectors induced during the initial phase of blade expansion reveal that the margin does not possess an unusually high frequency of cell division. Figure 5 illustrates the number of sectors in 25 mm^2 sections of the leaf blade in leaves irradiated at a length of 8 mm and 13 mm. In order to compare the relative frequency of sectors in different regions of the blade, the number of sectors in each

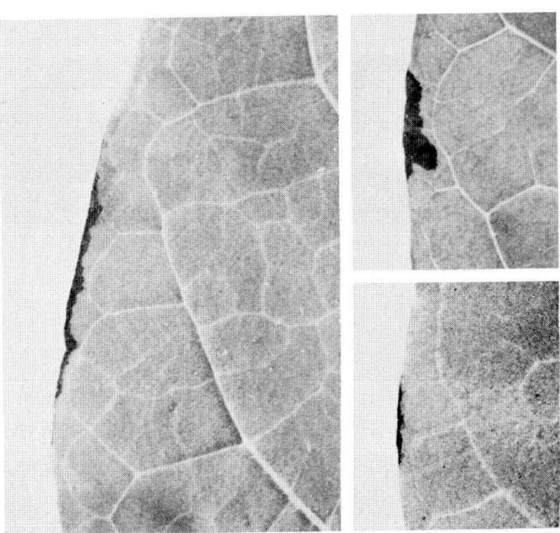

FIGURE 4. Submarginal sectors induced prior to or shortly after the initiation of the leaf blade.

square was divided by the average number of sectors per square in that transverse section of the blade. Squares with more than the average number are colored black. It is apparent from these data that regions containing relatively large numbers of clones occur more or less randomly throughout the blade early in leaf expansion. Since variation in the frequency of clones within a localized region of the lamina is correlated with variation in the frequency of cell division (Poethig, 1981) it can be concluded, therefore, that cell division occurs at about the same frequency within any transverse section of the leaf.

There is, on the other hand, significant variation in the

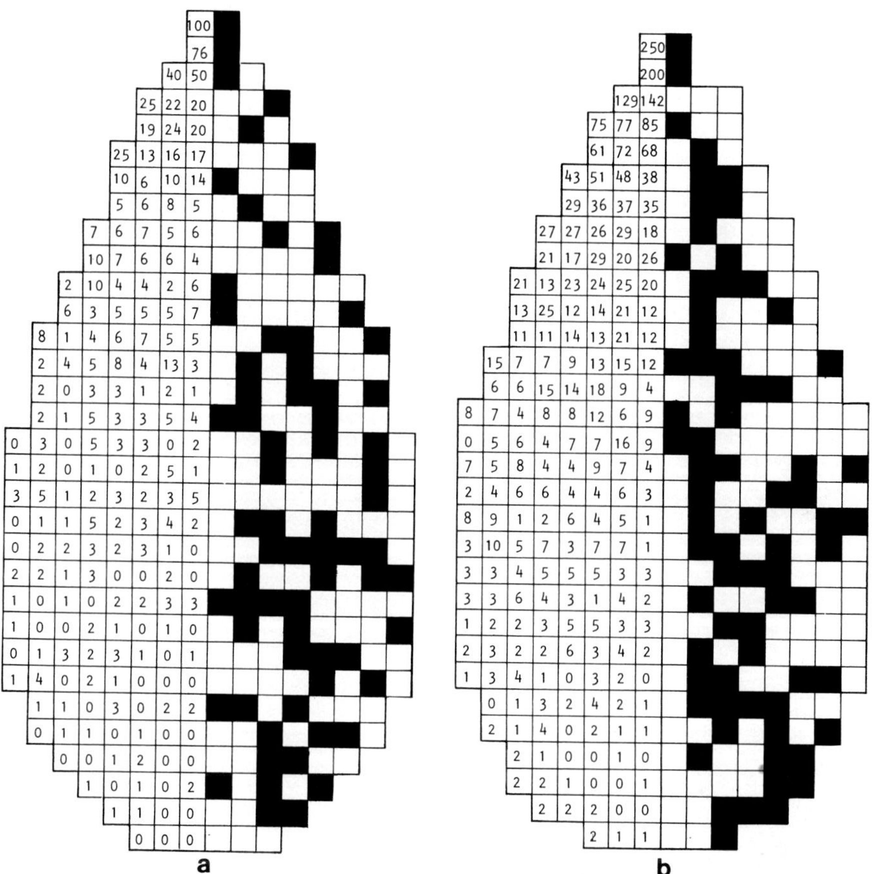

FIGURE 5. The distribution of sectors in leaves irradiated at a length of 8 mm (a) and 13 mm (b). These leaves were taken from successive nodes on the same plant, and differ in age by about 29 hrs. The number of sectors in corresponding sectors on both halves of a leaf were summed and is shown on the left of the diagrams. The black squares on the right half represent the sections with 1.1 or more times the average number of sectors in that transverse section of the lamina.

frequency of cell division along the length of the leaf. In a leaf irradiated during the period when it is entirely meristematic, the number of sectors per unit area decreases from the tip to the base of the leaf, while the average size of a sector increases (Figure 6). This phenomenon can be attributed, in part, to a gradient in the duration of cell division. As Avery (1933) originally noted, the tobacco leaf

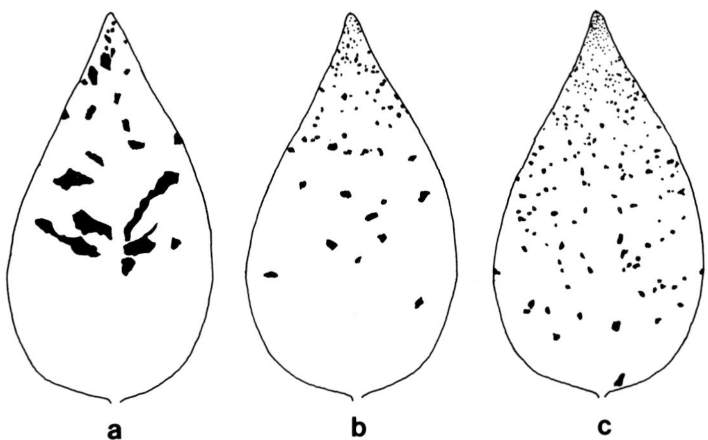

FIGURE 6. Composite diagrams of subepidermal sectors in leaves irradiated at a length of 0.7 mm (a) 1.5 mm (b) and 3 mm (c).

matures basipetally. Clonal analysis demonstrates that cell division stops at the tip of a leaf when it is about 1/10 its final size (here about 15-18 mm) and subsequently in progressively more basal regions of the blade. Thus, at any stage in development, a cell at the base of the leaf will contribute proportionately more to the growth of the blade than a cell at the tip. In terms of clonal analysis, this means that basal sectors should be larger than apical sectors and occur less frequently per unit area--as is, in fact, the case. Evidence for longitudinal variation in the frequency of cell division comes from an analysis of the sensitivity of cells to irradiation, a parameter known to be directly related to the rate of cell division. This parameter can be estimated by dividing the average number of sectors per unit area by the number of progenitor cells for that region--calculated as the reciprocal of average sector size. The sensitivity of cells to irradiation in a leaf 1.2 cm long is illustrated in Figure 7, while the frequency of mitosis in a leaf of the

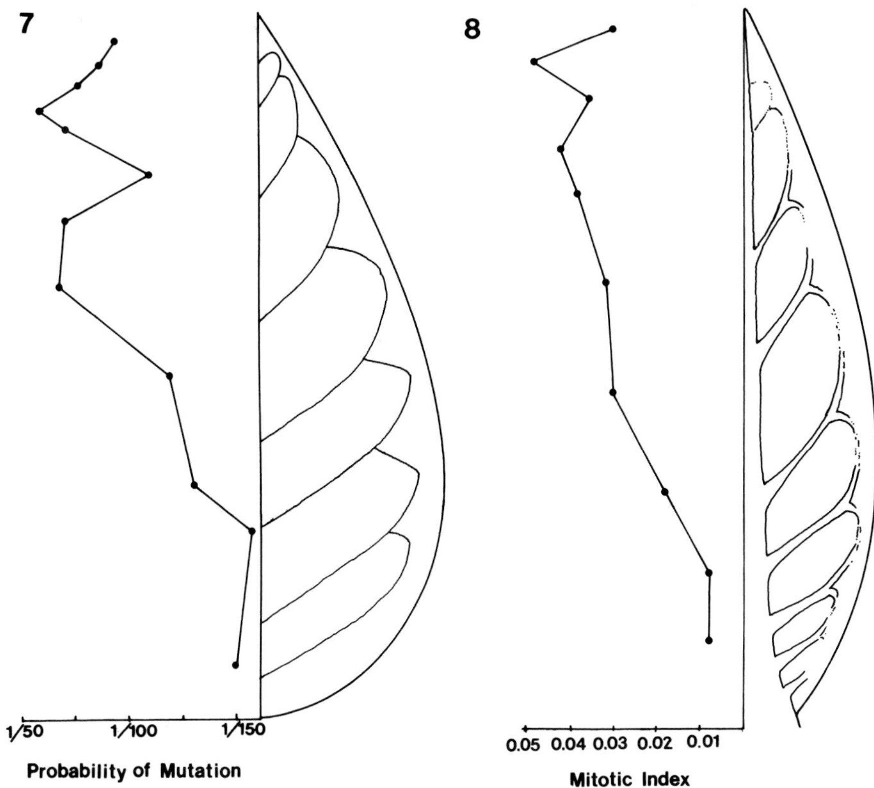

FIGURE 7. The probability of mutation in the palisade layer of a leaf irradiated at a length of 1.2 cm, calculated from the frequency and size of sectors at maturity. The leaf is 16 cm long.

FIGURE 8. The frequency of mitosis (mitotic index) in the palisade layer of a 1.2 cm long leaf, as determined from a cleared, Feulgen-stained specimen.

parameters indicate that the frequency (rate) of cell division is much lower at the base of the leaf than at the tip of the leaf at this early stage in development. After the cessation of cell division at the tip of the leaf the sensitivity of cells to irradiation increases dramatically at the base of the leaf. Throughout most of the post-emergent phase of leaf growth, the sensitivity of cells to irradiation (and presumably the frequency of cell division) is higher at the base of the leaf than in any other region. It is interesting that this increase in sensitivity to irradiation is correlated with an increase in the relative growth rate at the base of the leaf blade (Poethig, 1981).

In tobacco, variation in the rate and duration of cell division probably plays a more significant role in determining leaf shape than the orientation of cell division. There is some evidence that cell division is polarized early in the expansion of the lamina, but the degree of polarization is not great, and by the time the leaf is 1/10 final size, cell division no longer appears to be preferentially oriented.

IV. LEAF MORPHOGENESIS IN MAIZE

Histological observations suggest that the corn leaf arises from the outer two layers of the shoot apical meristem (Sharman, 1942). Clonal analysis confirms this conclusion, but demonstrates that the behavior of individual cells in these two layers is remarkably variable; sometimes LI cells contribute to a substantial portion of the mesophyll, and at other times the derivatives of this layer remain confined to the epidermis. Support for this conclusion is provided by sectors induced 1-2 plastochrons before leaf initiation. Four types of sectors have been observed (Figure 9):

a. Sectors running the length of the leaf blade and encompassing both epidermis and underlying mesophyll.
b. Sectors running the length of the leaf blade, but confined to the mesophyll.
c. Sectors in which both the epidermis and mesophyll are white from the leaf margin to a point in the middle of the leaf blade, beyond which only the epidermis is

white. The white sector in the epidermis may or may not reach the leaf sheath.

d. Mesophyll sectors extending from a point within the leaf blade into the leaf sheath or culm.

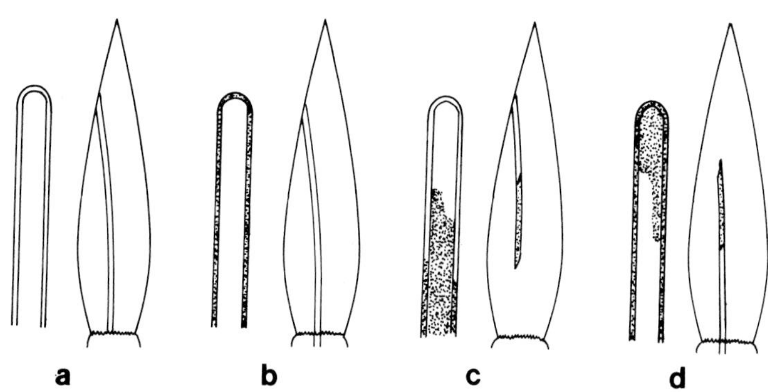

FIGURE 9. White sectors in leaves irradiated before initation. At the left of each leaf is a schematic diagram of a longitudinal section of the sector, showing the distribution of green (stippled) and white cells in the epidermis and internal layers of the leaf.

Sectors in which the epidermis and mesophyll are of different genotypes (Types b, c and d) can only arise if the leaf is derived from two cell layers (note: this conclusion depends on the fact that these sectors were generated before leaf initation). Type a and c sectors represent derivatives of LI cells while type b and d sectors represent derivatives of LII cells. The variation in the extent of these sectors can be attributed to variation in the pattern of cell division during leaf initation. Type a sectors, for example, arise when the cell lineage derived from a mutant LI cell undergoes extensive periclinal divisions during leaf initiation, thereby contributing to the internal tissue of the leaf blade. Type b sectors result when LI cells divide anticlinally and only contribute to the epidermis. Type c and d sectors represent cases in which the LI underwent periclinal divisions during leaf initation, but not to the extent that it replaced the LII.

Because the maize leaf completely encircles the shoot, the number of cells contributing to the width of the leaf equals the number of cells in the circumference of the shoot apical meristem just prior to leaf initiation (actually slightly more, because the leaf margins overlap). The circumference of the shoot at the time of irradiation can be estimated by dividing the circumference of the internode at maturity by the average width of sectors within that internode. On this basis, it is estimated that approximately 42 cells contribute to the transverse dimension of the leaf primordium. The longitudinal dimension of the primordium is more difficult to estimate accurately. Clones such as the one shown in Figure 9c can only arise, however, if at least 3 cells contribute to a longitudinal section of the LI, so the leaf must initially encompass at least 3 cells in the longitudinal dimension of the shoot apex.

Figure 10 illustrates the location and shape of clones induced before leaf initiation. As in tobacco, sectors in the middle of the leaf extend to its tip, while those at the margin of the leaf end in more basal regions. This pattern is not quite as obvious in corn as it is in tobacco because a large majority of the clones end in the distal part of the leaf due to the prolonged growth of the leaf base; if the leaf elongated uniformly, clones would end at equal intervals along its length. It is clear, nonetheless, that a precise correlation exists between the position of a founder cell in the apex and the part of the leaf to which it contributes. Cells located near the site of initiation contribute to the distal part of the leaf blade; cells recruited from the opposite side of the apex form the marginal and basal parts of the leaf

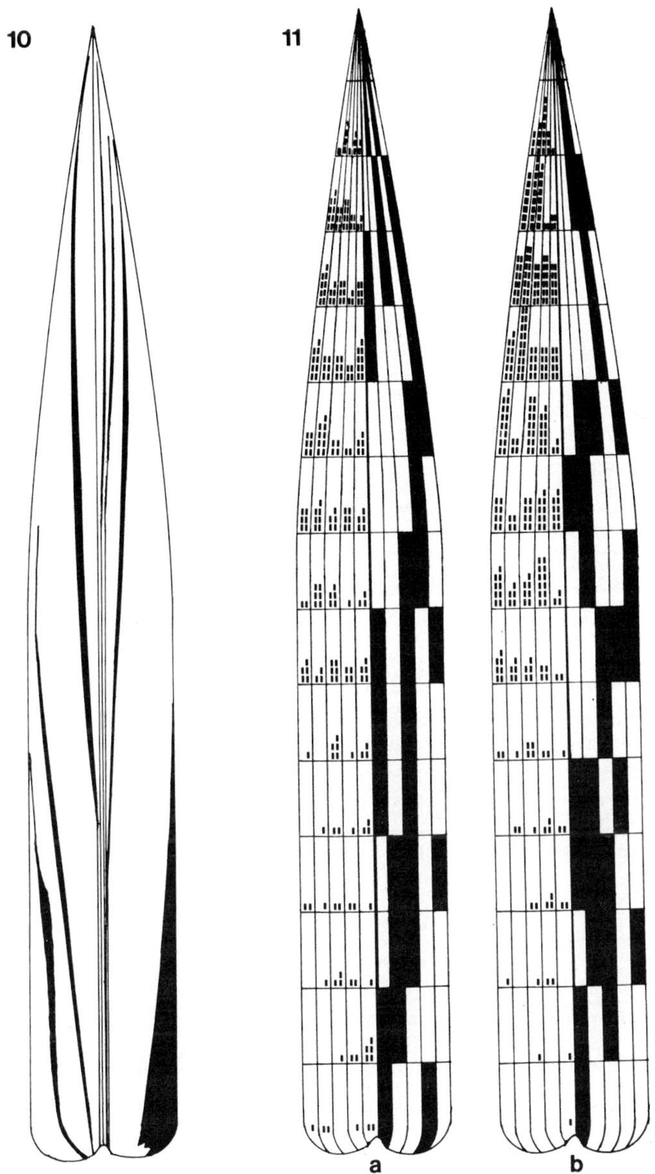

FIGURE 10. Sectors induced before leaf initiation.

FIGURE 11. The number (left) and relative frequency (right) of sectors in leaves irradiated at a length of 0.6-0.8 mm (a) and 1.2 mm (b). Black sections contain 1.1 times average number of sectors in that transverse section of the leaf.

blade. Clones occur more rarely in the basal margin of the leaf blade than in the central-apical region, and those that occur in this basal region tend to be unusually broad. This implies that the basal margin of the leaf blade is derived from relatively few cells, and that cell division is not as highly polarized in this region as it is in other parts of the leaf. Cell division appears to be most highly polarized near the tip of the leaf, since clones narrow considerably when they enter this region. Thus, the gradual increase in the width of the leaf blade from the tip to the base of the leaf can be attributed to two factors: 1) the addition of new founder cells to the leaf primordium as it extends around the apex, and 2) a decrease in the longitudinal polarization of cell division towards the base of the leaf. Unfortunately, it is not yet clear whether this spatial variation in the orientation of cell division persists throughout development, or is characteristic of a specific stage.

The distribution of sectors in leaves irradiated after initiation is shown in Figure 11. Two things should be pointed out. As in tobacco, sectors are smaller and more numerous at the tip of the leaf than at the base. This feature is consistent with the fact that the tip of the leaf matures long before the leaf base (Sharman, 1942). Of greater interest, perhaps, is the transverse distribution of sectors. Although the frequency of sectors is clearly elevated near the margin in some parts of the leaf, it is often just as great, if not greater, in intercalary regions. In the basal region of the blade, sectors occur more frequently near the midrib than near the margin. The data are not extensive enough to reveal a more subtle pattern in the distribution of sectors within the lamina--if such a pattern exists--but clearly suggest that the margin of the leaf does not possess an unusually high rate of cell division.

V. DISCUSSION

The ability to mark cell lineages genetically makes it possible to examine the number and fate of the cells that give rise to an organ, and the dynamics of cell division during its development. Two types of genetic mosaics have contributed to the analysis of cell lineage patterns in plant development: chimeras and radiation-induced mosaics. Chimeras were first used for cell lineage analysis by Satina and Blakeslee (1941) and have since been used extensively for this purpose, most notably by Stewart and Dermen (Dermen, 1960; Stewart, 1978; Stewart and Dermen, 1979). Though often ignored, the work of these investigators has added significantly to our understanding of the cellular basis of plant morphogenesis. Perhaps the most significant outcome of these studies is the observation that plant morphogenesis involves an indeterminate pattern of cell division. Cell lineages derived from the 3 layers of the apical meristem exhibit considerable variation in the amount of cell division they undergo and in their final differentiated state (Stewart and Dermen, 1975, 1979). The fact that this variation in cell behavior has no effect on the overall morphology of an organ means that plant morphology does depend on determinate patterns of cell lineage. The fate of a cell is apparently determined by its position, not its history. This conclusion does not mean that cell behavior is inconsequential in morphogenesis. Although it is often stated that the shape of a plant organ is independent of cell-specific behavior, the extent to which morphogenesis depends on specific spatially and temporally regulated features of cell behavior remains unclear. To assess the role that these parameters might play in morphogenesis, it is essential to have an accurate picture of the pattern of cell division and cell expansion. It was with this goal in mind that a clonal analysis of leaf development in maize and tobacco was undertaken.

The patterns of variegation in chimeras represent the fate of the three fundamental cell layers in the apical meristem. To visualize the behavior of individual cells in a primordium, a different method of marking cell lineages must be used. Radiation-induced somatic mutations offer the advantage of being relatively easy to generate at any stage of meristematic growth, and of providing unequivocal evidence of cell lineage. Moreover, because these mutations occur at a low frequency, somatic sectors can be assumed to arise from a single cell and can therefore be used to estimate the cell number of a primordium.

The results of this study of leaf morphogenesis in maize and tobacco confirm Dulieu's (1968) conclusions about the cell lineage of angiosperm leaves. As he reported, there is no evidence that cells at the apex of a leaf primordium serve as initials. Instead, each of the founder cells of a leaf contributes to a unique section of the primordium, and only a small fraction of the original population resides at the apex. This feature may be characteristic of determinate organs in general, since it is also typical of shoot growth in determinate plants such as maize (Coe and Neuffer, 1978; Poethig, unpublished observations) and tobacco (Dulieu, 1969). The fact that apical cells contribute relatively little to the growth of the leaf is consistent with the observation that cell division ceases at the leaf apex early in development (Avery, 1933; Sharman, 1942), and implies that the apex plays a minimal role in establishing the physical organization of the leaf. Green (Green and Brooks, 1978; Green and Lang, 1981) has shown that the development of a new axis in Graptopetalum, whether it be a shoot or a leaf, involves a global reorganization of the pattern of cell division within a population of founder cells. Cells whose polarity is appropriate to the new axis retain their original polarity, while those with inappropriate polarity experience a change in their orientation of division and expansion. The cell lineage patterns of maize and tobacco leaves suggest that a similar phenomenon occurs during leaf initiation in these species.

The cellular dynamics of leaf morphogenesis in maize and tobacco are remarkably similar. In both species the expansion of the blade is governed primarily by intercalary cell division. Neither the orientation of cell division at the leaf margin, nor the frequency of cell division in this region suggest that it has a major role in blade expansion. In both species cell division is initially suppressed at the base of the leaf, and later exhibits a dramatic increase in frequency. In both, cell division ceases first at the tip of the leaf and is prolonged at the leaf base. The principal difference between the pattern of cell division in these species lies in its orientation. In the leaf blade of maize, cell division is highly polarized (with cell walls perpendicular to the midrib), while in the tobacco leaf blade cell division is for the most part unpolarized. As a consequence, the rapid and prolonged growth of the leaf base in maize only leads to the elongation of the leaf, whereas in tobacco it results in an increase in the width of the base relative to the leaf tip.

The results of this study add to a growing body of information about the orientation, distribution and frequency of cell division during leaf morphogenesis. In all of the species examined so far the shape of the leaf blade is closely correlated with local variations in the frequency, duration or orientation of cell division (Fuchs, 1975, 1976; Jeune, 1972; Thomasson, 1970). But further studies will be needed in order to determine whether these parameters directly determine leaf shape or, in Haber and Foard's (1963) words, "play a secondary . . . role in influencing organ form." To solve this problem it will be necessary to carefully examine the relationship between the growth dynamics of the leaf and its pattern of cell division and cell expansion, as has been done in Graptopetalum (Green and Poethig, 1982). Attention must also be given to the cellular correlates of experimentally induced and genetic variations in leaf shape. With the help of clonal analysis and other more direct methods of following cell behavior we should be able to learn how plant morphogenesis is regulated at a cellular level.

ACKNOWLEDGEMENTS

I am grateful to I. M. Sussex, E. H. Coe and J. H. Duesing for their interest in and helpful criticism of this research. This research was supported by an NSF Predoctoral Fellowship and an NIH Training Grant (HD 07180-01) at Yale University, and by a postdoctoral fellowship (funded by the USDA) at the University of Missouri.

REFERENCES

Avery, G.S. (1933). Structure and development of the tobacco leaf. Amer. J. Bot. 20, 565-592.

Coe, E.H., Jr. and Neuffer, M.G. (1978). Embryo cells and their destinies in the corn plant. In "The Clonal Basis of Development" (S. Subtelny and I. Sussex, eds.), pp.

113-129. Academic Press, New York.

Dermen, H. (1960). The nature of plant sports. Am. Hort. Mag. 39, 123-173.

Dubuc-Lebreux, M.A. and Sattler, R. (1980). Développement des organes foliacés chez Nicotiana tabacum et le problème des méristèmes marginaux. Phytomorphology 30, 17-32.

Dulieu, H. (1968). Emploi des chimères chlorophylliennes pour l'etude de l'ontogénie foliare. Bulletin Scientifique de Bourgogne 25, 1-60.

Dulieu, H. (1969). Mutations somatiques chlorophylliennes induites et onotgénie caulinaire. Bull. Sci. Bourgogne 1-84.

Dulieu, H. (1974). Somatic variations on a yellow mutant in Nicotiana tabacum L. (a_1^+/a_1 a_2^+/a_2). I. Non-reciprocal genetic events occurring in leaf cells. Mut. Res. 25, 289-304.

Dulieu, H. and Bugnon, F. (1966). Chiméres chlorophylliennes et ontogénie foliaire chez le tabac (Nicotiana tabacum L.). C. R. Acad. Sci. Paris 263(D), 1714-1717.

Foster, A.S. (1936). Leaf differentiation in angiosperms. Bot. Rev. 2, 349-372.

Fuchs, C. (1966). Observations sur l'extension en largeur du limbe foliaire due Lupinus albus L. C. R. Acad. Sci. Paris D 263, 1212-1215.

Fuchs, C. (1975). Ontogenèse foliare et acquisition de la forme chez le Tropaeolum peregrinum L. I. Les premiers stades de l'ontogenese du lobe médian. Ann. Sci. Nat. (Botanique) 16, 321-390.

Fuchs, C. (1976). Ontogenèse foliare et acquisition de la forme chez le Tropaeolum peregrinum L. II. Le developpement du lobe après la formation de lobules. Ann. Sci. Nat. (Botanique) 17, 121-158.

Green, P.B. and Brooks, K.E. (1978). Stem formation from a succulent leaf: its bearing on theories of axiation. Amer. J. Bot. 65, 13-26.

Green, P.B. and Lang, J.M. (1981). Toward a biophysical theory of organogenesis: birefringence observations on regenerating leaves in the succulent Graptopetalum paraguayense E. Walther. Planta 151, 413-426.

Green, P.B. and Poethig, R.S. (1982). Biophysics of the extension and initiation of plant organs. In "Developmental Order: Its Origin and Regulation" (Eds. S. Subtelny and P.B. Green), pp. 485-509. A. R. Liss, Inc., New York.

Haber, A.H. and Foard, D.E. (1963). Nonessentiality of concurrent cell divisions for degree of polarization of leaf growth. II. Evidence from untreated plants and from chemically induced changes of the degree of polarization. Amer. J. Bot. 50, 937-944.

Jeune, B. (1972). Observations et expérimentation sur les feuilles juveniles du Paulownia tomentosa. Bull. Soc. Bot. France 119, 215-230.

Maksymowych, R. and Erickson, R.O. (1960). Development of the lamina in Xanthium italicum represented by the plastochron index. Amer. J. Bot. 47, 451-459.

Poethig, R.S. (1981). The cellular parameters of leaf development in Nicotiana tabacum L. Ph.D. thesis, Yale University.

Satina, S. and Blakeslee, A.F. (1941). Periclinal chimeras in Datura stramonium in relation to the development of leaf and flower. Amer. J. Bot. 28, 862-871.

Sharman, B.C. (1942). Developmental anatomy of the shoot of Zea mays L. Ann. Bot. 6, 245-282.

Steffensen, D.M. (1968). A reconstruction of cell development in the shoot apex of maize. Amer. J. Bot. 55, 354-369.

Stein, O.L. and Steffensen, D. (1959). Radiation-induced genetic markers in the study of leaf growth in Zea. Amer. J. Bot. 46, 485-489.

Stewart, R.N. (1978). Ontogeny of the primary body in chimeral forms of higher plants. In "The Clonal Basis of Development" (Eds. S. Subtelny and I. Sussex), pp. 131-159. Academic Press, New York.

Stewart, R.N. and Burk, L.G. (1970). Independence of tissues derived from apical layers in ontogeny of the tobacco leaf and ovary. Amer. J. Bot. 57, 1010-1016.

Stewart, R.N. and Dermen, H. (1975). Flexibility in ontogeny as shown by the contribution of shoot apical layers to leaves of periclinal chimeras. Amer. J. Bot. 62, 935-947.

Stewart, R.N. and Dermen, H. (1979). Ontogeny in monocotyledons as revealed by studies of the developmental anatomy of periclinal chloroplast chimeras. Amer. J. Bot. 66, 47-58.

Thomasson, M. (1970). Quelques observations sur la repartition de zones de croissance de la feuille du Jasminum nudiflorum Lindl. Candollea 25, 297-340.

ALTERNATIVE MODES OF ORGANOGENESIS IN HIGHER PLANTS[1]

Donald R. Kaplan

Department of Botany
University of California
Berkeley, California

This article examines the question of alternative modes of organ development in higher plants, taking its examples from convergences in dissected leaf form that are produced by different morphogenetic programs. In Araceae leaf dissection occurs either by conventional marginal lobing, as in the pinnate leaf of *Zamioculcas*, or by localized tissue death as in the perforated leaf of *Monstera*. In the palms, both palmate and pinnate leaves originate by a process of intralaminar corrugation followed by a cleavage of the resultant pleats into the free lobes or leaflets. The origin of plications has been the most controversial aspect of palm leaf development. Some workers suggested that this pleating arises by differential growth or meristem folding while others have supported a tissue cleavage process. By a detailed, comparative morphological and histogenetic analysis of plication inception and early growth in the pinnate leaf of *Chrysalidocarpus lutescens* and the palmate leaf of *Rhapis excelsa*, it could be shown that plication origin in both taxa occurs by differential growth rather than by tissue cleavage. It is concluded that the principal reason for this difference in past interpretation was the reliance of previous workers on matters of plication shape for the determination of the developmental

Contemporary Problems in
Plant Anatomy

ISBN 0-12-746620-7

mechanism involved. By using a quantitative evaluation of the cellular basis of plication origin we were able to transcend matters of shape and show that plication configuration in the respective species is a consequence of physical constraints within the leaf itself or between leaves in the bud rather than different morphogenetic mechanisms. It is emphasized that in future researches greater attention should be directed toward the determination of the common denominators of such different developmental programs rather than the mere cataloging of their superficial differences.

I. INTRODUCTION

Flowering plants are known for their enormous structural diversity, representing adaptations to a wide range of habitats and life styles. For the plant biologist interested in the developmental basis of variation in plant form, such morphological variety raises certain basic questions. For example, how much of this morphological diversity is merely the result of quantitative variations in growth distribution on a common developmental theme and how much is the result of qualitatively unique developmental programs which are distinctive for individual plant groups? Answers to such broad questions can be of significance to both evolutionary and developmental biologists. Evolutionary biologists have shown an increasing interest in the developmental mechanisms underlying structural innovation. The recent publications of Gould (1977) and Bonner (1982), for example, underscore this increasing integration of developmental biology with evolutionary thought. Similarly, developmental biologists can learn a great deal about fundamental mechanisms of growth and development by studying markedly divergent structural features and assessing their relationships. The results of such studies not only reinforce the most basic developmental themes, but they also lead to discoveries of new developmental phenomena and relationships.

While we have acquired considerable understanding of the range of developmental mechanisms operating in animal ontogeny, especially those of Metazoans (Wessels, 1982), our information is far less extensive for plants. Part of the problem is that there are very few practicing plant developmental morphologists today, especially compared to their zoological counterparts. But the problem goes deeper than that. Most structural botanists, particularly in the United States, have been concerned more with the anatomical-cell biological elements of plant construction than with the relationships of those elements to the general levels of morphological organization. Where they have been morphological in outlook, they have tended to be interested more in systematic than in developmental-mechanistic approaches. This narrowness of focus has led to a gap between the anatomical (histological) and morphological levels of analysis. By studying plant developmental phenomena from the organ level down to the anatomical-cellular levels of organization my colleagues and I are attempting to achieve an appreciation of the contribution of each of these organizational levels to the generation of plant form.

My own developmental studies have focused on changes that occur during the growth of an individual plant axis, *i.e.*, on what has been termed "heteroblastic development" (Kaplan, 1973, 1980; Bruck and Kaplan, 1980). I would now like to consider developmental mechanisms in organisms exhibiting distinctly different ways of producing apparently similar end products, for example dissected or lobed leaves. Dissected leaves are foliar appendages which have their blade surfaces cut into either lobes or segments, termed leaflets or pinnae. The developmental mechanisms of leaf dissection are quite diverse, especially in monocotyledonous flowering plants and most particularly in the aroids (Araceae) and palms (Palmae). I shall focus on the distinctiveness of these developmental programs as well as their common denominators.

II. DEVELOPMENTAL MODES OF LEAF DISSECTION IN THE ARACEAE

The large and vegetatively diversified tropical family Araceae is noted for its range of leaf morphology, including leaves with elaborately dissected blade regions (Ertl, 1932; Goebel, 1932; Troll, 1939). I have chosen two genera, *Zamioculcas* and *Monstera*, to illustrate two distinctly different morphogenetic bases of leaf dissection in this family.

Zamioculcas zamiifolia is a rhizomatous, succulent member of subfamily Pothoideae (Engler, 1905). It exhibits a typical sympodial growth habit with a single foliage leaf formed on each sympodial segment after a succession of scale leaf and scale-foliage leaf transitions (Engler, 1905). Adult foliage leaves are once-pinnately compound, bearing 4-8 leaflet pairs in a paripinnate configuration (Fig. 1). The leaf is succulent, stands erectly (Fig. 1A) and is supported

FIGURE 1. Mature leaf morphology in *Zamioculcas zamiifolia*. A, Mature pinnate leaf. B-I, Transections of various sectors of the leaf showing the transectional symmetry of each part and the configuration of vascular bundles at each level in the leaf. In each bundle xylem is black and phloem clear.

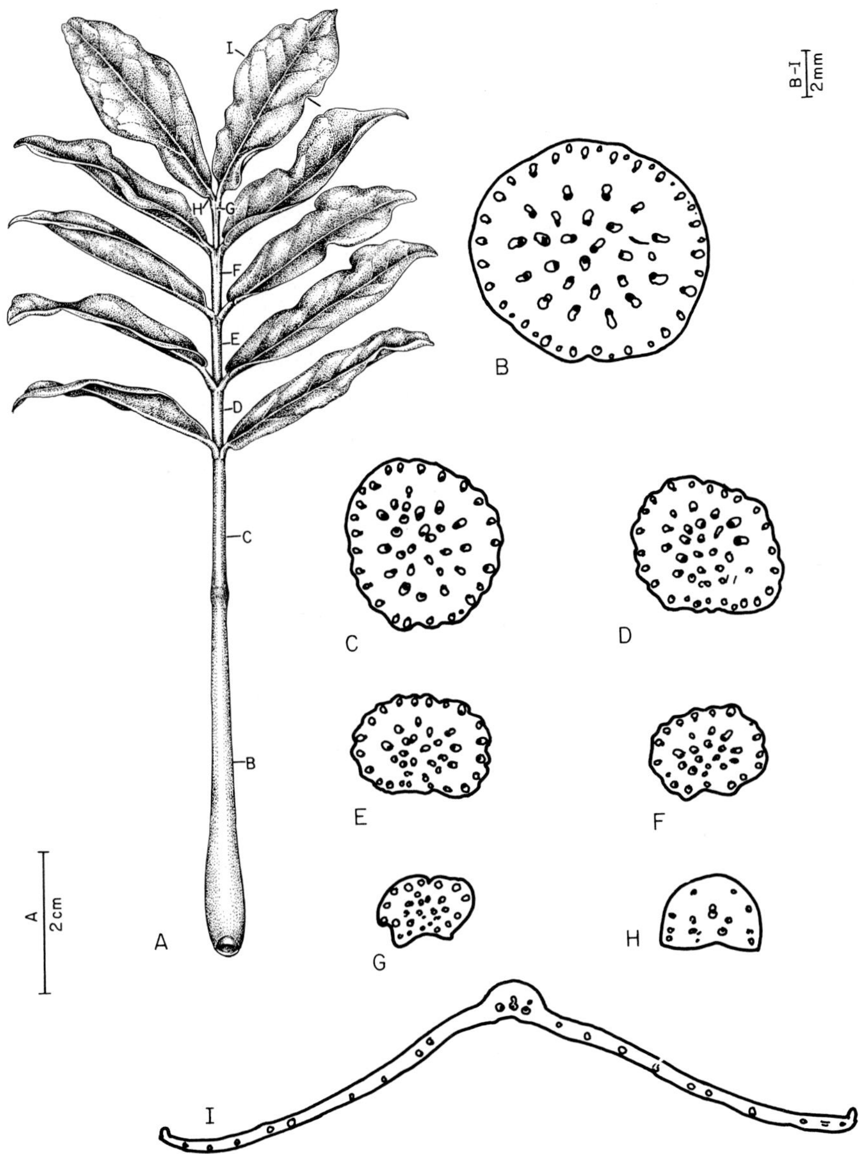
I
H
G
F
E
D
C
B
A
2 cm
B–I
2 mm
B
C
D
E
F
G
H
I

by a long, unifacial petiolar zone (Fig. 1B,C). A pulvinar articulation joint approximately 2/3 the length of the petiole permits some orientation of the leaf axis (Fig. 1A).

Our interest in the leaf of Zamioculcas is that its pinnae originate by a pattern of morphogenesis characteristic of the majority of compound leaf types. Soon after the foliage leaf is initiated its primordium appears hood-shaped and completely enshrouds the shoot apical meristem (Fig. 2A,B). The primordial pinnae first appear as bump-like protuberances along each of the two margins in a close basipetal sequence (Fig. 2B,C). Their initiation is by a process of differential growth that has been termed "meristem fractionation" (Hagemann, 1970). In this process regions of growth (the pinnae) alternate with zones of no growth (the sinuses or future rachis internodes) to give a freely lobed result (Fig. 2C). Following their initiation each pinna enlarges significantly as a result of surface growth (Fig. 2D,E) and becomes vertically oriented to fit into the space formed by the surrounding scale leaves (Fig. 2E,F). Early in development the leaf primordium becomes differentiated into a distal lamina and a basal, meristem-encircling leaf base with the dissected lamina comprising the greatest proportion of total leaf length (Fig. 2C,D). However a distinct petiole is intercalated between the leaf base and lamina (Fig. 2D,E) and this petiole ultimately elongates to form the greatest proportion of total leaf length (Fig. 1A). Finally, the rachis units elongate to separate the points of leaflet insertion (compare Fig. 2F and 1A).

The pattern of initiation of leaflets as free-lobe-like outgrowths from the embryonic leaf margins as described above is the most widespread mechanism of leaf dissection found in higher plants. Not only is it characteristic of the majority of aroids with lobed or dissected leaves, such as *Gonatopus*, *Amorphophallus*, *Syngonium* and *Helicodiceros* (Troll, 1939), but it also is the developmental pattern responsible for leaf dissection in the ferns, cycads and virtually all dicotyledonous angiosperms (Troll, 1939).

Troll (1939) emphasized that one can find all manner of intermediates between compound leaves, where the leaflets appear as fully individualized units bearing their own petioles, and simple, lobed leaves borne along the length of an individual plant. Hagemann (1970) has shown that the developmental difference between pinnately compound and pinnately lobed leaves in *Pelargonium* depends on the relative timing of lobe initiation and growth in blade surface. In the pinnately compound leaf of *P. alternans* pinna primordia appear as marginal lobes prior to any marked growth in breadth of the central portion of the blade. Conversely, in the development of the pinnately lobed leaf of *P. betulinum*, the central portion of the lamina grows in surface *before* the lobes are initiated at its margins. Recent studies by Fuchs (1975, 1976) and Jeune (1982) have provided a detailed characterization of the anatomical basis of the differential growth involved in marginal lobe initiation in the leaves of *Tropaeolum peregrinum* and *Castanea sativa*. Thus we have a fairly comprehensive appreciation of both the range of expression and the cellular basis of this mode of leaf dissection in plant species.

A markedly different mechanism of leaf dissection is exhibited by the aroid genus *Monstera* (subfamily Monsteroideae). Looking at the adult foliage leaf of the root climber *Monstera deliciosa*, one can easily deduce the developmental basis of the dissection. Not only is the blade

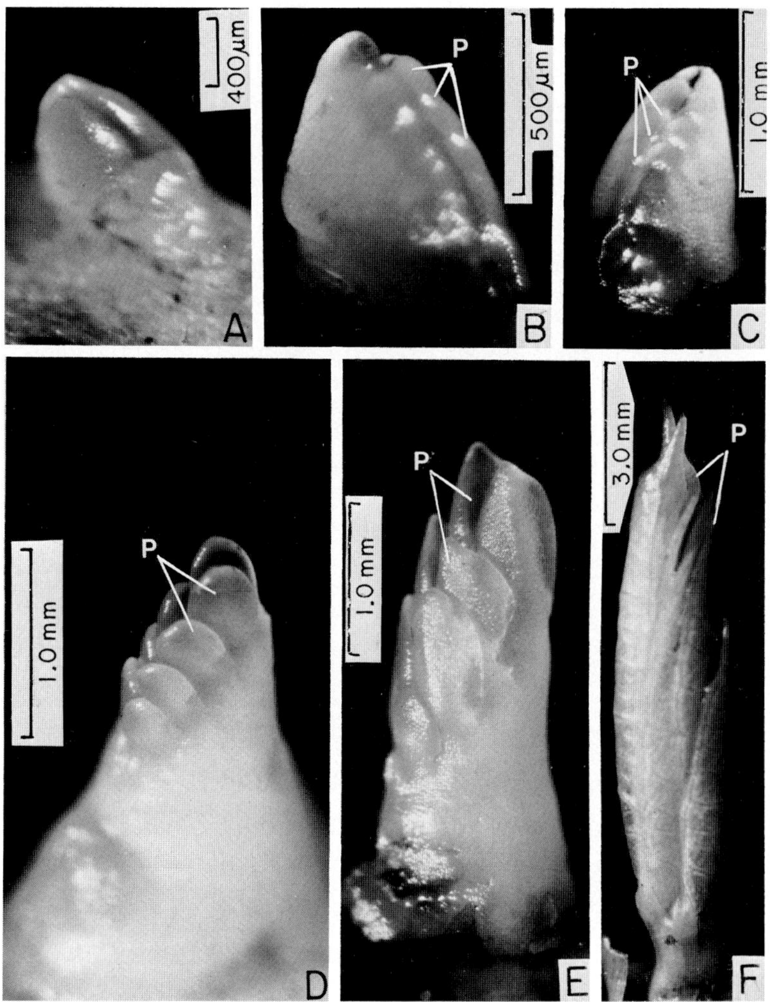

FIGURE 2. Developmental stages of the foliage leaf of *Zamioculcas zamiifolia*. A, Recently initiated leaf primordium 400 um in length. B, Leaf 900 um in length showing inception of pinnae (P). C, Leaf 1500 um in length. D, Leaf 2 mm in length showing leaflet expansion. E, Leaf 3 mm in length. F, Leaf 11 mm in length.

surface cut into a series of horizontal to obliquely oriented peripheral lobes, but there also are perforations in the intercostal sectors of the blade (Fig. 3S). The shape of the perforations differ, depending upon the sector of the lamina, reflecting differences in growth distribution in the different regions of the blade. Those perforations located at the lamina periphery, where the greatest amount of lateral expansion occurs, tend to be long and slit-like whereas those adjacent to the midrib, where there is less lateral growth, are smaller and elliptical to circular (Fig. 3S).

The developmental origin of laminar lacunae in *Monstera* has been known for over a century and was described by Trécul (1854), Schwarz (1878), and Melville and Wrigley (1969). All lacunae, whether they are peripheral and break through the margin to form a lobed blade or remain as holes toward the lamina center, originate as localized regions of tissue necrosis. In scanning electron micrographs (SEM) of young leaves of *M. deliciosa* it can be seen that the future necrotic areas differentiate after the major lateral veins have been defined (Fig. 4A) (leaf length > 5 mm) and appear initially as ovoid to elliptical indentations in the lamina surface (Fig. 4B). They appear indented because the dying cells have less volume than the living cells around them. Ultimately the dying tissue dries up and falls out leaving a hole in the blade surface (Fig. 4C,D). According to Schwarz (1878) a secondary epidermis is differentiated from the mesophyll cells that border the hole and it blends so well with the preexisting surface layer that it is difficult to distinguish its limits in the mature leaf. Melville and Wrigley (1969) have demonstrated that the first perforations to develop arise at the periphery of the blade, and if additional ones are initiated, they arise in a centripetal sequence with the newest being closest to the midrib.

While blade lacunae can develop in leaves of other Monsteroideae and Lasioideae of Araceae (e.g., Dracontioides and Anchomanes) as well Aponogeton fenestrale (Aponogetonaceae), our interest is in the range of adult leaf morphology exhibited by members of the genus Monstera. In contrast to species such as M. adansonii, (Fig. 3K) M. acacoyaguensis (Fig. 3O) and M. oreophila (Fig. 3G) whose adult leaves are perforated but retain a simple outline, there are species such as M. subpinnata (Fig. 3M), M. dilacerata (Fig. 3R) and M. tenuis (Fig. 3V) whose leaves are distinctly pinnately lobed without any central lacunae. These leaves exhibit a morphology which converges with that of Zamioculcas but which originate by a vastly different developmental process. While none of the pinnatifid species of Monstera cited above have received the requisite developmental study, their mature morphology suggests that blade dissection comes about through the initiation of a single row of submarginal perforations that secondarily break through the margin just as they do in the leaves of M. deliciosa (Compare Fig. 3R and 3V with 3S). Support for this interpretation comes not only from the morphological linkage to the other perforated leaves in the genus (Fig. 3) but also from the observation that lobes in these leaves have attenuated apices suggestive of an origin by laminar splitting (Fig. 3R and V).

Thus in members of a single family, we have examples of two radically different developmental processes operating to produce similar looking, dissected foliar organs.

III. DEVELOPMENTAL BASIS OF LEAF DISSECTION IN THE PALMS

Palms are noted for and characterized by their large and elaborately dissected foliar appendages (Moore and Uhl,

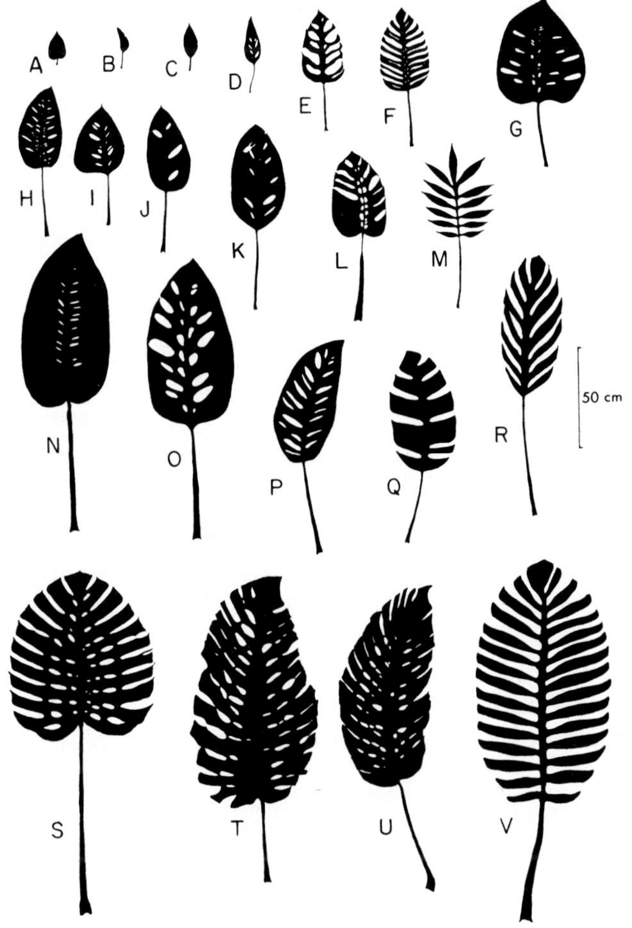

FIGURE 3. Adult foliage leaves of _Monstera_ species (reproduced from Madison, 1977 with permission). A, _M. tuberculata_. B, _M. pittieri_. C, _M. obliqua_. D, _M. xanthospatha_. (E) _M. gracilis_. (F) _M. epipremoides_. (G) _M. oreophila_. H, _M. siltepecana_. I, _M. adansonii_ var. _laniata_. J, _M. adansonii_ var. _adansonii_. K, _M. adansonii_ var. _klotzchiana_. L, _M. membranaceae_. M, _M. subpinnata_. N, _M. lechleriana_. O, _M. acacoyaguensis_. P, _M. acuminata_. Q, _M. supruceana_. R, _M. dilacerata_. S, _M. deliciosa_. T, _M. punctulata_. U, _M. dubia_. V, _M. tenuis_.

1982). In fact the fronds of some palms are the largest leaves in the plant kingdom. All leaves of palms are regionally differentiated into a proximal, stem-encircling leaf base, a middle petiole and a distal blade. Depending upon the distribution of growth when the leaflets arise, the blade of a palm leaf can either be palmate or pinnate. In fan (palmate) leaves, leaflet initiation is not accompanied by rachis elongation; hence in the mature frond all of the leaflets are inserted at a single locus at the tip of the petiole. By contrast, in feather (pinnate) palms, leaflet inception is accompanied by rachis extension so that at maturity the pinnae are distributed laterally along the length of an elongated rachis axis. There also tends to be a correlation between blade shape and petiole length (Troll, 1949); palmate leaves tend to have long petioles whereas those of pinnate fronds tend to be short.

Our interest in the leaves of palms is that they represent a third and distinctly different mechanism of leaf dissection, one that in some ways is a combination of the patterns described for *Zamioculcas* and *Monstera*. In contrast to the more conventional mode of origin of leaflets as free lobes, the pinnae in a feather palm, for example, appear as a series of folds *within* the lateral blade surface and at some distance from the leaf's margin (Fig. 6A). In sections cut tangential to the blade margin these pleats (termed

FIGURE 4. SEMs of developing leaves of *Monstera deliciosa*, showing stages of leaf perforation. A, Leaf 1.5 mm in length, prior to perforation. (B) Leaf 11.5 mm in length, (C) Leaf 17 mm in length. (D) Leaf 17 mm in length when perforation has broken through completely.

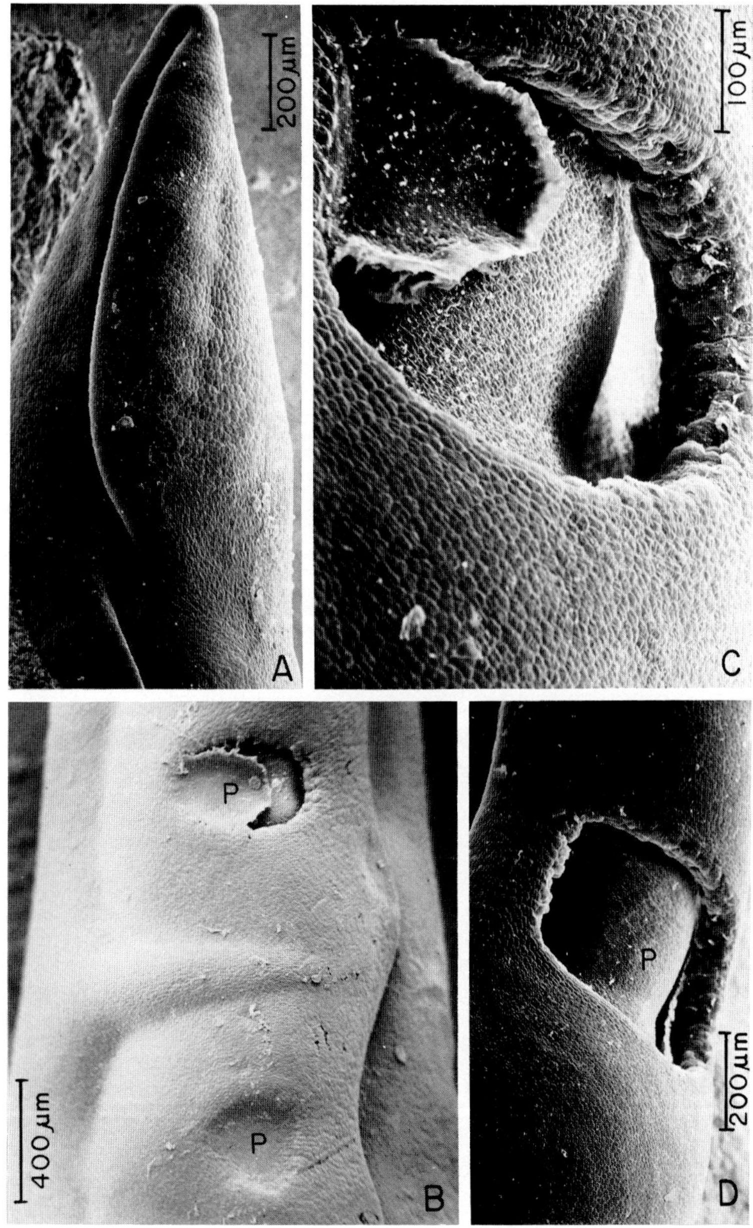
200 µm
A
100 µm
C
P
P
400 µm
B
P
200 µm
D

plications) resemble the foldings of a camera bellows (Fig. 5A-C). Ultimately the pleats in each lamina half will be cleaved by tissue separation into the free pinna segments of the mature frond (Fig. 5D-I).

Plications in pinnate fronds are initiated in either a basipetal or bidirectional sequence, reflecting the polarity of longitudinal differentiation in the leaf. Pinnate leaves have the long axis of their plications oriented at right angles to the long axis of the leaf margin and hence appear as horizontal slits when viewed from the outer or abaxial surface of the leaf (Fig. 6A). Palmate leaves, on the other hand, have the long axis of their plications oriented parallel to the long axis of the blade (Fig. 6B).

Once initiated, the plications grow not only in length, *i.e.*, parallel to a given blade surface (Fig. 6A,B) but also in depth; *i.e.*, at right angles to the blade's flat surface (Fig. 5A-E; Fig. 6A, B). In their growth in depth the plications actually increase in surface by intercalary growth of the intercostal sector between the adaxial and abaxial plication ridges; alternate ridges are pushed apart by the horizontal extention of the laminar sheets between them (Fig. 5C-E).

Subsequent stages of palm leaf development entail not only physical separation of the pleats into distinct leaflets or lobes but also their extrication from the blade surface. Depending upon the species studied, leaflet delimitation occurs by an actual abscission of the folds from one another (Kaplan, Dengler and Dengler, 1982a, Fig. 5D-I). This separation typically takes place either through the abaxial or adaxial plication ridges, although in some species (*e.g.* *Rhapis excelsa*, Kaplan et al., 1982a), it can occur across the intercostal zone as well. If the cleavage occurs through the abaxial ridges, then when the resultant V-shaped leaflets are reflexed in the final phases of frond expansion and the

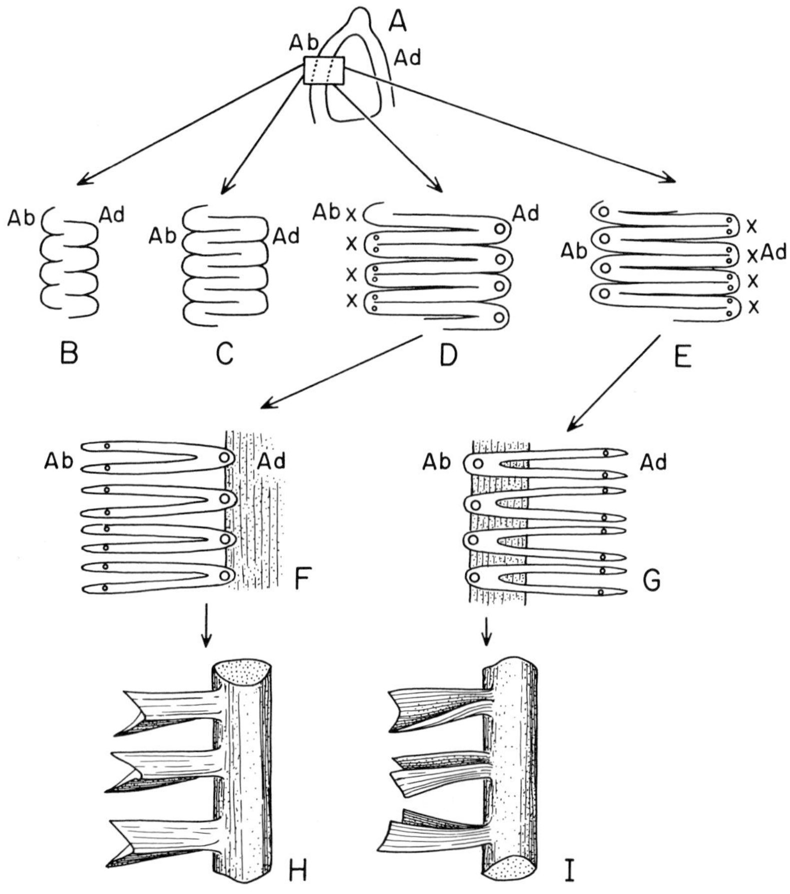

FIGURE 5. Diagrams illustrating an adaxial view of a hood-shaped pinnate palm leaf primordium (A) with the process of plication having been initiated (B) and growth in surface occurring (C) in a submarginal locus at right angles to leaf margin of A. D,E illustrate plications that have undergone intercostal, intercalary growth but which will abscise through either the abaxial ridges (x in D) to form reduplicate leaflets (F and H) or the adaxial ridges (x in E) to form induplicate leaflets (G and I). (From Kaplan et al 1982a with permission).

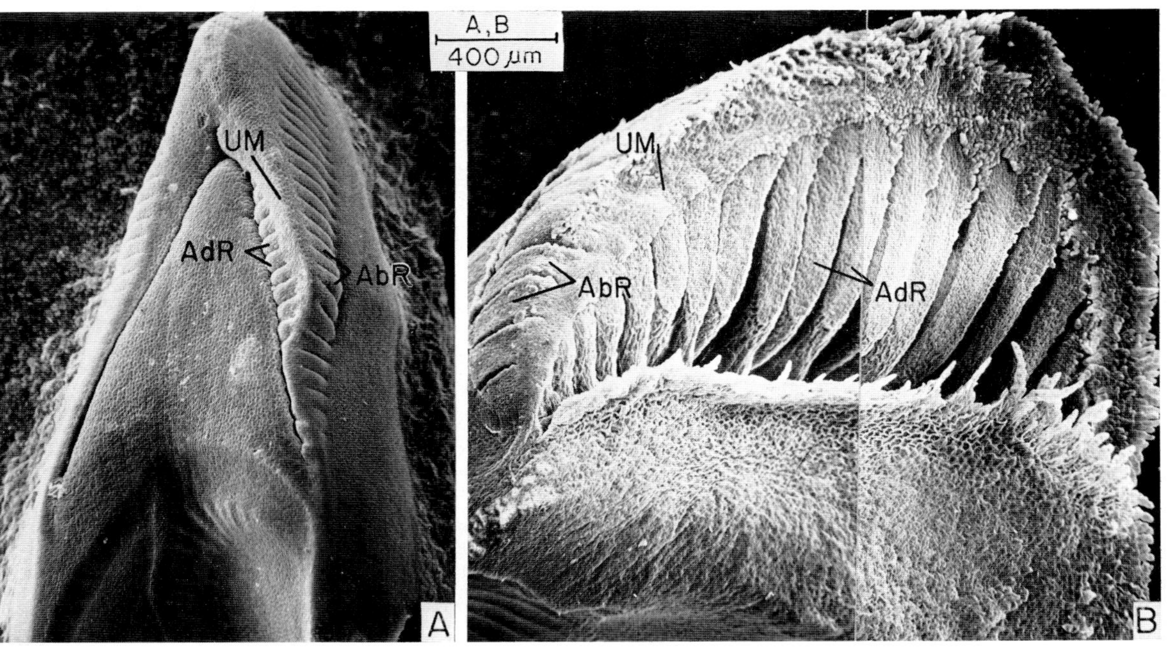
A, B
400 μm
UM
AdR
AbR
A
UM
AbR
AdR
B

opening of the V will be upside down with the point away from the morphological undersurface of the frond (Δ Fig. 5D,F and H). This leaflet orientation is termed reduplicate. If, however, the cleavage occurs through the adaxial ridges then the V will point downwards and this orientation is termed induplicate (V Fig. 5E,G and I). Once such leaflet delimitation has occurred, the pinnae become separated along their length as a result of elongation of the rachis internodes (Fig. 7) and expansion of pulvinar tissue in each pinna axil and they finally assume a near-horizontal plane of flattening (Fig. 5H and I).

Since plication occurs within the lamina surface and at some distance from the margin, in many species of palms the leaf margin remains unpleated and is retained until late in development as a strip of tissue attached to the pinna apices, like a curtain rod supporting a series of pleated drapes (Corner, 1966) (Fig. 7). Thus the final phases of palm leaf morphogenesis involves either the breakthrough or the shedding of these marginal strips or "reins" (Eames, 1953) as intact units. Once the reins have been abscised, the mature palm leaf resembles dissected leaves in taxa where the leaflets have originated by more conventional means.

It should be emphasized that, since plication and leaflet delimitation are temporally separate developmental events, it is possible to have blade corrugation without the pleats being cleaved into pinnae or lobes. The latter

FIGURE 6. A, SEM of a 3 mm long pinnate leaf of *Chrysalidocarpus lutescens*. B, SEM of 3.75 mm long palmate leaf of *Rhapis excelsa*. Adaxial ridges (AdR) and abaxial ridges (AbR) and unplicated margin (UM) indicated.

typically occurs in juvenile leaves of dissected--leaved palm species and in the adult leaves of some taxa that retain a simple but plicate blade morphology (Goebel, 1926; Moore and Uhl, 1982). Conversely, some palm species can exhibit even higher orders of blade dissection with bipinnate leaves developed in genera such as Caryota (Moore and Uhl, 1982). In Caryota not only is the primary lamina plicated and cleaved into segments but the pinnae themselves also undergo plication and cleavage to form pinnules.

While the foregoing account of the overall aspect of palm leaf development has met with general agreement, the mechanism of plication origin has remained a source of controversy for over 150 years. Two radically different morphogenetic models have been proposed to explain the origin of plications in palm leaves. On the one hand, Hofmeister (1868), Goebel (1884, 1926), Eichler (1885), Deinega (1898), Hirmer (1919), Arber (1922), Periasamy (1962, 1977), Corner (1966), and Bugnon (1980) have indicated that plication simply is the product of differential growth or meristem folding in a localized region of the lamina wing (Fig. 8A-C). By contrast, Von Mohl (1845), Trécul (1853), Naumann (1887), Yampolsky (1922), Eames (1953), Venkatanarayana (1957), Padmanabhan (1963, 1967a, 1967b, 1969, 1973) and Padmanabhan and Veerasamy (1973, 1974) have interpreted the zig-zag pattern as a result of some differential growth plus alternating lines of cell separation (proceeding from either the outside or the inside of the respective blade surfaces) that ultimately break through the abaxial and adaxial leaf surface to give rise to the pattern of alternating ridges and furrows comprising the plicate configuration (Fig. 8D-F).

One consequence of the tissue cleavage model is that, not only would the original protodermal (epidermal) layers on both leaf surfaces be cut up into isolated patches (Fig. 8E,F),

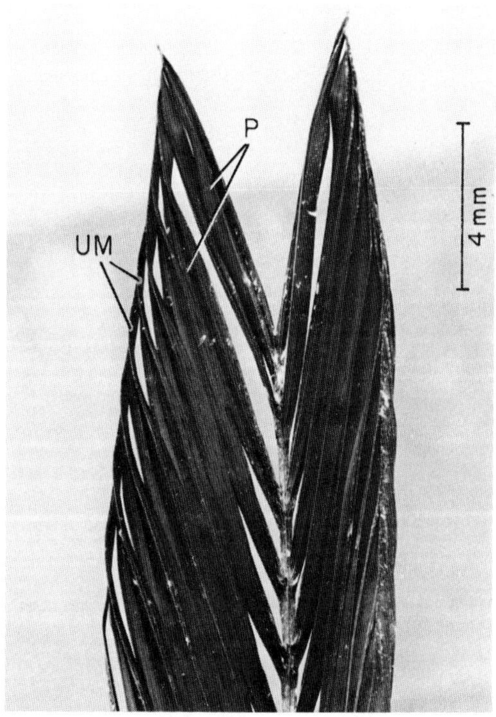

FIGURE 7. Partially opened leaf of <u>Chamaedorea</u> sp. showing the plications separated along their lengths as pinnae (P) but still attached at their apices by the unplicated margin or reins (UM).

but also subsequent epidermal continuity around the base of the plication furrows would have to come from a redifferentiation of epidermal cells from mesophyll tissue surrounding the base of each cleft (Fig. 8E,F). This would be in marked contrast to the behavior of the epidermis in the meristem folding model where it simply follows the changes in contour of the folded blade surface (Fig. 8A-C). Furthermore, according to the tissue separation model, the interior of each of the furrows would not correspond to the adaxial and abaxial leaf surfaces, as they would in the folding model (Fig. 8A-C); rather they would represent new surfaces derived from cleavage of the internal part of the lamina; only the plication ridges would represent the original abaxial and adaxial surfaces of the leaf (Fig. 8D-F).

Since the original goal of our research program was to characterize the range of organogenetic processes in higher plants, the tissue splitting model (Fig. 8D-F) was particularly intriguing to us because, if it proved to be valid, then it would represent a unique developmental mechanism.

An extensive review of the literature on palm leaf development (Kaplan, et al., 1982a) showed that while each mechanism had its proponents, no one had carried out the research required to discriminate effectively between these two developmental models. Furthermore, most workers had based their conclusions on superficial observations of plication shape rather than on a more basic developmental analysis. Those plications whose ridges were closely abutted and separated by narrow, incision-like furrows were judged to have originated by a tissue separation mechanism, whereas those plications whose ridges were separated by wide, trough-like indentations were considered to have originated by meristem folding. It was clear from our review that if we were going to resolve this long-standing problem that we would have to make an analysis that went beyond matters of plication shape.

In order to properly orient ourselves with regard to the three-dimensional aspects of palm leaf development and avoid the problems of improper section orientation that had plagued the work of some previous researchers, we did an extensive SEM study of the overall morphogenesis of the leaves of two feather palm species, Chrysalidocarpus lutescens and Chamaedorea seifrizii and a fan palm, Rhapis excelsa. From this survey not only did we gain the three-dimensional perspective that was a necessary prerequisite for our histogenetic analyses, but we also were able to determine the relative timing of the inception of plications in these species and to select those best for our anatomic studies (Kaplan et al, 1982a). For example, the pinnate leaf of Chrysalidocarpus initiates a greater number of plications (40 vs. 20) than Chamaedorea over a greater span of leaf lengths (up to 5.5 mm in Chrysalidocarpus vs. 1.5 mm for Chamaedorea). Thus Chrysalidocarpus supplies a greater number of developmental stages for the study of plication histogenesis. Furthermore, the relatively short plastochron interval in the shoots of Chrysalidocarpus meant that we would have to sacrifice fewer specimens in order to get an abundance of leaf developmental stages. We also were particularly interested in analyzing plication histogenesis in Chrysalidocarpus because its folds have a shape that previously had been interpreted as having resulted from a tissue cleavage process (Eames, 1953). Since the palmate leaf of Rhapis exhibited shoot development properties similar to that of Chrysalidocarpus, it was selected for a detailed study of a fan type of leaf ontogeny.

To increase the resolution of cell lineages in our histogenetic analysis of plication development, we used plastic embedded material fixed for electron microscopy and cut 1.2-2.0 um in thickness. Since plications in different positions of the blade could exhibit developmental differences, we followed the ontogeny of a particular plication position with time and then compared its development with those in other parts of the leaf rather than assuming that the development of successively initiated folds would be

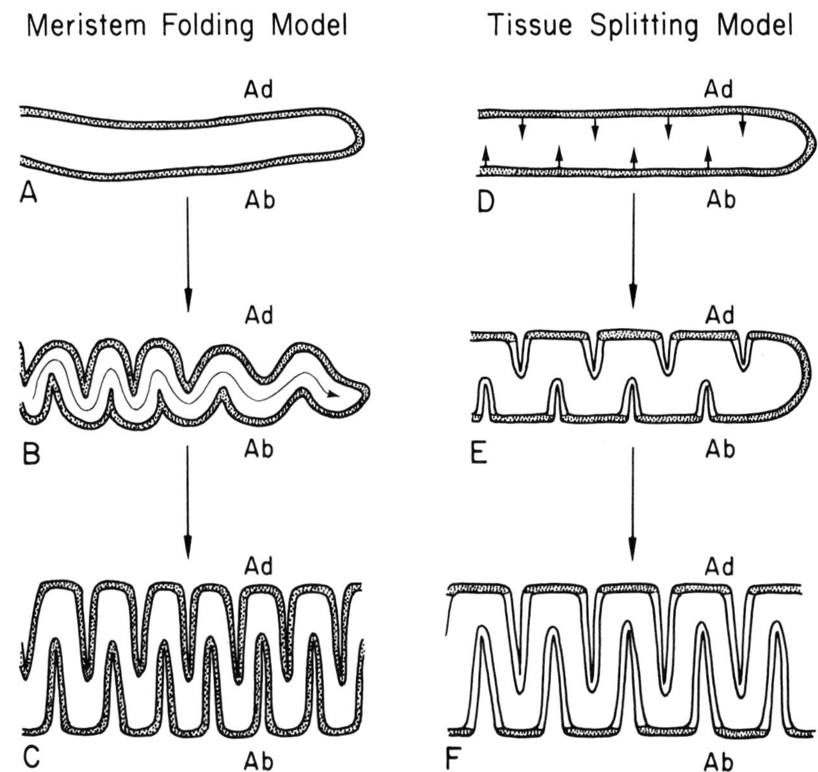

FIGURE 8. Comparison of different models of plication origin in palm leaves. A-C, Meristem folding model. D-F, Tissue splitting model. Original leaf epidermis stippled and that redifferentiated from mesophyll cells, clear.

equivalent to that of a single fold. Moreover, since the shape of an individual pleat varies along its length (from unplicated leaf margin to the unplicated rachis (Fig. 6A,B), we made all measurements and histological analyses along the complete length of a fold at all stages of development.

The earliest stage of plication inception in *Chrysalidocarpus* is marked by the upgrowth of slightly separated adaxial mounds associated with periclinal cell division in the middle layer of a 5-layered lamina (Fig.

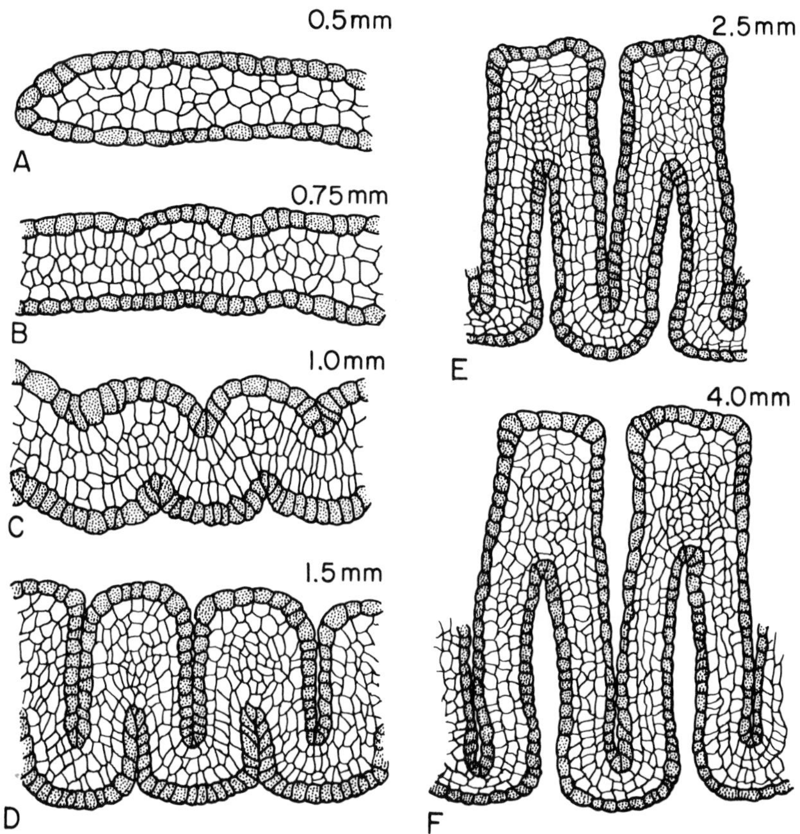

FIGURE 9. Transections taken through the middle of a plication's length in successively longer pinnate leaves of Chrysalidocarpus lutescens. The adaxial surface faces up and ad- and abaxial protodermal layers are stippled (Modified from Dengler et al, 1982).

9A,B). With continued periclinal division the adaxial ridges become more protuberant (Fig. 9C). At the same time the earliest signs of an equivalent abaxial deformation become evident indicating the abaxial ridges that alternate with their adaxial counterparts (Fig. 9C,D). Once a pattern of alternating ridges is established (Fig. 9D), growth is shifted from the ridges to the intercostal sections between the ridges (Fig. 9E). Continued intercalary plate meristem activity results in the progressive displacement of the adaxial and abaxial ridges from one another as the intercostal sectors of the folds continue to expand in surface (Fig. 9F). Ultimately, at a leaf length of 18-20 mm, a tissue separation occurs through the thickness of the abaxial ridges, resulting in the formation of a series of V-shaped reduplicate pinnae (Fig. 5D, F and H).

Since one of the correlates of the tissue cleavage model was that the original protodermal layer would be cut into isolated patches and epidermal continuity around the plication folds would have to come from the redifferentiation of new epidermal cells out of mesophyll derivatives (Fig. 8D-F), one means of evaluating this mechanism would be to examine the protoderm during plication development. By examining a closely-graded developmental series we demonstrated that the protoderm is continuous over both leaf surfaces at all stages of development (Fig.9). The epidermis simply follows the contours of the folds as predicted by the differential growth model (Compare Fig. 9 with Fig. 8A-C). At no stage is there evidence of the original protoderm being subdivided or new lineages being produced by mesophyll redifferentiation. Furthermore our TEM observations showed a continuous cuticle over the surfaces of all plication epidermal cells regardless of their location (Dengler, Dengler and Kaplan, 1982). If a true epidermal splitting were to occur, then it should be evident as a disrupted cuticular surface particularly in the cells at the base of a furrow where separation was purported to occur. We never observed such a cuticular discontinuity at any stage of plication development. The most conspicuous properties of cells at the base of the furrows is their cuneate or wedge shape in contrast to the more tabular shape of their neighbors lining the furrows (Fig. 9D). It is assumed that their wedge or even bulbous shape is a result of being compressed between adjacent, vertically extending ridges (Fig. 9C,D).

Beyond these more qualitative histological arguments, it is possible to design quantitative tests to discriminate further between the two principal models of plication proposed. By making measurements of plication dimensions during development it can be shown that the furrows between the ridges progressively increase in length (Dengler et al, 1982). However, the mode of increase in furrow length would differ depending upon the mechanism of plication involved. According to a meristem folding concept, increase in furrow length would be a result of upgrowth of adjacent mounds above or below the basal part of the furrow. Conversely, according to a tissue separation model, the furrows themselves would elongate, penetrating toward the interior of the leaf as cells at their base progressively separate from one another (Fig. 8D-F).

By using the cuneate epidermal cells as a marker of the furrow base and counting the number of contiguous cells in both the ab- and adaxial ridges opposite their respective furrow bases through the earliest phases of plication inception and growth, it was possible to discriminate between tissue cleavage and differential growth as two possible mechanisms of plication inception. If, for example, a tissue cleavage process were operating, we would predict that the number of contiguous cells between the furrow base and top of the opposite ridge would decrease because the cells would be pulled apart to cause the furrow to elongate. If, however, furrow length increased as a result of upgrowth of adjacent ridges, cell number in the ridge above the furrow should either increase or stay the same but not decrease.

Cell counts taken through the adaxial and abaxial ridges and across the width of the intercostal sector and plotted as a function of total leaf length for the pinnate leaf of *Chrysalidocarpus* show unequivocally that plication inception is by differential growth and not by schizogenous splitting. Cell numbers in the ridges either increase (adaxial ridge) or stay about the same (abaxial ridges Fig. 10A,B). At no time do they ever decrease as would be indicative of a cell separation process.

At this juncture it was also of interest to make similar observations and measurements of plication development in the palmate leaf of *Rhapis excelsa* to compare with our data on *Chrysalidocarpus*. Since plication orientation in fan palm leaves is vertical rather than horizontal (Fig. 6B), the spatial relationships in the developing folds are quite different from those of pinnate fronds. In the leaf of *Rhapis* the oldest folds occur in the center of the blade and the newest arise at the basalmost periphery of both margins of the lamina (Fig. 6B). As a result of this more lateral-basal initiation site, young plications in *Rhapis* have a slightly different transectional shape than those in *Chrysalidocarpus*. Instead of appearing as closely appressed mounds (Fig. 9C,D), primordial folds in *Rhapis* appear more widely separated from one another and sinusoidal in shape (Fig. 11A-C; 12A-C). Through all stages of development and the formation of some 22 plications, newly initiated folds have the same shape because they arise at the same relative

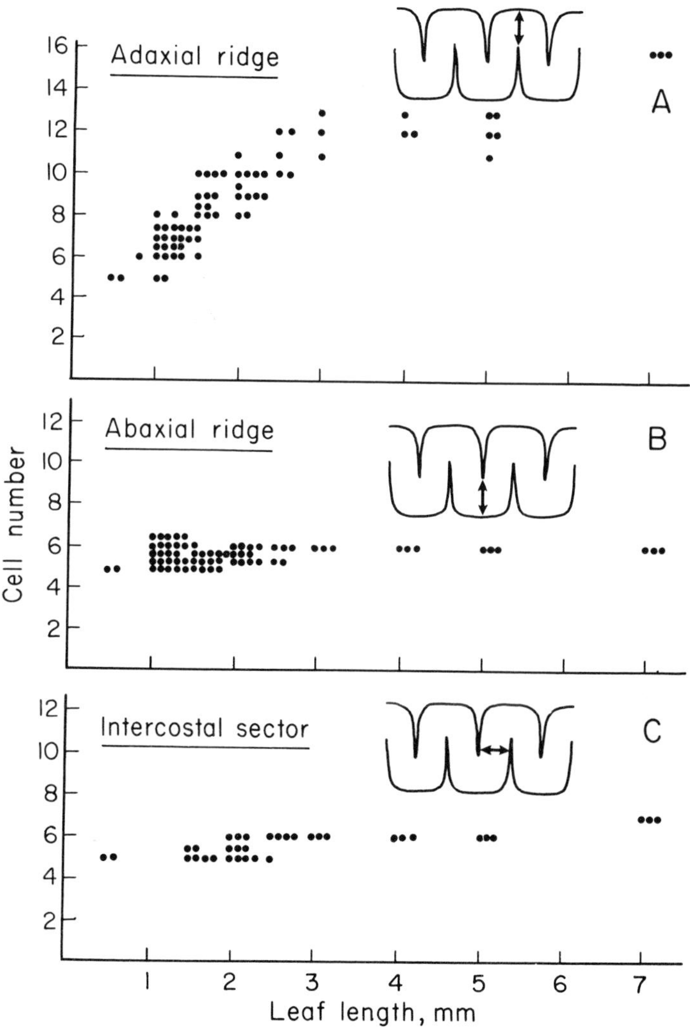

FIGURE 10. Cell numbers in (A) adaxial, (B) abaxial, and (C) intercostal sectors of median sections of plications in leaves of Chrysalidocarpus lutescens plotted as a function of leaf length. (From Dengler et al 1982 with permission)

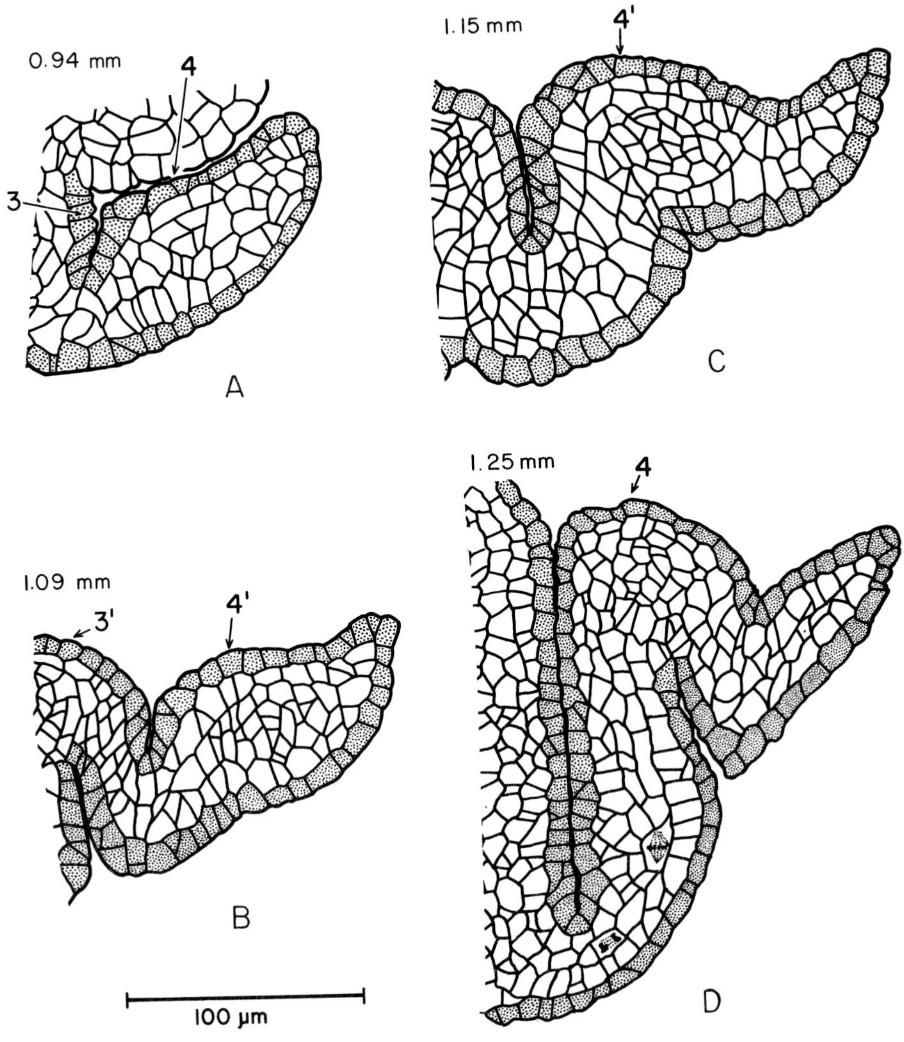

FIGURE 11. Transections of developmental stages of plication 4 in the palmate leaf of Rhapis excelsa taken through leaves of successively greater lengths. The adaxial surface faces upward and the protodermal layers have been stippled. (Modified from Kaplan et al 1982b).

position in the leaf (Fig. 11 and 12). As in Chrysalidocarpus adaxial ridge upgrowth in the leaf Rhapis is correlated with periclinal divisions in the middle layers of the embryonic blade (Fig. 11, Fig. 12A-D). Similarly, abaxial mound protuberance is a result largely of cell enlargement (Figs. 11 and 12). The difference is that the most recently initiated folds in Rhapis have broad, shallow troughs separating adjacent adaxial ridges rather than the narrow furrow lines characteristic of the pleats in Chrysalidocarpus (compare Fig. 11B and C and Fig. 12B and C with Fig. 9C and D). It is only the addition of new pleats laterally that crowds the more central folds together and converts the shallow depression between the ridges into a sharply incised furrow (Figs. 11 and 12).

When cell counts and leaf measurements are made during plication development in Rhapis, the results are virtually identical to those for Chrysalidocarpus (Kaplan et al, 1982b). Counts of cell numbers in adaxial and abaxial ridges and the intercostal sector show that cell number either increases (adaxial ridge) or stays about the same (abaxial ridge and intercostal sector; Fig. 13A-C). Thus like the plications of Chrysalidocarpus, blade pleating in Rhapis is a result of differential growth and not tissue cleavage. Furthermore, like leaves in the other genus, there never are signs of epidermal disruption and redifferentiation; at all stages of development and along the length of each fold the protodermal layer is continuous over the pleat surface and merely follows changes in shape that the plications undergo (Figs. 11 and 12).

Aside from demonstrating the generality of our results, this comparison of plication development between pinnate and palmate species has also helped to eliminate plication shape

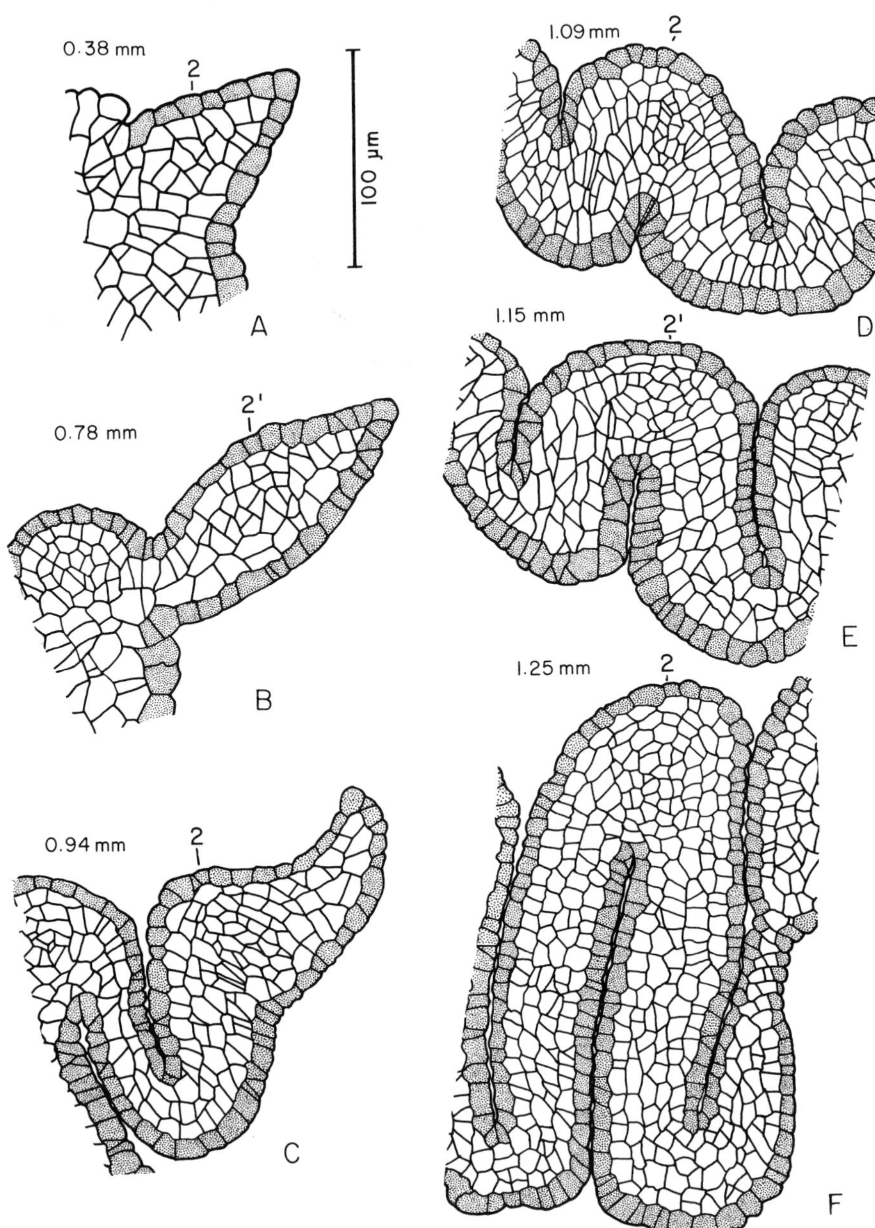
0.38 mm
2
100 μm
A
0.78 mm
2'
B
0.94 mm
2
C
1.09 mm
2
D
1.15 mm
2'
E
1.25 mm
2
F

as a basic indicator of the type of morphogenetic mechanism operating. If, as we have demonstrated, the basic cellular patterns of plication initiation and early growth are the same in these two different leaf types, regardless of differences in plication orientation, then matters of shape merely reflect spatial or physical relationships within or between leaves at the site where the plication arises rather than suggesting a fundamentally different morphogenetic process. Plication initiation in the palmate leaf of *Rhapis* appears more obviously as a folding process because its pleats arise in a free lateral portion of the blade that is unconstrainted relative to the other, earlier-initiated folds. By contrast, plications in *Chrysalidocarpus* have a shape more suggestive of a tissue cleavage mechanism than folding because of their intercalation between a rigid leaf apex, a thick rachis and rigid, unplicated leaf margin (Dengler et al, 1982), in addition to their being enclosed more tightly by surrounding older leaves in the bud (Kaplan et al, 1982a). Thus it would seem to be the physical circumstances in which plication occurs that affects the

FIGURE 12. Transections of developmental stages of plication 2 in the palmate leaf of *Rhapis excelsa* taken through leaves of successively greater lengths. The adaxial surface faces upward and the protodermal layers have been stippled. (Modified from Kaplan et al. 1982b).

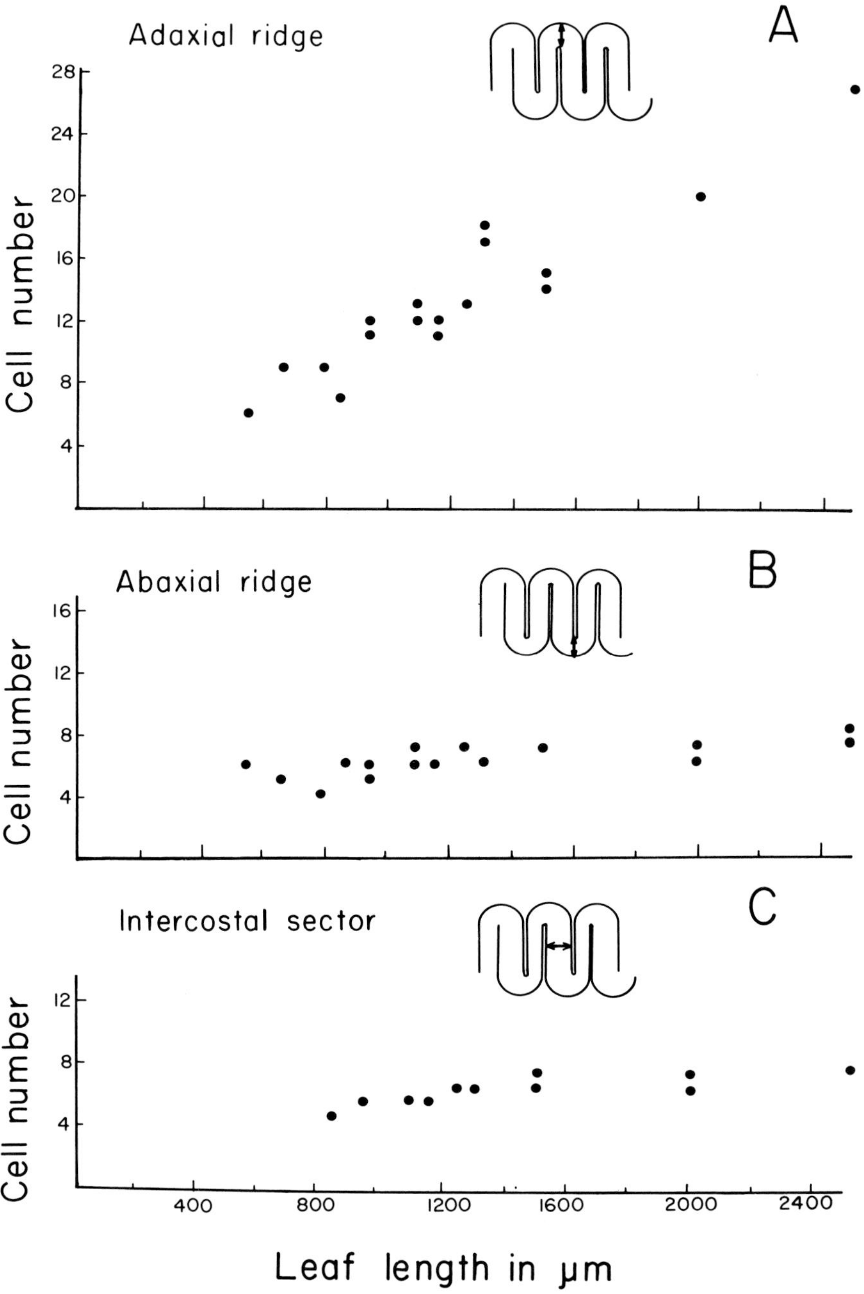
Number of cells at median section of plication
Adaxial ridge
A
28
24
20
16
12
8
4
Cell number
Abaxial ridge
B
16
12
8
4
Cell number
Intercostal sector
C
12
8
4
Cell number
400
800
1200
1600
2000
2400
Leaf length in μm

shape of the folds rather than fundamentally different developmental processes.

Our original reason for studying palm leaf development was to determine whether the leaves of palms initiate their plications by a morphogenetic mechanism that is unique for the plant kingdom, as had been claimed by some, or whether it is the result of a more conventional growth process. While there can be little doubt that the overall mechanism of palm leaf development is complex, involving processes of both differential growth and actual splitting of tissues to delimit the leaflet component, palm leaf morphogenesis is not as aberrant as it might seem at first glance. If we compare palm leaf development with the more conventional mode of leaf dissection we have described for *Zamioculcas*, in both cases the pinnae arise as a product of differential growth. What differs is the locus of their initiation and pattern of subsequent leaflet expansion. In the case of *Zamioculcas*, and most other species with compound leaves, differential growth occurs at the free margins of the blade so that pinnae develop as free outgrowths from the start. By contrast, in the palms, differential growth occurs within the lamina surface and as a consequence abscission of the folds from one another and abscission of the marginal strips are essential steps in producing free leaflets. When viewed in this perspective, the developmental complexities of palm leaves

FIGURE 13. Cell numbers in (A) adaxial, (B) abaxial, and (C) intercostal sectors of median sections of plications in leaves of *Rhapis excelsa* plotted as a function of leaf length. (From Kaplan et al. 1982b with permission).

seem less abnormal and more a product of specific developmental constraints, particularly the position of leaflet inception.

While our studies of plication in palms have failed to substantiate a tissue cleavage mechanism in plication origin, it is clear that tissue splitting does occur in the delimitation and extrication of the leaflets, processes unique to the palms and worthy of greater attention than they have received. The only paper which deals with leaflet delimitation and extrication in any detail is the classic publication of Eichler (1885). According to Eichler leaflet delimitation, _i.e._, plication cleavage, can range in mechanism from gross mechanical tearing to very precise separation of cells, presumably along the middle lamella. Unfortunately, despite the careful nature of his investigations, Eichler's 19th century studies were documented only by drawings and thus are open to interpretation. Some of the plication separations attributed to localized cell death may have been artifacts of his preparative or microscopic methods. We presently have access to a range of sophisticated light and electron microscopical techniques that can give us a more accurate picture of how this separation process is being effected in leaves of different palm taxa.

IV. GENERAL CONCLUSIONS AND DIRECTIONS FOR FUTURE RESEARCH

At the beginning of this article I raised the question of the existence of developmentally distinctive programs in plants and especially those that represent developmental convergences in form. While we have described in representatives of two families of monocotyledons morphogenetic patterns that are quite different, our goal shouldn't be limited to the mere cataloging of developmental pathways. This should only be the first step in a series of investigations designed to determine whether such disparate ontogenetic sequences have any underlying common denominators or whether they are indeed indicative of different systems of developmental control at a more fundamental biochemical level. To answer some of these broad questions it will be necessary to investigate leaf morphogenesis in *Monstera* and the palms at more basic cellular and subcellular levels. For example, in *Monstera* not only will it be of interest to determine which cell organelles first react to the necrotic stimulus but also how it spreads within the leaf.

Despite the seeming obvious differences in morphogenetic mechanisms operative in a leaf like that of *Monstera* and those of palms it is possible that their patterns of development may have more in common than would be predicted from first considerations. If, for example, we consider that cell separation phenomena involve the action of hydrolytic enzymes such as pectinases and/or cellulases, then it is possible that tissue necrosis in *Monstera* is also a manifestation of such hydrolytic activity but at the cytoplasmic level. Similarly, as emphasized by Addicott (1982) in his monograph on abscission in plants, in those species, such as some palms (Eichler, 1885) where abscission processes involve a mechanical component in their tissue separation, partial breakdown of wall components may occur but it is not extensive enough to result in a complete separation without some mechanical aid.

The key to future work in this area will lie not only in attempting to integrate more effectively knowledge from different levels of biological organization but also in studying these phenomena comparatively between representatives of these different organismal groups. Only by such comparative perspectives will we be able to decide whether we are dealing with truly distinctive developmental patterns or variations on a basic biochemical theme.

ACKNOWLEDGEMENTS

The research reported on palms was done collaboratively with Professors Nancy and Ronald Dengler of the University of Toronto. Special thanks are expressed to Mrs. Emily Reid who supplied the artwork and to Mrs. Eleanor Crump for her help with manuscript editing. Work supported in part by NSF grant DEB 74-02823 A01.

REFERENCES

Addicott, F.T. (1982) Abscission in Plants. Univ. Calif. Press, Berkeley.

Arber, A. (1922). On the development and morphology of the leaves of palms. _Proc. Roy. Soc. Ser. B_, 93: 249-261.

Bonner, J.T. (1982). Evolution and development. Springer-Verlag, Berlin, Heidelberg, New York.

Bruck, D.K. and Kaplan, D.R. (1980). Heterophyllic development in _Muehlenbeckia_ (Polygonaceae). _Amer. J. Bot._ 67: 337-346.

Bugnon, F. (1980). Quelques aspects fondamentaux de l'organogenèse foliaire chez un Palmier (_Washingtonia filifera_ Wendl.) _Rev. Cytol. Biol. Végét. Bot._ 3: 157-166.

Corner, E.J.H. (1966). The natural history of palms. Univ. Calif. Press, Berkeley.

Deinega, V. (1898). Beiträge zur Kenntnis der Entwickelungsgeschichte des Blattes und der Anlage der Gefässbündel. Flora 85: 439-498.

Dengler, N.G., R.E. Dengler, and Kaplan, D.R. (1982). The mechanism of plication inception in palm leaves: histogenetic observations on the pinnate leaf of Chrysalidocarpus lutescens. Can. J. Bot. 60: 2976-2998.

Eames, A.J. (1953). Neglected morphology of the palm leaf. Phytomorphology 3: 172-189.

Eichler, A.W. (1885). Zur Entwickelungsgeschichte der Palmenblätter. Abhandl. K. Akad. Wissen. Berlin I, 1-28.

Engler, A. (1905). Das Pflanzenreich IV 23B Araceae-Pothoideae.

Ertl, P.O. (1932) Vergleichende Untersuchungen über die Entwicklung der Blattnervatur der Araceen. Flora 126: 115-248.

Fuchs, C. (1975). Ontogenèse foliaire et acquisition de la forme chez le Tropaeolum peregrinum L. I. Les premiers stades de l'ontogenèse du lobe median. Ann. Sc. Nat. Bot., Paris ser. 12, 16: 321-390.

_____ (1976). Ontogenèse foliare et acquisition de la forme chez le Tropaeolum peregrinum L. II. Le développement du lobe après la formation des lobules. Ann. Sc. Nat., Bot. Ser. 12, 17: 121-158.

Goebel, K. (1884). Vergleichende Entwicklungsgeschichte der Pflanzenorgane. Schenk's Handbuch der Botanik. Verlag von Edward Trewendt, Breslau 3: 99-432.

____. (1926). Die Gestaltungsverhältnisse der Palmenblatter. Ann. Jard. Buitenzorg. 36: 161-185.

_____. (1932). Organographie der Pflanzen. Dritte Auflage Teil III. Samenpflanzen. VEB Gustav Fischer Verlag, Jena.

Gould, S.J. (1977). Ontogeny and phylogeny. Belknap Press, Harvard Univ. Press, Cambridge.

Hagemann, W. (1970). Studien zur Entwicklungsgeschichte der Angiospermenblätter. *Bot. Jb.* 90: 297-413.

Hirmer, M. (1919). Beiträge zur Morphologie und Entwicklungsgeschichte der Blätter einiger Palmen und Cyclanthaceen. *Flora* 113: 178-189.

Hofmeister, W. (1868). Allgemeine Morphologie der Gewächse. Verlag von Wilhelm Englemann, Leipzig.

Jeune, B. (1982). Morphogenese des feuilles et stipules de *Castanea sativa* Miller. *Bull. Mus. Natn. Hist. Nat, Paris ser. 4 section B, Adansonia* 1-2: 85-101.

Kaplan, D.R. (1973). Comparative developmental analysis of the heteroblastic leaf series of axillary shoots of *Acorus calamus* L. (Araceae) *Cellule* 69: 251-290.

_____. (1980). Heteroblastic leaf development in *Acacia*; morphological and morphogenetic implications. *Cellule* 73: 135-203.

_____ Dengler, N.G. and Dengler, R.E. (1982a). The mechanism of plication inception in palm leaves: problem and developmental morphology. *Can. J. Bot.* 60: 2939-2975.

_____, _____, and _____. (1982b). The mechanism of plication inception in palm leaves: histogenetic observations on the palmate leaf of *Rhapis excelsa*. *Can. J. Bot.* 60: 2999-3016.

Madison, M. (1977). A revision of *Monstera* (Araceae). *Contr. Gray Herb.* 207: 3-100.

Melville, R., and Wrigley, F.A. (1969). Fenestration in the leaves of Monstera and its bearing on the morphogenesis and colour patterns of leaves. Bot. J. Linn. Soc. 62: 1-16.

Moore, H.E., Jr. and Uhl, N.W. (1982). Major trends of evolution in palms. Bot. Rev. 48: 1-69.

Naumann, A. (1887). Beiträge zur Entwickelungsgeschichte der Palmenblätter. Flora 70: 193-202, 209-218, 227-242, 250-257.

Padmanabhan, D. (1963). Leaf development in palms. Curr. Sci. 32: 537-539.

_____. (1967a). Direct evidence for schizogenous splitting in palm leaf lamina. Curr. Sci. 36: 467-468.

_____. (1967b). Some aspects of the histogenesis in the leaf of Phoenix sylvestris L. Proc. Ind. Acad. Sci. Sect. B 65: 221-229.

_____. (1969). Leaf development in Phoenix sylvestris L. In: "Recent advances in the anatomy of tropical seed plants" (K.A. Chowdhury, ed) Delhi Hindustan Pub. Corp. (India) Delhi, India. Pp 165-177.

_____. (1973). Direct and indirect evidence for the occurrence of schizogenous splits in the young palm-leaf lamina. Indian Biologist 1: 34-42.

_____ and Veerasamy, S. (1973). Late "splitting" in the juvenile leaf of Phoenix dactylifera L. Curr. Sci. 42: 470-472.

_____ and _____. (1974). Ontogenetic studies on the juvenile leaves of Phoenix dactylifera L. Aust. J. Bot. 22: 689-700.

Periasamy, K. (1962). Morphological and ontogenetic studies in palms I. Development of the plicate condition in the palm-leaf. Phytomorphology 12: 54-64.

_____. (1977). Morphological and ontogenetic studies in palms VI. On the ontogeny of plication in the palm leaf. Proc. Ind. Acad. Sci. 85: 269-273.

Schwarz, F. (1878). Über die Entstehung der Löcher und Einbuchtungen an dem Blätte von Philodendron pertusum Schott. Sitzber. K.K. Akad. Wiss. Wien. Abt. I, 77: 267-274.

Trécul, A. (1853). Mémoire sur la formation des feuilles. Ann. Sci. Nat. Bot. Sér. 3, 20: 235-314.

_____. (1854). Notes sur la formation des perfortions que presenteent les feuilles de quelques Aroidees. Ann. Sci. Nat. Ser. 4, Bot. 1: 37-40.

Troll, W. (1939). Vergleichende Morphologie der höheren Pflanzen. Band I Vegetationsorgane Teil 2. Gebrüder Borntraeger Berlin.

_____. (1949). Die Stiel-Spreiten-Relation als Ausdruck des Prinzips der variabeln Proportionen. Naturwiss. 36: 333-338.

Venkatanarayana, G. (1957). On certain aspects of the development of the leaf of Cocos nucifera L. Phytomorphology 7: 297-305.

Von Mohl, H. (1845). Vermischte Schriften Botanisches Inhaltes. L.F. Fues, Tübingen.

Wessels, N.K. (1982). A catalogue of processes responsible for metazoan morphogenesis. In "Evolution and Development" (Ed. J.T. Bonner), pp. 115-154. Springer Verlag, Berlin, Heidelberg and New York.

Yampolsky, C. 1922. A contribution to the study of the oil palm Elaeis guineensis Jacq. Bull. Jard. Bot. Buitenzorg. Ser. 3, 5: 107-174.

MORPHOLOGICAL ASPECTS OF LEAF DEVELOPMENT IN FERNS AND ANGIOSPERMS[1]

Wolfgang Hagemann

Institut für Systematische Botanik
und Pflanzengeographie
der Universität, Heidelberg, Germany

A survey of the morphogenesis and possible phylogenesis of the foliar organs of ferns and angiosperms including the fertile structures is given. The construction of the sporangia-bearing fern leaf reflects a primitive state of a thallose land plant. The most important morphogenetic processes occur in its marginal meristem which reaches its maximum thickness only a short time after primordium initiation. Dilation of the marginal meristem leads to fractionations which produce the pinnae acroplasticly, periplasticly, or basiplasticly in an alternate, ternate, or binate order. During this period, thickness of the marginal meristem declines until in the last growth period the pinnae blades are modelled. The longitudinal articulation of fern leaves leads to a loose differentiation of a leaf base, petiole, and leaf blade. All these processes are also observed in angiosperms, but diversity is enlarged by the capacity of meristem incorporation and the development of true intercalary growth zones. Gamophylly, interpetiolar and median stipules, peltate blades, and pinnae were possible by incorporation and fusion of marginal meristems.

[1]Dedicated to Professor Dr. Werner Rauh on his 70th birthday.

Contemporary Problems in
Plant Anatomy

ISBN 0-12-746620-7

The common basipetal type of blade development in angiosperms is the result of a transgression of the leaf petiole intercalary growth zone into the base of a basipetally growing leaf blade. It is suggested that these processes, which initially occurred in the sori of dennstaedtioid ferns might have been transmitted to the reproductive phyllomes of angiospermous ancestors and eventually to their vegetative phyllomes. Angiospermy was reached by the formation of an ascidiate carpel. Transgression of the carpel blade margins onto the axis brought about all types of coenocarpous gynoecia. Adelphic microsporophylls of angiosperms are interpreted as the primitive pollen organs, the stamens being synangia situated on the under face of reduced phyllomes. Organ fusion to ringlike androecia and basipetal or acropetal stamen production demonstrates the capacity of meristem incorporation and intercalary growth of these organs.

I. INTRODUCTION

Eichler's (1861) studies on leaf development were most impressive and continue to influence the basic concepts of leaf morphology to the present day. But after him, for more than a century, developmental studies were based nearly exclusively on the cell theory. Plant cells were believed to be the fundamental elementary units of the plant body. Cell genealogy seemed to be the key to all major information necessary for developmental theories and as a consequence of this, modern experimental research is biased to a large extent by histological concepts. However, in my opinion, the cell theory prevents a proper comprehension of the plant body as the living unit. This was the reason I proposed the concept of the phragmoblastem, which is envisioned as the differentiating substrate in all higher land plants (Hagemann 1978, 1982). Differentiation starts in the terminal meristems by sculpturing the plant body as-a-whole, is followed by the zonal differentiation of the tissue systems in the established plant organs, and terminates with the differentiation of single cells. If this idea is correct, morphogenetical research should begin with the description of organ formation.

Following the theory of evolution, the complex construction of modern plants is the result of a long phylogenesis which started with relatively simple organisms. However, the original plant body had to fulfill all major functions of a plant. Hence, evolution of higher plants must have started from plants with relatively small, only slightly differentiated but multifunctional plant bodies.

Plants are principally sessile organisms requiring large surfaces for their photosynthetic activities. The indefinitely growing, ribbonlike, frondose liverworts or fern gametophytes spreading over the soil provide excellent models for primitive plant construction (Figs. 1, 2), even if this contradicts the phylogenetic theories since the times of v. Goebel (1930), v. Wettstein (1935), and Zimmermann (1959). Some of the fundamental processes of differentiation already existed before the differentiation of shoot organs was established. These processes were mainly the differentiation of marginal meristems and the finished nongrowing parts of the thallus producing carbohydrates and gonidia. The cormophytes with their highly elaborated, rooted shoot systems extended from the soil into the atmosphere. The leaves of cormophytes, especially those of ferns, resemble mostly frondose land plants in construction (Fig. 3 A,B). Compared to the leaf, stems and roots are newly developed, derived organs with special functions that have evolved under terrestrial conditions. Hence, leaves are the most conservative plant organs of the plant body.

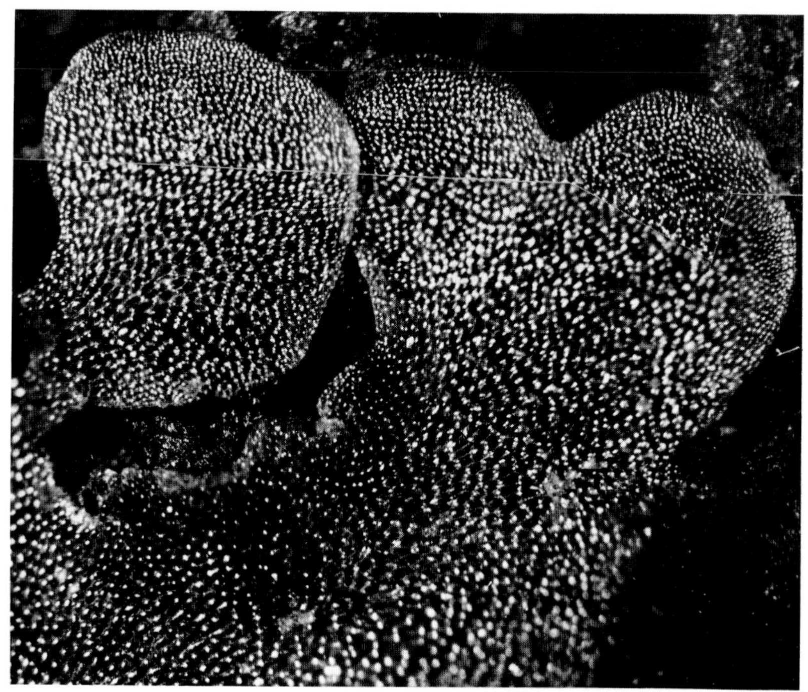

FIGURE 1. Gametophyte of a *Vittaria* species.

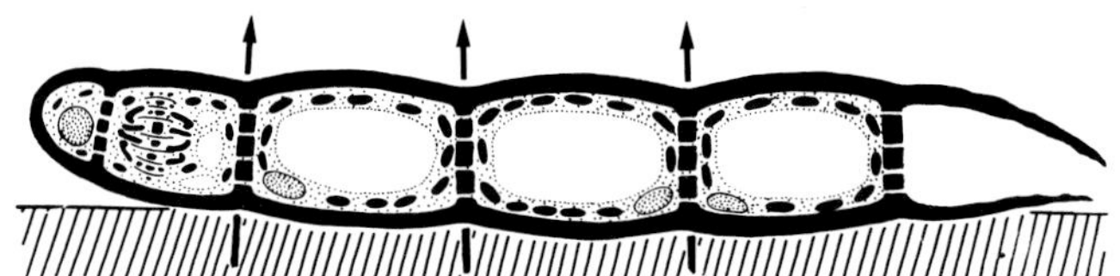

FIGURE 2. Model of a simple frondose plant body growing continuously on the soil and dying from behind. The arrows show the direction of water transport passing the phragmoblastem from the soil through the apoplast into the atmosphere.

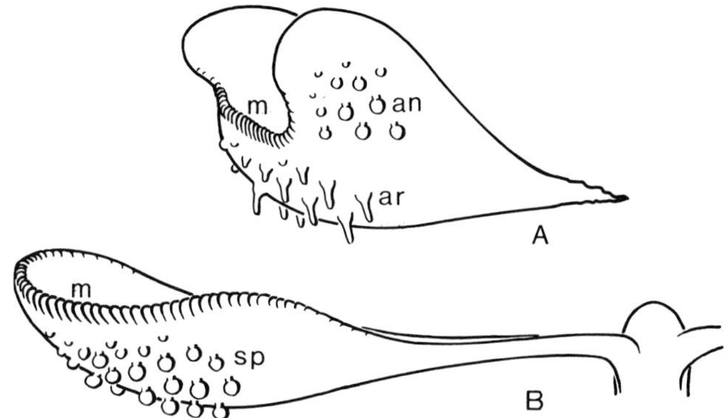

FIGURE 3. A, Sketch of a fern gametophyte growing with a marginal meristem (m) and producing gametangia (an, ar) on its under side. B, Sketch of a fern leaf growing by a marginal meristem (m) and producing sporangia (sp) on its lower face.

II. FERN LEAVES

Fern gametophytes as well as fern leaves grow by means of marginal meristems. The dorsiventral construction with midribs and the production of gametangia or sporangia on the lower surface are common primitive features. Important differences are (1) the gametophyte is an independent plant growing indefinitely, dying from behind, and bearing rhizoids on its lower surface, and (2) the leaf with definite growth is developed as an outgrowth of the stem, is devoid of rhizoids, and is detached as-a-whole from the remaining shoot axis. These differences can be understood as a consequence of the position of leaves in the free space of the atmosphere which makes indeterminate growth impossible, rhizoids superfluous, and which favors greater width and the longitudinal articulation of the leaf.

Leaves are inserted in a transverse position on the shoot apex. Their upper face is turned toward the apex of the growing shoot, and their dorsiventral symmetry seems to be the consequence of the transverse position of the young leaf meristem in the differentiation gradient of the shoot apex (Fig. 4 A,B). The early leaf primordium, therefore, is subjected from its beginning to an integrated system of differentiation gradients, one of which is the newly initiated leaf's own gradient between the growing leaf apex and the leaf base, and the other is the gradient perpetuated from the stem, extending from the lower to the upper face of the leaf. The latter gradient controls the bifacial leaf construction inserted on a radial stem. The short-celled embryonic tissue on the upper face of the leaf primordium and the more elongated tissue on its lower surface are well known features of all leaf primordia exhibiting acrovergent coiling (Fig. 5 A,B). The peripheral zone of such a primordium has the form of a marginal meristem. In ferns this meristem has a striking zonation due to the conspicuous marginal initials followed by a small-celled, quickly growing submarginal zone of meristematic tissue (Hagemann, 1965). The marginal initials are established in the very young leaf primordium, the first being a two-sided apical initial which, in most fern leaves, after a short period of growth is transformed into an equal marginal initial between its own segments (Fig. 6).

In many ferns leaf growth is restricted to the margin, which is believed to be a primitive feature. Leaves of this type have an open venation, because veins can only differentiate below the growing margin (Fig. 7). Regular bifurcation

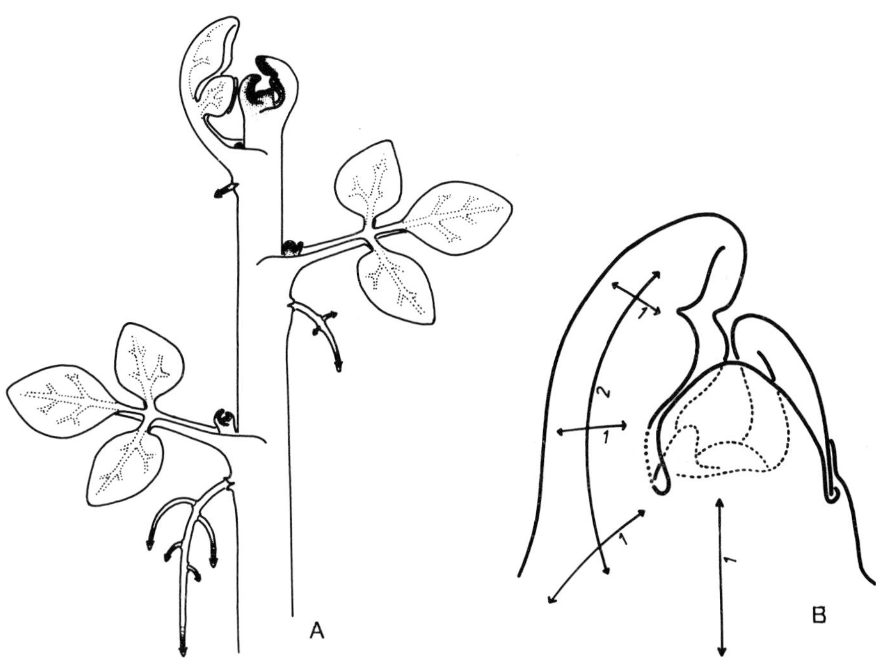

FIGURE 4. A, Diagram of the cormus of a seed plant with stem, ternate leaves, axillary buds, and roots arising endogenously beneath the leaves. B, Diagram of a shoot apex with three leaves in successive developmental stages, the largest showing an ensheathing base, stipules, petiole, and a ternately divided blade. Bifaciality is due to a crossed gradient system: (1) the differentiation gradient of the stem and (2) the leaf's own gradient between the leaf apex and the leaf base.

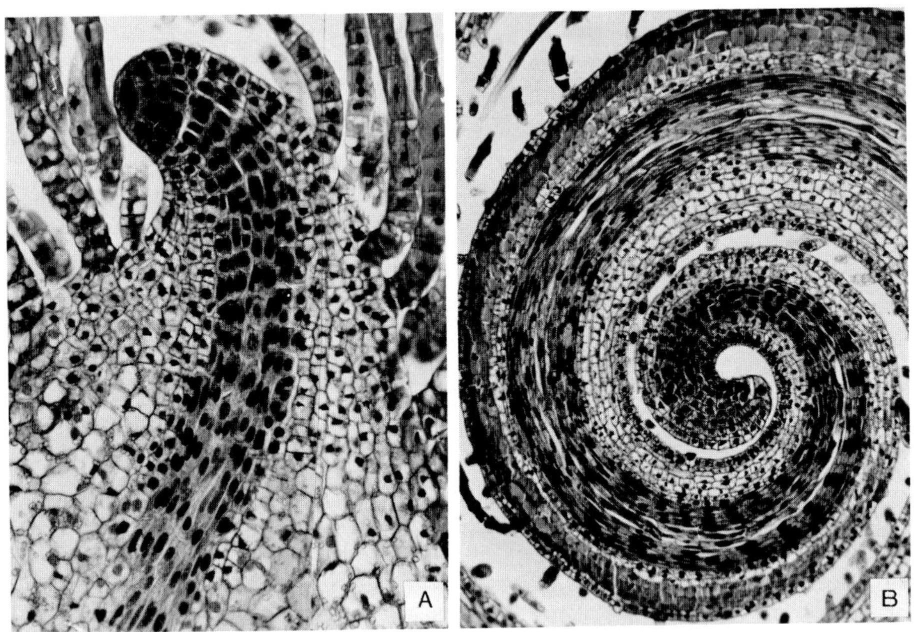

FIGURE 5. A, Young leaf of *Adiantum cuneatum*. B, Older coiled leaf of *Adiantum edgeworthii* in median longisection.

occurs in response to the dilation of the marginal meristem due to the pattern regulating the specific distances between the veins. If growth continues for a longer time in the submarginal leaf areas, growth is intensified, and under extreme conditions a regular meshwork of veins is blocked out (Fig. 8). As a result, venation patterns may reflect the type of growth the leaf has undergone, though there are histological differentiations (Goebel, 1922). As shall be discussed later, the differentiation of veins may influence morphogenesis, but

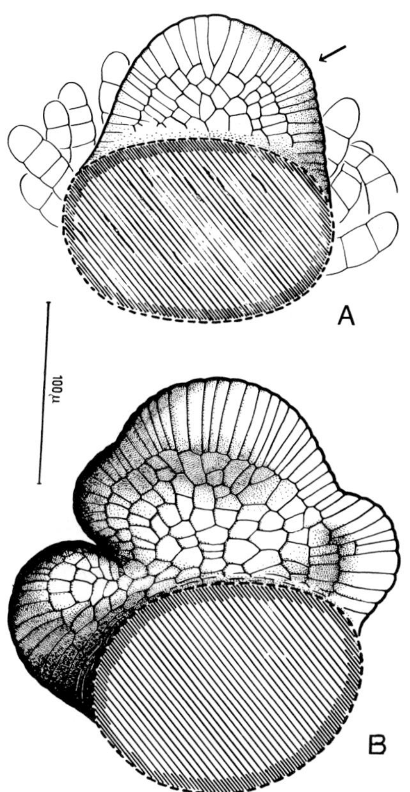

FIGURE 6. A, Leaf primordium of Adiantum cuneatum with a two-sided apical cell and the first pinna just initiated (arrow). B, Older leaf without apical initial.

this will occur only in advanced organs. Simpler leaves with open venation provide better insight into growth processes which result in different leaf forms. This is so because primary morphogenesis--not influenced by histogenetic differentiations-- is the dominant influence controlling future leaf form.

The most important developmental processes occur in the marginal meristem. The great majority of fern leaves are

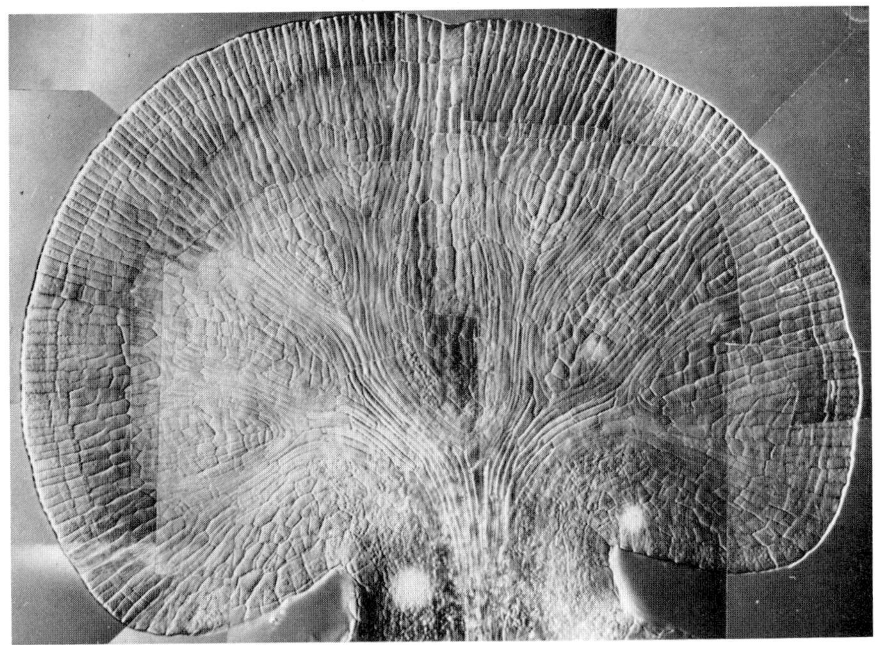

FIGURE 7. Growing pinna of *Adiantum formosum*.

pinnately divided. Only a few are smooth edged. The latter seem to be derived forms which result from the suppression of pinna formation. The formation of pinnae is based on the fractionation of the marginal mersitem, the division into small portions being directly related to meristem elongation (Fig. 9). This process can be seen in connection with the acrovergent coiling of the fern leaf, because coiling must result in an extremely folded or crumpled leaf margin. Marginal subdivision allows the resulting pinnae to be piled up in the bud in an orderly manner (Fig. 10A,B). This seems to be an elegant solution to the mechanical problems of coiling.

FIGURE 8. Part of a growing pinna with a network of procambial strands from *Acrostichum aureum*.

Fractionation of the marginal meristem always occurs successively following its prolongation. The division process occurs alternately in the flanks of the leaf apex resulting in an alternate, pinnate leaf (Figs. 11A, 12A). This means that the resulting portions of the marginal meristem are unequal. In some ferns one can observe a certain degree of independence of the fractionation along the two flanks of the leaf apex (Fig. 11B). The fractionation process in one flank of the apex may overtake that on the other side. Synchronization will result in the ternation of the leaf margin. Ternate fern

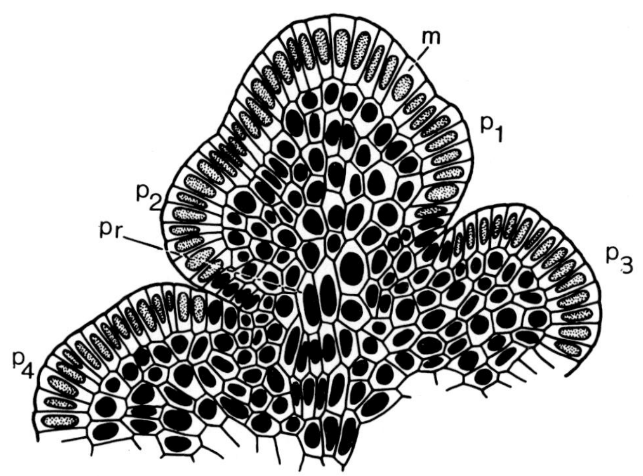

FIGURE 9. Fractionating marginal meristem of a pinna of _Adiantum cuneatum_. p_1-p_4, pinnae; m, marginal initials; pr, procambial strand.

leaves are particularly common in the Gleicheniaceae (Fig. 11C). Finally, division of the marginal meristem may occur in the middle of the leaf apex. The equal fractionation results in a binate or bifurcated leaf (Fig. 11D). Binately furcated leaves are very rare in modern ferns, and all existing examples are regarded as considerably specialized forms. Consequently, bination seems to be a derived condition. In summary we can separate three types of leaf division in ferns--alternation, ternation, and bination. Alternation appears to be both the most frequent and the most primitive condition.

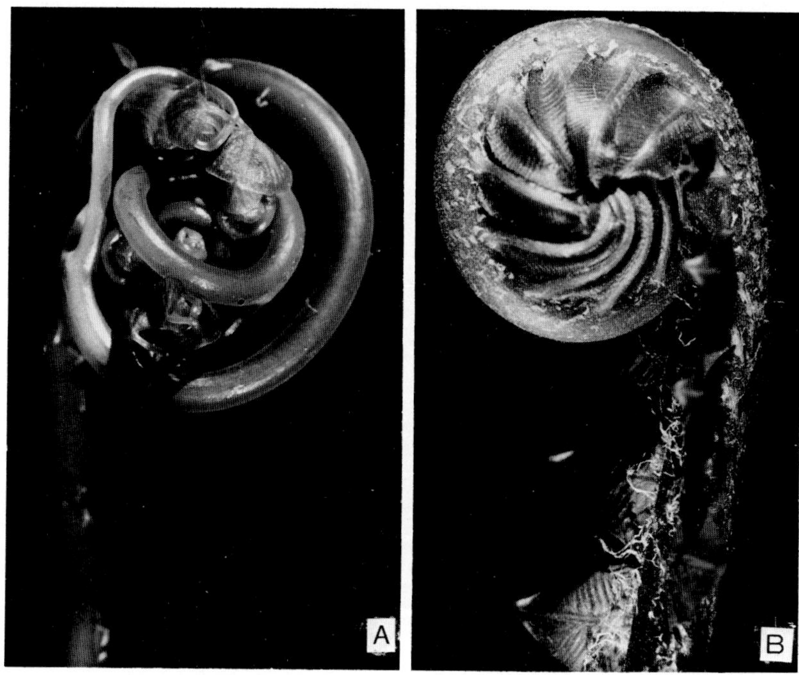

FIGURE 10. A, Coiled leaf of Adiantum cuneatum. B, Coiled leaf of Blechnum brasiliense.

Just as in fractionations, the distribution of active growth in the leaf marginal meristem allows for the observation of three basic growth types which are important for the comprehension of leaf forms. One can probably assume that the acroplastic mode of development is the most primitive of these types, because frondose gametophytes grow in this manner. Leaf growth continues only in or near the apex (or apices of the pinnae respectively), whereas in older and more basal parts of the leaf growth ceases (Fig. 13B). Consequently new pinnae are formed only beneath the leaf apex. The resulting leaf form is a prolongated one (Matteuccia struthiopteris, Nephrolepis cordifolia, Jamesonia cinnamonea (Fig. 13). In some cases fern

FIGURE 11. A, Adiantum cuneatum. Periplastic basitonous leaf. B, Hypolepis repens first order pinna. The pinnae of the second order develop in the upper part alternately, but in the base some pinnae are paired (arrows). C, Dicranopteris pectinata. The leaf is ternately divided, but the terminal pinna is inhibited resting in a terminal bud (adapted from Troll). D, Peltapteris flabellata. Binately divided leaf.

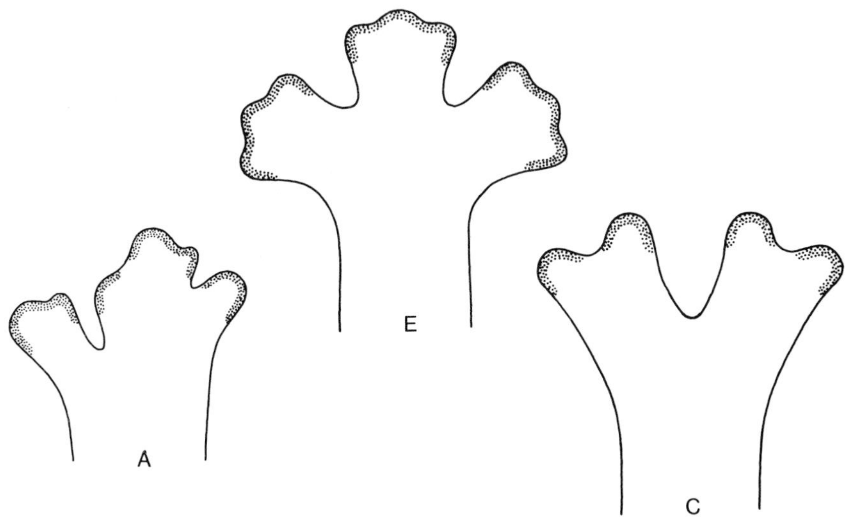

FIGURE 12. Diagrams for alternate (A), ternate (B), and binate (C) leaves.

leaves may have unlimited growth. In a leaf with limited growth, wide leaves will be advantageous instead of very long ones. Therefore, basitonic deltoid leaf forms will result if the marginal meristem continues to grow all along the leaf margin (Fig. 11A). This is the periplastic mode of blade development which is represented in the leaf forms of *Davallia* species or in *Adiantum cuneatum*. Finally, the mode of growth in fern leaves may even be basiplastic, if the activity of the marginal meristem continues only near the basal parts of the blade and produces new pinnae in a basipetal sequence. The leaf form then is pedate, as occurs in *Pyrrosia polydactylis*

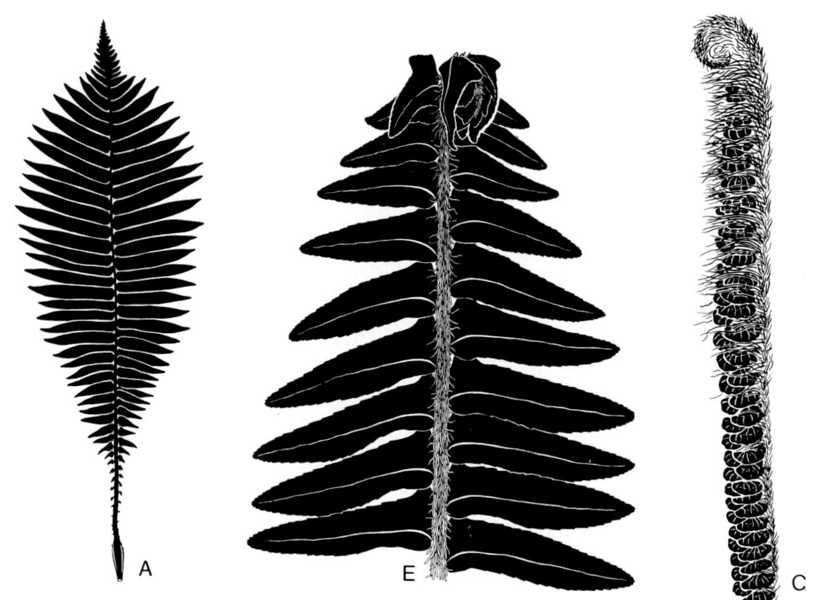

FIGURE 13. Acroplastic fern leaves. A, Leaf of Matteucia struthiopteris as an example for a mesotonous leaf. The pinnae of the first order which are pinnatifid in reality are drawn as entire. The leaf has a scalelike leaf base, a petiole zone with rudimentary pinnae, and a large pinnate blade. B, Nephrolepis cordifolia with a coiled apex. The more basal pinnae are finished. C, Jamesonia cinnamomea with the permanently curled leaf apex.

and Adiantum pedatum (Fig. 14 A,B). Certainly in Adiantum pedatum each pinna itself continues to grow acroplasticly.

During leaf development one can distinguish three growth periods. The first is the initiation of the primordium and the resulting establishment growth. Since nearly all fern leaf primordia have a two-sided apical initial, one may conclude that this first period of growth is characterized by an accelerated apical growth (Fig. 6A). The next stage

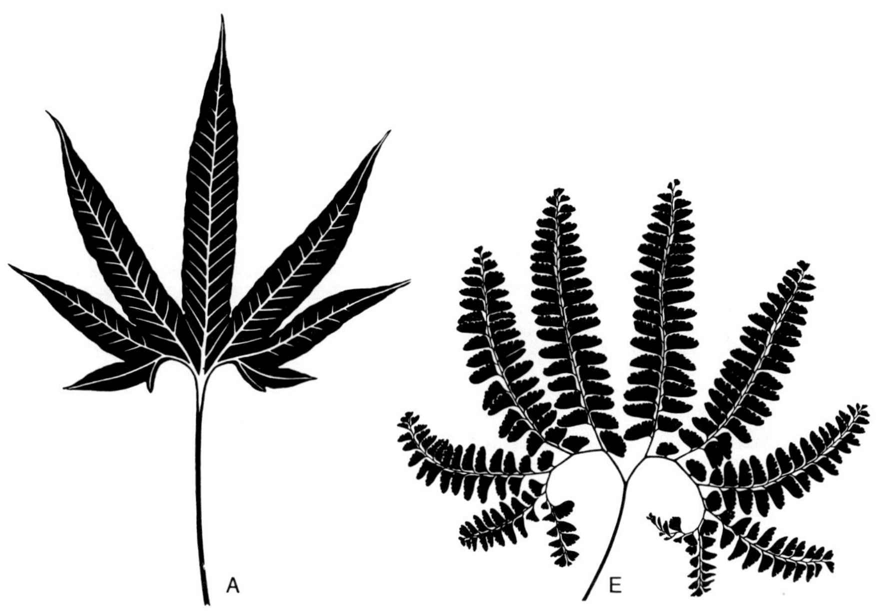

FIGURE 14. Pedate fern leaves. A, Pyrrosia polydactilis. B, Adiantum pedatum.

begins with fractionations of the marginal meristem which continues in an acroplastic, periplastic, or basiplastic manner. In the beginning of this phase the marginal meristem has its maximum thickness. During this second period, meristem thickness declines slowly until further fractionation stops. In the third growth phase, the marginal meristem continues to growth without fractionation thereby modelling the laminae of the pinnae (Fig. 7). Only in periplastic leaves do all three growth periods proceed simultaneously over the margin of the blade. In acroplastic and basiplastic leaves growth periods change in the acropetal or basipetal direction. Different parts of the developing leaves are in different developmental stages.

Having discussed some fundamental processes of primary morphogenesis, it is necessary to discuss some processes of histogenesis which may influence the future leaf form. The histogenetic differentiation of the leaf blade becomes obvious at the time of occurrence of costal and intercostal areas. As can be observed in Fig. 7, the differentiation of procambial strands has some influence on the growth of the marginal initials of a pinna, because the initial lying above the intercostal area no longer divides anticlinally. The intercostal area, therefore, doesn't take part in the further dilation of the margin. In some *Adiantum* species, suppression of marginal growth of the intercostal initial is intensified resulting in more or less shallow incisions in the margin of the pinna. Costal areas are not necessarily identical to vascular bundles, because more than one vascular strand may be observed in large costae. The costal system of the blade also functions as a skeleton. If one assumes that differentiation of procambium depends on the transport of anabolic substances into the growing meristematic zones, and that growing procambial tissues for their part will stimulate growth in thickness in the surrounding parenchyma, then quantitative differences in apical growth will affect the thickness of the resulting costae. A growing periplastic pinna, for example, with promoted apical growth will differentiate a prominent midrib. This can clearly be seen in the pinnae of the genus *Pellaea* (Fig. 15 A,B).

Independent from the developmental processes discussed above, leaves undergo a longitudinal articulation which produces the leaf base connecting leaf and stem. The leaf

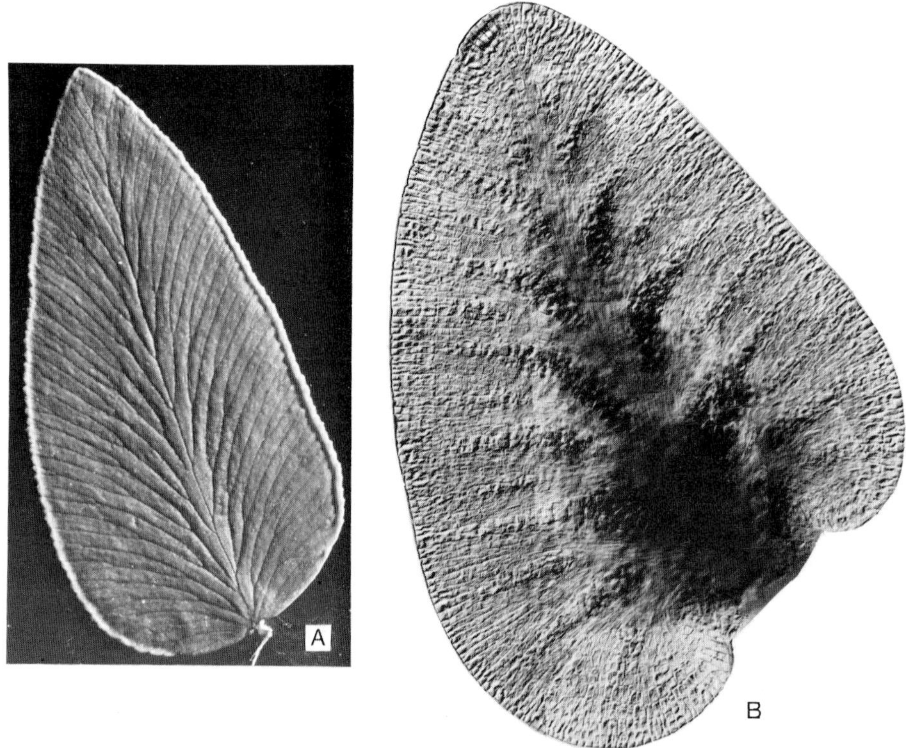

FIGURE 15. A, Pellaea viridis. Periplastic pinna with accelerated apical growth. The open venation system develops a clearly visible midrib. B, Pellaea flexuosa. Growing pinna with developing procambial system showing the midrib.

base is mechanically the most stressed region of the leaf, because it supports the entire leaf. The petiole exposes the leaf blade and originates between the leaf base and the blade. In ferns the relationship of the petiole to the leaf base and blade shows clearly that petiole differentiation is to some degree independent from other developmental steps, for example, the fractionation of the marginal meristem producing the pinnae. In most periplastic leaf forms the petiole develops beneath the first pinna (Fig.11 A). In Marattiaceae, however, one pair of stipule-like pinnae remains in the region of the leaf base. The intervening rachis link develops into the petiole (Fig. 16). Matteuccia struthiopteris is an example of a plant with a prolongated petiolar zone with numerous rudimentary pinnae (Fig. 13A).

III. LEAVES OF ANGIOSPERMS

All previously described developmental processes occuring in fern leaves also are observed in angiosperms. However, two developmental processes are added. The first is the ability of marginal meristems to spread out by incorporation of neighboring meristematic tissues, the second is the capacity to form intercalary meristems. These abilities increase considerably the diversity of leaf forms (Hagemann, 1970, 1973). The capability of meristem incorporation in marginal leaf meristems is seen throughout the angiosperms, in which actively growing free ends of leaf marginal meristems are adjoining other meristematic tissues (Fig. 17). Such tissues are stimulated by the margin to grow in the direction of the marginal meristem, thus prolongating it. One might say that they are "marginalized." In ferns, marginal meristems can elongate only by the anticlinal divisions of marginal initials. In angiosperms marginal meristems lack clearly differentiated marginal initials and, hence, look like normal small-celled meristems which they may incorporate. The incorporating ends of the marginal meristems may meet one another opposite the median portion of the leaf embracing the shoot apex. Since they marginalize the last remaining tissues between them they will fuse. This

FIGURE 16. *Marattia fraxinea*. Leaf of a young plant showing the basal pair of stipules and the long pinna-free petiole.

process can be interpreted as an actual fusion of organs occurring in angiosperm leaves. Since organ fusion seems to be absent in ferns, we have a strong argument against the telome-theory of Zimmermann (1959), because there is no real basis for what he has called "webbing" ("Verwachsung") of telome-like entities in connection with the phylogenetic derivation of megaphyllous leaves (Bower, 1963, p. 83; Bierhorst, 1971, p. 212).

In angiosperms, meristem incorporation is possible wherever the ends of marginal meristems are in close

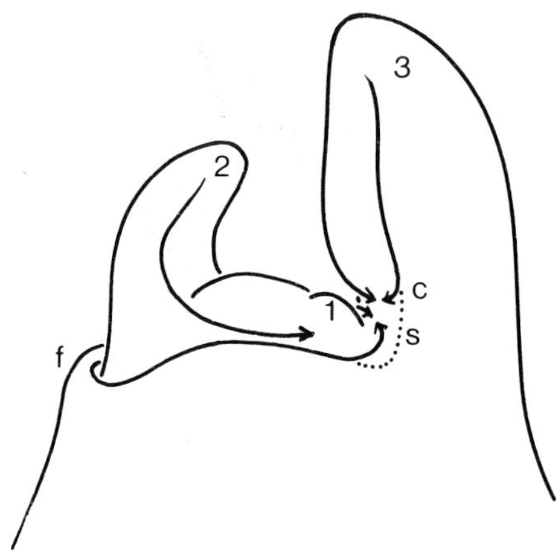

FIGURE 17. Shoot apical meristem with three leaves (1-3) with incorporating marginal meristems (arrows). f, fusion zone of the sheath of leaf 3; c, cross zone of the peltate blade; s, cross zone of the median stipule of leaf 3.

proximity to other meristematic tissues. This process is fundamental for the comprehension of gamophylly, peltation of the leaf blade, pinnae peltation, and the formation of interpetiolar and median stipules (Fig. 18). Indeed, there exists diverse leaf constructions of which the leaves of pitcher plants are only the most spectacular. In the following figures the main events in the ontogeny of a peltate leaf are illustrated (Fig. 19). At first the young, totally bifacial leaf primordium is surrounded by a continuous marginal meristem. The longitudinal articulation of the leaf base, petiole, and blade causes the interruption of the marginal meristem in the future petiolar zone, with the consequence that the resulting free ends of the marginal

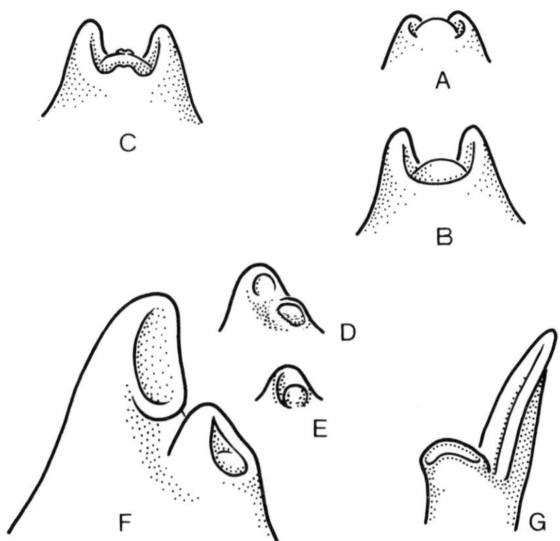

FIGURE 18. Diagrammatic figures of young leaf primordia resulting from meristem incorporation. A-C, two opposite primordia (A) producing interpetiolar stipules (C) after gamophyllous fusion of their marginal meristem (B). D-F, primordium (D) producing a closed sheath (E) surrounding the shoot apex with median stipule and a peltate blade (F). G, Young leaf of Polygonum sachalinense with its ochrea which arises from marginal fusion opposite to and in the axil of the leaf.

meristems will creep over the short distance of the embryonic upper face of the leaf and fuse in the so-called cross-zone. The resulting ring-shaped marginal meristem constitutes the peltate blade primordium at the same time as the median stipule at the upper end of the leaf base is established.

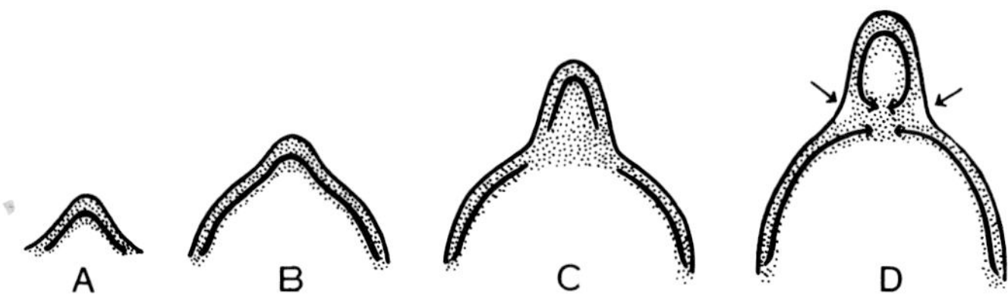

FIGURE 19. Diagram for the development of a peltate leaf with median stipules. For explanation see text.

The notion of intercalary growth zones in angiospermous leaves needs a clear definition of intercalary meristems: they are growing zones between non-growing, differentiated parts of the shoot (Fig. 20). The elongating zone between the apical meristem and the mature parts of the shoot does not satisfy this definition even if this zone may be very long, for example in a coiled young leaf of *Adiantum edgeworthii* (Fig. 5B). Typical examples of intercalary meristems are those at the base of strap-shaped monocotyledonous leaves, or the growing upper ends of most leaf petioles, for example, *Cyclamen coum* (Fig. 20B,C). Better known are the intercalary meristems in the internodal bases of grasses or horsetails (Fig. 20A).

As a rule, fern leaves are described as acroplastic, angiospermous leaves as basiplastic. Leaf growth is characterized in this way in most text books since the times of Eichler (1861). However, it has already been pointed out that in fern leaves this generalization is not tenable. However, no one can deny that basiplastic growth of the leaves is a very common feature in angiosperms. This is

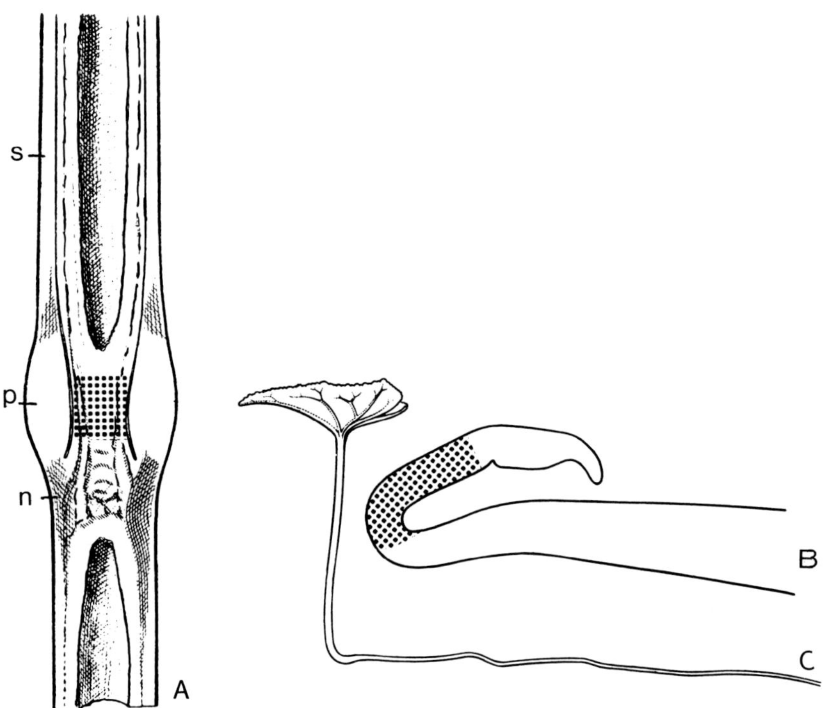

FIGURE 20. A, Construction of a grass node. n, node; p, pulvinus; s, basal part of the leaf sheath (adapted from Troll). B-C, Growing and mature leaves of Cyclamen coum. The intercalary meristems stippled.

especially striking in basipetal pinnate leaves. Polemonium caeruleum which was chosen by Troll (1939, p. 1523), may serve as an example of a general type of development in angiospermous pinnate leaves (Fig. 21 A,B). Polemonium has an intercalary meristem in the base of the blade combined with an active fractionating marginal meristem in the flanks, although the leaf tip clearly ceases to produce new leaf parts early in development. The phylogeny of this unique, simple looking but elegant construction is difficult to

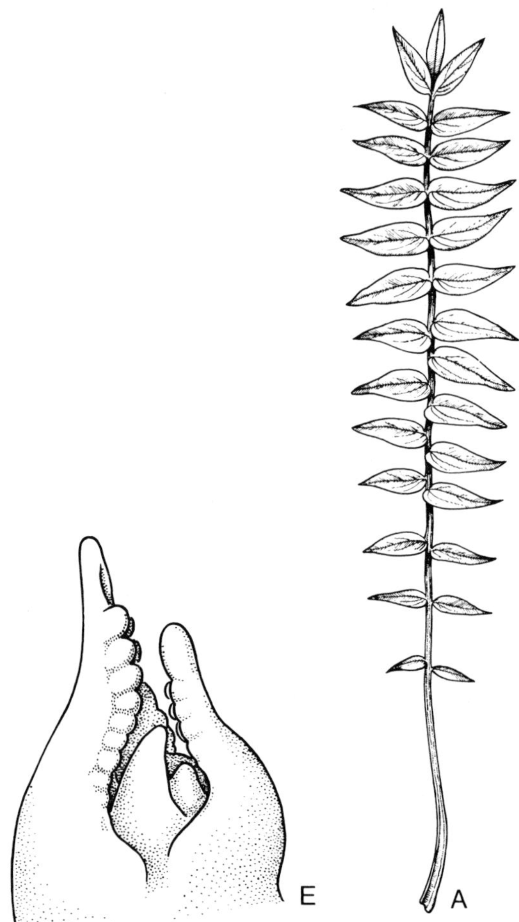

FIGURE 21. Polemonium caeruleum. A, Basipetal pinnate leaf. B, Shoot apex with young leaves showing the basipetal appearance of the pinnae primordia.

explain. An explanation is possible, however, if one compares the basipetal type with more primitive leaves in angiosperms. Periplastic and acroplastic leaves are also represented in this group of plants. A nearly ideal example for a periplastic developing, ternate pinnate leaf is that of Thalictrum aquilegifolium (Fig. 22 A,B). Multiple ternations produce pinnae of successively higher orders. In contrast to fern leaves, the development of the rachis segments between pinnae is by means of intercalary meristems, the leaf petiole being only the most basal one. Acroplastic leaf blades are common in Leguminosae (Fig. 23 A-C). It is very easy to understand the acropetal developing leaf if one keeps in mind the polyternate one. If the ternations are strictly limited to the terminal pinna, an acropetal leaf will result. In comparison with the polyternate leaf, all pinnae along the rachis must be pinnae of successively higher orders (Fig. 23 C).

The same condition can be observed in a pedate leaf, for example, the leaf of *Helleborus foetidus* (Fig. 24 A,B). The leaf of this plant is constructed in the same way as the pedate fern leaf except that the former is developed from a ternately growing leaf. But only the first fractionation can really be a ternate one. Because of the extreme basiplastic condition of the marginal meristem the fractionations in the flank pinnae must be limited to the basiscopic halves, the real basal ends of the marginal blade meristem. As a result, pinnae must arise in a strong basipetal sequence. Just as in acropetal leaves, the younger pinnae are of successively higher orders (Fig. 24B).

To understand how the basipetal pinnate leaf of the *Polemonium*-type develops, it is most convenient to refer to the intercalary meristems of the petiole zones in the growing pedate leaf, especially the leaf petiole itself. To shift the growth zone upwards from the petiole into the base of the blade is an operation which transforms the pedate leaf into a basipetal pinnate leaf, because the base of the blade producing the pinnae in basipetal sequence is elongated. Hence, in the basal zone of the blade, terminal and intercalary growth

A

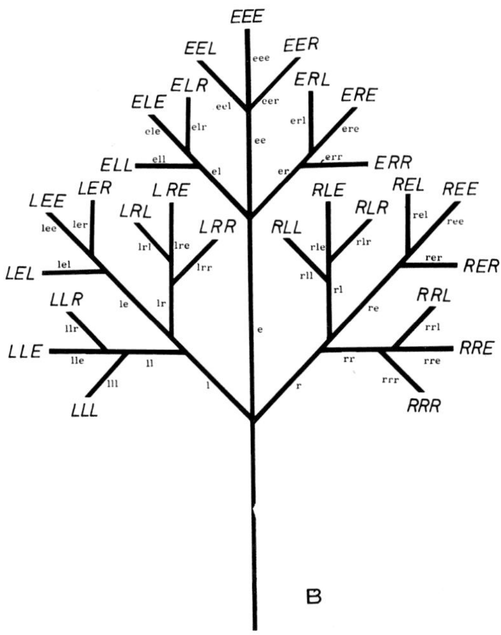
EEE
EEL
EER
ELR
ERL
ELE
ERE
ELL
ERR
LEE
LER
LRE
LRL
LRR
RLE
RLR
RLL
REL
REE
RER
LEL
LLR
RRL
LLE
RRE
LLL
RRR
eee
eel
eer
ele
elr
ee
erl
ere
ell
el
er
err
lee
ler
lrl
lre
lrr
rle
rlr
rel
ree
lel
rll
rl
rer
le
lr
re
e
rrl
llr
lle
ll
rr
rre
lll
l
r
rrr
B

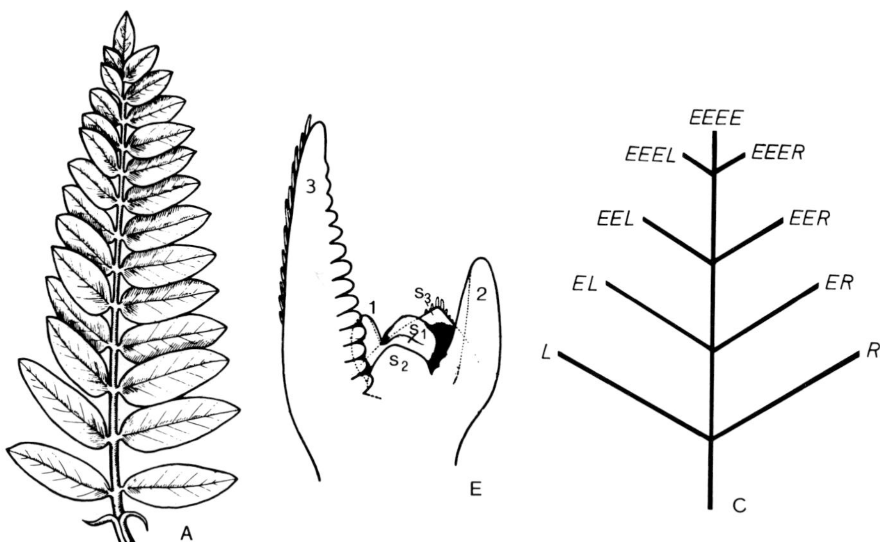

FIGURE 23. A, Acropetal leaf of Astragalus cicer. B, Shoot apex with three leaf primordia (1-3) with their stipules ($s_1 s_3$), the largest one in the foreground being omitted. New pinnae appear beneath the apex of leaves 2 and 3 (adapted from Troll, 1937). C, Diagram showing the pinnae of successive higher orders. For labelling see Fig. 22B.

FIGURE 22. Leaf of Thalictrum aquilegifolium (A) with a diagram (B) depicting the orders of pinnae and pinnae petioles until the third order. In the strict periplastic ternate system all pinnae are of the same order. The capitals E, R and L stand for Terminal (E), Right (R), and Left (L) pinna and the combination of three capitals signals the position of pinnae of the third order. The corresponding small letters indicate the pinnae petioles.

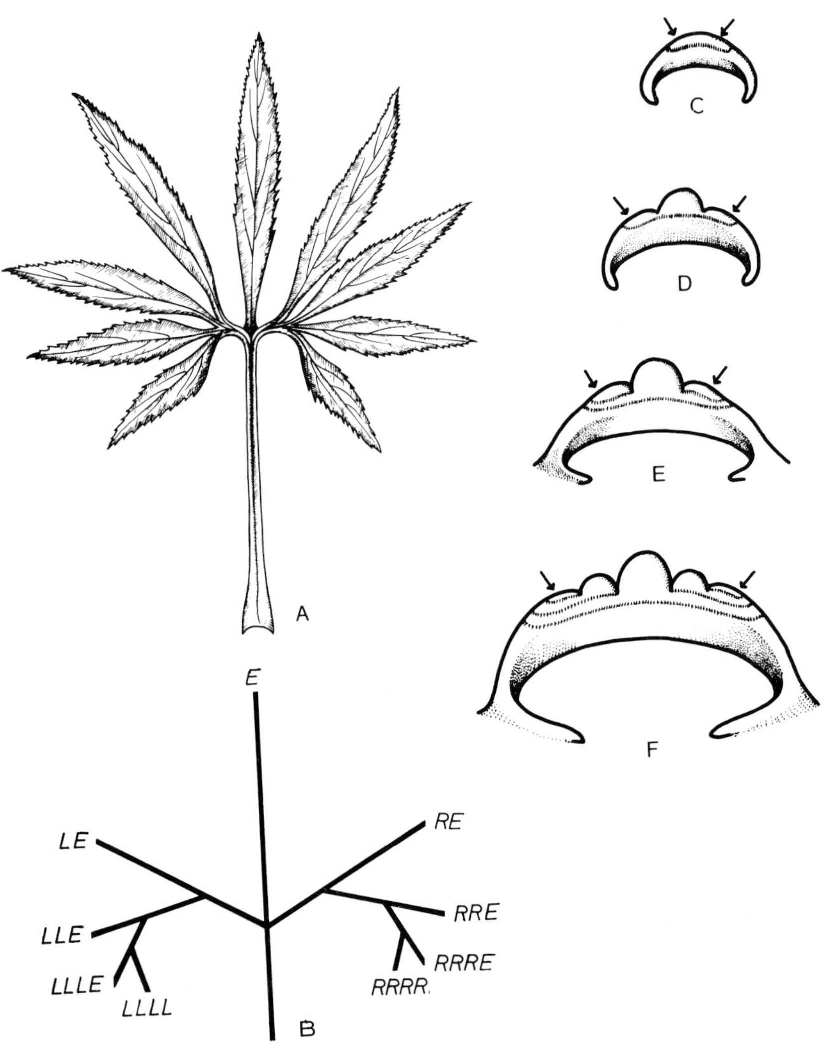
A
E
RE
LE
RRE
LLE
RRRE
LLLE
RRRR.
LLLL
B
C
D
E
F

zones are combined. This has important consequences insofar as the distribution of intercalary growth zones changes totally in comparison with the primitive polyternate leaf. The following diagram shows the position of the intercalary meristems in the pedate condition together with the intercalary meristems of the pinnate condition of the basipetal leaf (Fig. 25A). As can be observed, the intercalary meristems in the pedate leaf are homologous with the petioles of the blade and pinnae petioles (Figs. 24C-F, 25 A,B). In the pinnate leaf intercalary growth ceases in the vicinity of leaflet pairs and continues between the pairs. The resulting intercalary meristems are new ones compared with the petiole meristems of the pedate blade. In spite of the homology of the pinnae in the pedate and the pinnate leaf, the petioles of the pinnae and the links of the rachis are by no means homologous growth zones. Hence, the shifting of the intercalary growth zones produces a new construction which is typical for most angiosperms (Fig. 25A,C,D). This seems to be an important fact in the phylogeny of angiospermous leaves.

Undivided leaf forms are observed in all major groups of angiosperms. If smooth-edged leaf blades grow in a periplastic manner, a reniform blade will result, for example that of _Caltha palustris_. There are also extremely acroplastic smooth-edged leaf blades like those of _Drosophyllum_

FIGURE 24. A, Pedate leaf of _Helleborus foetidus_. B, Diagram showing pinnae of successively higher orders. For labelling see Fig. 22B. C-F, Successive stages of leaf primordia showing the basipetal fractionation of the marginal meristem. The arrows show the next fractionation zone. The stippled lines demarcate the zones of the future leaf petiole and pinnae petioles.

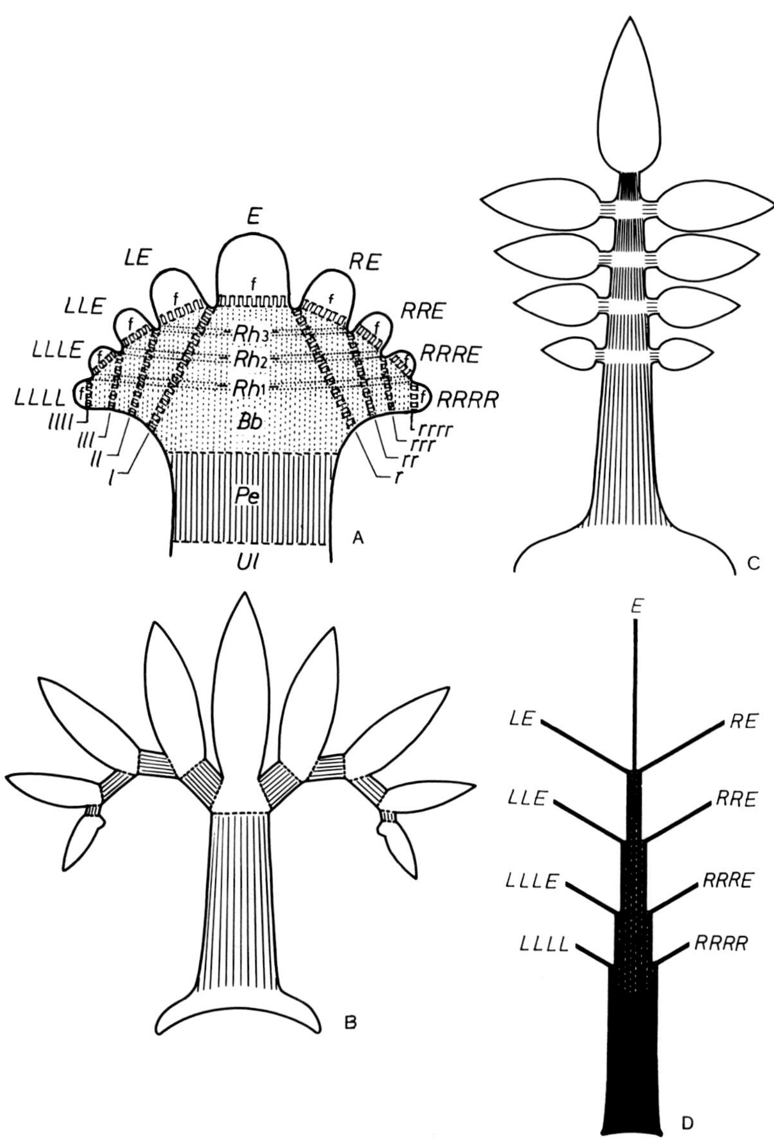
E
LE
RE
LLE
RRE
LLLE
RRRE
LLLL
RRRR
f
Rh3
Rh2
Rh1
Bb
Pe
Ul
llll
lll
ll
l
rrrr
rrr
rr
r
A
B
C
D

lusitanicum, and a very spectacular example of a basiplastic leaf blade is Streptocarpus wendlandii. Since all major types of growth exist in smooth-edged angiosperm leaf blades, I suggest a polyphyletic origin of these forms. As in ferns, it seems reasonable that pinnate leaves are more primitive in angiosperms, and smooth-edged leaf blade types have developed from all types of leaves by suppression of marginal meristem fractionation. If this is the case, the type of leaf construction in magnoliaceous plants cannot be considered as a primitive one. It seems more likely that the Paeonia-type of leaf may serve as a better model for a primitive angiosperm leaf.

FIGURE 25. A, Diagram showing a leaf primordium with basipetal pinnae labeled as in Fig. 24B. The future intercalary growth zones are marked for pedate and pinnate construction. Ul, Under leaf; pe, petiole; Bb, base of the blade; l-llll and r-rrrr, successive pinnae petioles of the pedate leaf; Rh_1Rh_3, successive intercalary growth zones of the rachis links in the pinnate leaf; f, petiole zones of the pinnae of the pinnate leaf. B, Diagram of the pedate leaf with prolongated intercalary zones hatched. C, The same for the pinnate leaf. D, Diagram of the basipetal pinnate leaf showing lack of homology of the rachis links with those of the acropetal pinnate leaf (See Fig. 23C).

IV. REPRODUCTIVE LEAF STRUCTURES

The foregoing discussion has examined some fundamental processes of development in vegetative leaf structures. I think it would now be useful to have a further look at the development of the sporophylls and especially the reproductive organs of angiosperms. According to the classical theory of the flower, floral organs are of phyllomic nature (Eichler, 1975; Troll, 1957, p. 4; Eames, 1961, p. 86; Weberling, 1981, p. 11). If this is correct, one may argue that knowledge of vegetative structures must be a prerequisite for successful research on floral organs. But I think this is not necessarily so. We have seen that vegetative and reproductive structures are combined in primitive fern leaves. In the course of phylogeny these two functions were separated on specialized leaf organs. I cannot dismiss the impression that some of the significant abilities of angiospermous leaves may have occurred first in the evolution of reproductive structures. Once established they may have been transmitted to vegetative structures. Meristem incorporation by marginalization and intercalary meristems may have had such an origin.

Fern leaves are lacking these capacities and this is true for all vegetative leaf structures in this group of plants. But in reproductive structures we can find exceptions. I cannot agree with the views of Bower (1963, p. 221), Goebel (1930, p. 1296), Zimmermann (1959, p. 147), Bierhorst (1971, p. 324), and others, that the marginal position of reproductive organs is a primitive feature. I mentioned already the homology between sporangia and gametangia and their position on the under side of leaves and gametophytes respectively. The structural entities for the two main functions, photosynthesis and reproduction, are produced simultaneously in the growing zones. It seems to be nonsequential that a functional division and hence a specialization with respect to timing and space will occur in evolution. It is commonly accepted that the differentiation of green pinnae and reduced spore producing pinnae, or the differentiation of trophophylls and sporophylls in ferns, are examples of such specializations. But it is less obvious that the production of submarginal or marginal sori signifies a differentiation in the same direction, because this observation is contradicted by the hypothesis of the phyletic slide of the sori from the marginal into the superficial position (Bower, 1963, p. 221). However, the submarginal or marginal position of sporangia is undoubtedly the more specialized condition, because the production of vegetative and reproductive structures is separated into two different developmental phases (Fig. 26). In ferns these phases are clearly separated by the acrovergent and revergent coiling of the leaf margin.

It is of special interest to follow the course of development of marginal, gradate sori of dennstaedtioid ferns, because their highly developed and complex sori show for the first time in ferns developmental processes of significant importance, namely organ fusion and intercalary meristems.

In Dennstaedtia the sorus occupies a marginal position which is clearly shown by scanning electron micrography (Fig. 27). The first sporangia are initiated by the marginal initials and further sporangia are produced in basipetal sequence (Bower, 1963, p. 221). The submarginal indusium on the lower face, seen also in Pteridium, is initiated first, and the young sorus is bent simultaneously in a reversed

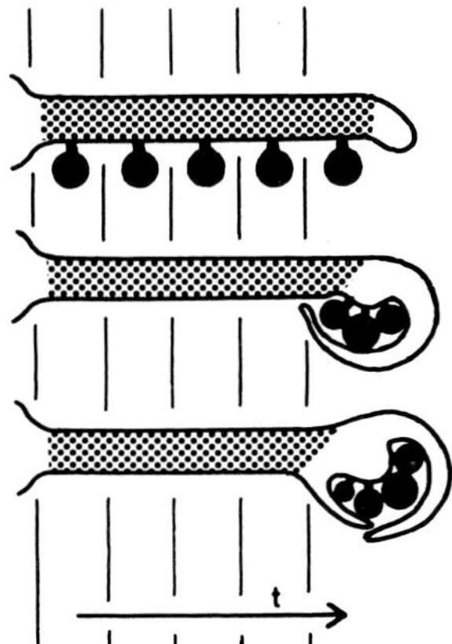

FIGURE 26. Diagram showing the stepwise differentiation in the development of superficial, submarginal, and marginal sporangia in relation of time.

direction. Somewhat later a dorsal indusium begins to develop. A short time later the two indusia are united across the marginal meristem of the leaf and this involves a true incorporation and fusion process. However, this is not realized by the marginal leaf meristem but by the marginal meristems of the indusia which fuse across the leaf marginal meristem. Tube-like indusia are typical for some highly developed fern families. The basipetal sequence of sporangial development produce the gradate type of sorus. The most elaborated gradate sorus occurs in the Hymenophyllaceae (Fig. 28). In this case the basipetal production of sporangia continues nearly indefinitely, because the receptacle is provided with a true intercalary meristem. An important question is whether it is accidental or not that these two striking developmental processes occur together in the

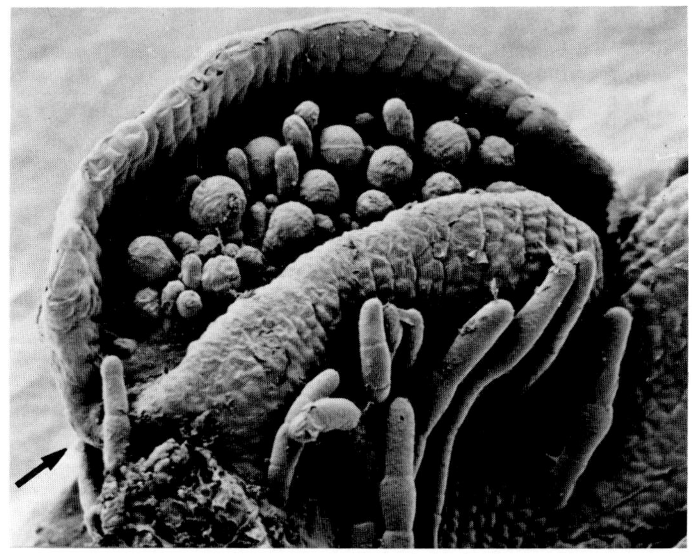

FIGURE 27. Young marginal sorus of *Dennstaedtia dissecta* showing the fusion of the upper and lower indusia across the leaf margin (arrow).

phylogenetic course of the reproductive structures in ferns, but not in the development of their vegetative leaf structures.

In angiosperms both processes are of importance for the construction of floral organs. Since the floral organs are regarded as leaves, the two growth processes possibly could have been transmitted to the marginal meristem, and the developing leaves as a whole, in connection with their reproductive functions. However, it may be that carpels and pollen organs owe their typical constructions to their capacity for meristem incorporation and intercalary growth. This is more obvious in carpels than in pollen organs, and, therefore, I shall start with a view of carpel development.

Troll (1934), Baum (1952a, 1952b), Leinfellner (1969a, b), Tucker and Gifford (1964), Van Heel (1981) proposed the

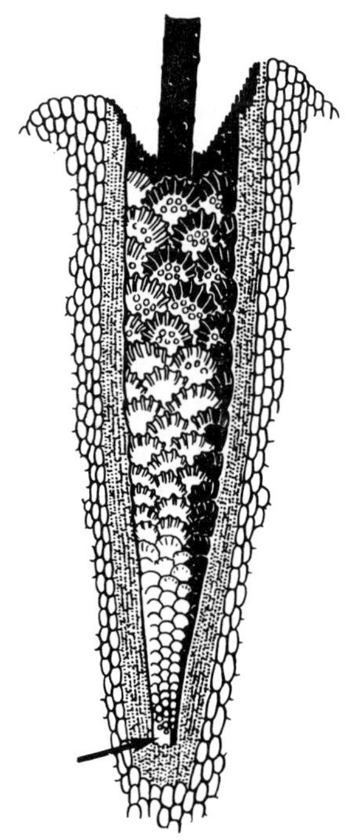

FIGURE 28. Opened gradate sorus of <u>Trichomanes membranaceum</u>

peltate carpel as a basic type and which can be observed within the ranalian complex. If one takes winteraceous carpels or those of <u>Helleborus</u> as an example, peltation is obvious (Fig. 29A,B). Like normal leaves carpel development begins with the initiation and differentiation of the leaf tip--the future stigma. Carpel development continues by the basal fusion of marginal meristems in the cross zone. The intercalary growth of the carpel is extremely obvious in the floral development of Betulaceae, where the stigma may be

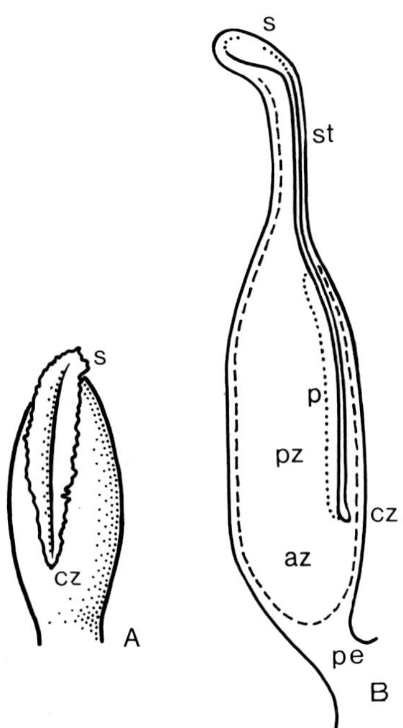

FIGURE 29. A, Peltate winteraceous carpel. B, Construction of the carpel of Helleborus niger. S, Stigma; st, style; pz, plicate zone; az, ascidiate zone; cz, cross zone; p, placenta; pe, petiole.

pollinated before an ovule in the basal part of the carpel develops.

Because the carpel has lost its vegetative functions it no longer needs to be exposed, and hence, the leaf base and leaf petiole may be extremely reduced. As a result, the leaf blade comes into close contact with the shoot apical meristem. This, together with peltation, is the starting point for the development of the syncarpic gynoecium. If more than one carpel is initiated around the floral meristem, and if each incorporates a part of the apical dome, then the carpels must be united by the incorporated stem tissue (Fig. 30 A-C). Thus, the syncarpous condition is reached by transgression of the carpels onto the flower axis (Hagemann, 1975). This process is described by Eames (1961, p. 195 "A primordium that is at first crescent-shaped may become ring-shaped at the base..."), and is clearly illustrated in many recently published photographs. Good examples are the pistillate flowers of Phoenix dactylifera (Demason & Stolte, 1982) and Ochna atropurpurea (Pauzé & Sattler, 1979) (Fig. 31A-C).

The fusion of carpels may occur in two steps. The first is gamophylly, perhaps continued with a primary curvature of the fusing margins in a central direction (Fig. 30D). The second step is the fusion of the united carpel margins in the middle of the floral apex, thereby constituting the septate of the syncarpous gynoecium (Fig. 30E). If the hitherto described events represent the original condition of syncarpous gynoecium development, then all other types can easily be derived. If one supposes that development remains incomplete, peltation will not be reached. Such a situation will result in the conduplicate carpel in the case of apocarpy which Eames (1961, p. 194) thought to be primitive. But his arguments are not convincing, because from the viewpoint of carpel construction the basal fusion of the blade marginal meristems favors the closure of the young carpel. Meanwhile Leinfellner (1969 a,b) and others have shown that many of the carpels which Eames thought to be conduplicate in reality are peltate. In monospermous carpels, like those of Ochna atropurpurea (Pauzé & Sattler, 1979), the ovule seems to have an axial origin (Fig. 31A-C). I would suppose in this case, that the region between the two horseshoe-shaped edges

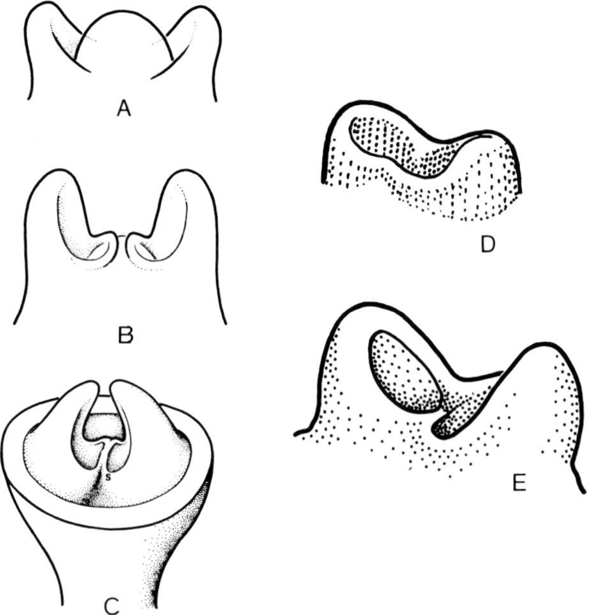

FIGURE 30. A-B, Two opposite carpel primordia invading the flower apex with their cross zones (B). The carpels are united by axial tissue. C, The invading carpel margins have fused gamophyllous and afterwards the united margins fuse on the flower apex constituting a common septum of the syncarpous gynoecium. D-E, Gamophyllous fusion of two carpels (D) showing afterwards the common cross zone in the middle (E). The margins of the carpels and the two halves of the septum are clapped one against the other around the cross zone.

of one carpel is marginalized, and, thus, the potential nongrowing cross-zone may develop into an ovule, while gamophyllous fusion between the neighboring carpels is

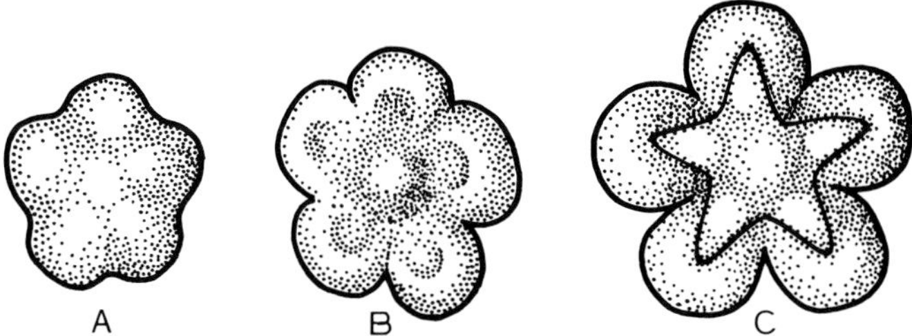

FIGURE 31. *Ochna atropurpurea*, flower apex with five carpel primordia. A, The apices of the carpels. B, U-shaped carpel marginal meristems fusing gamophyllous and marginalizing apical tissue in front of them. C, Gamophylly is completed and in the marginalized central zone the ovular mounds appear (adapted from Pauzé & Sattler, 1979).

occurring. If marginalization doesn't occur, fusion of the carpels may produce a paracarpous construction of the gynoecium with parietal placentation (Fig. 30 D). Paracarpy becomes comparable in this respect with the conduplicate condition of single carpels in a gamophyllous structure. I think that these few remarks on gynoecium development may indicate that a new approach, from a morphological point of view, by means of comparative developmental studies promises a coherent theory of the angiospermous gynoecium.

There are more problems in developing a theory of the construction of the androecium. I agree with the classical concept of the flower only to the extent that the microsporangia are produced by a microsporophyll. However, this cannot be identical with the stamen. The concept of the leaf requires a strict usage. In any case the leaf is a unit which arises on the flank of the apical meristem. The leaf may be compound or not. However, a single pinna or any part of a leaf will never be comparable for an entire foliar organ. Indeed many stamens seem to arise in the site of one leaf, but it is a common phenomenon that stamens may be augmented by so-called dedoublement. In doing so, groups of stamens, so-called adelphia, result and, hence, one of these stamens cannot represent the leaf (Fig. 32A). Adelphic stamens arise from a common primordium which in most cases produces the stamens in the basipetal direction, in others development occurs in an acropetal direction, as Leins (1975, 1979) and Gemmeke (1982) have clearly shown (Fig. 32). The common primordia of adelphia may or may not fuse to a ringlike swelling around the flower apex. This is parallel with gamophylly (Fig. 32B). Basipetal stamen production demonstrates clearly the intercalary growth of a leaf in an early stage of development. Only stamens producing primordial swellings as whole organs are leaf homologues, and as such they consist nearly entirely of the lower leaf face. So it seems convincing that stamens arising in the dorsal position on a leaf primordium are identical with synangial organs placed on the under side of a reduced sporophyll. Having in mind the marginal position of ovules in the carpel, the superficial position of sporangial clusters in the microsporophylls seems anomalous. But in cycads the same phenomenon is clearly seen.

In angiosperms the stamen number on each sporophyll may be reduced. If only one stamen per microsporophyll remains, this will occur in a leaf-like position on the flower axis. Sporophyll reduction may proceed so far that the leaf primordium is only determined on the flanks of the floral apex without further growth. In this case the determined leaf area

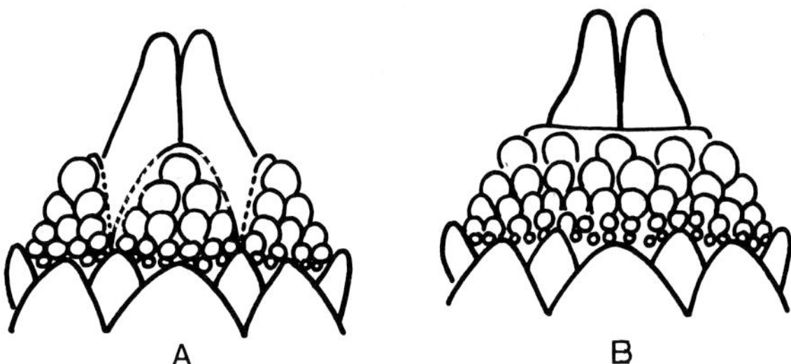

FIGURE 32. A, Diagram of a flower with nonfused basipetal adelphia. B, with gamophyllous fused basipetal adelphia.

comprises only the lower leaf surface, but this is sufficient for the production of a synangial stamen. A comparable case exists in the carpels of *Ochna atropurpurea* with the single ovule arising directly from a marginalized zone on the apical meristem.

Since the single stamen occurs in the site of a leaf organ as the result of extreme reduction in the angiosperm flower, the adelphic construction of the microsporophyll proves to be primitive. The brushlike or penicilliform microsporophylls which turn their under face above and stretch their sporangial clusters upwards by their prolongated intercalary-growing receptacles are most suitable for unspecialized flying animals as bees, beetles, birds and bats (Figs. 33, 34). Hence, from a morphological point of view recent speculations by taxonomists on the primitiveness of the dilleniaceous flower constructions seem to be tenable (Stebbins, 1974, p. 287; Leins, 1975, 1979; Vogel, 1978).

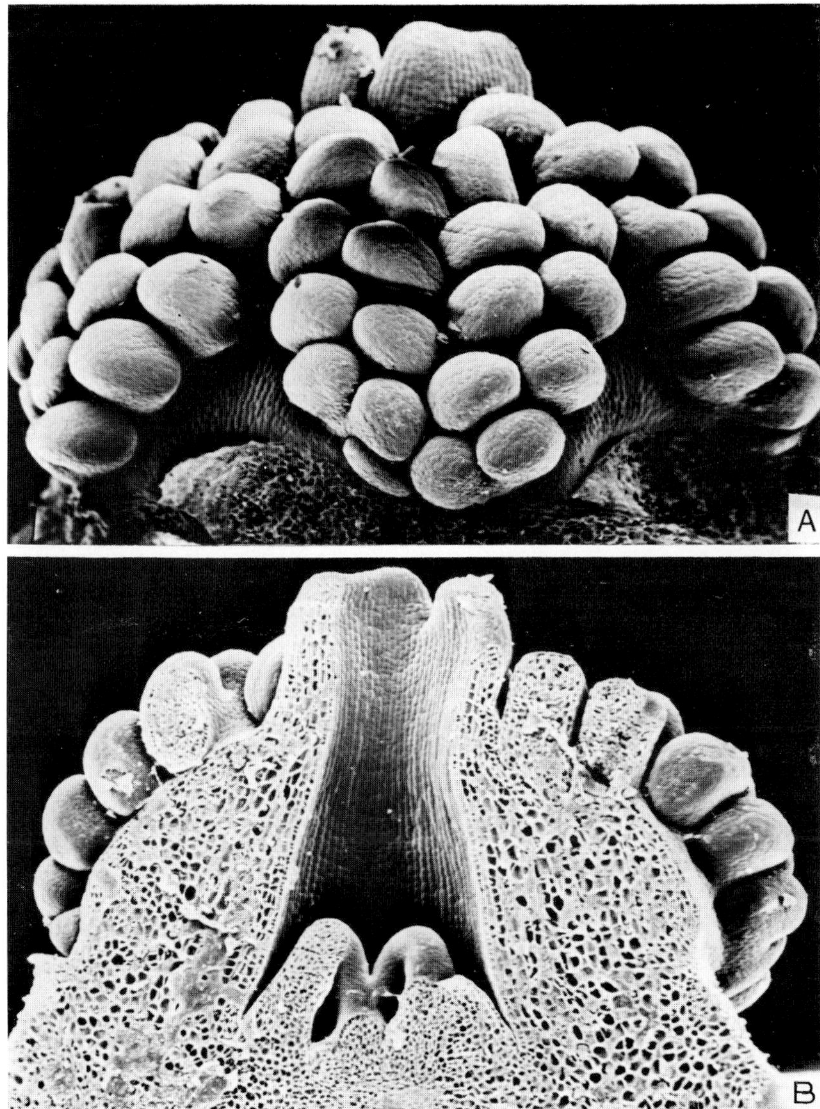

FIGURE 33. A, Young adelphic androecium with tube-like fused adelphia of Hibiscus syriacus. B, The same in median longisection. The apices of the microsporophylls are sterile on the top of the androecial tube. Two already fused carpel primordia are seen at the bottom of the androecial tube.

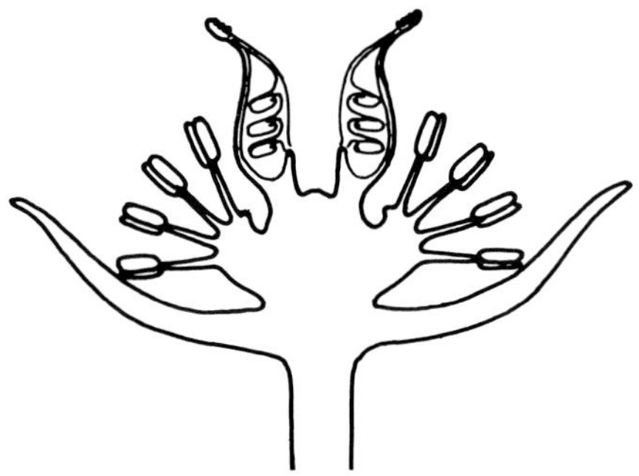

FIGURE 34. Diagram of a primitive angiospermous flower with adelphic microsporophylls.

The *Paeonia* type of flower is more primitive than the magnoliaceous type.

However, the crucial growth processes involved in the construction of primitive angiosperm floral phyllomes are doubtless the capacity for marginalization and intercalary growth. In my opinion it is not merely accidental that these processes are observed in the development of reproductive structures in some ferns. Consequently their occurrence in the vegetative phyllomes of angiosperms may be secondary.

With these remarks I have attempted to develop a comprehensive theory of the morphological, developmental, and phylogenetic aspects of the cormophytean leaf. I hope that criticism may stimulate its improvement.

ACKNOWLEDGMENTS

I am very grateful for the kind help of Dr. Jack B. Fisher, Miami, in reading the English translation of the manuscript, his kind suggestions improved the text considerably.

REFERENCES

Baum, H. (1952a). Die Querzonenverhältnisse der Karpelle von Helleborus foetidus und ihre Bedeutung für die Beurteilung der epeltaten Karpelle. Österr. Bot. Z. 99: 402-404.

_____. (1952b). Über die "primitivste" Karpellform. Österr. Bot. Z. 99: 632-634.

Bierhorst, D.W. (1971). Morphology of Vascular Plants. MacMillan, New York.

Bower, F.O. (1963). The Ferns. t l. Reprint Edition. Today and Tomorrow's Book Agency, New Delhi.

DeMason, D.A. and Stolte, K.W. (1982). Floral development in Phoenix dactylifera. Can. J. Bot. 60: 1439-1446.

Eames, A.J. (1961). Morphology of the Angiosperms. McGraw-Hill Book Company, New York.

Eichler, A.W. (1961). Zur Entwicklungsgeschichte des Blattes mit besonderer Berücksichtigung der Nebenblattbildungen. Marburg.

_____. (1975). Blüthendiagramme I. Engelmann, Leipzig.

Gemmeke, V. (1982). Entwicklungsgeschichtliche Untersuchungen an Mimosaceen-Blüten. Bot. Jb. Syst. 103: 185-210.

Goebel, K. (1922). Gesetzmäßigkeiten im Blattaufbau. Bot. Abh. H l Jena.

Goebel, K. (1926, 1930, 1933). Organographie der Pflanzen. G. Fischer, Jena.

Hagemann, W. (1965). Vergleichende Untersuchungen zur Entwicklungsgeschichte des Farnsprosses II. Die Blattentwicklung in der Gattung Adiantum L. Beitr. Biol. Pflanzen 41: 405-468.

_____. (1970). Studien zur Entwicklungsgeschichte der Angiospermenblätter. Ein Beitrag zur Klärung ihres Gestaltungsprinzips. Bot. Jb. 90: 297-413.

_____. (1973). The organization of shoot development. Rev. Biol. (Lisboa) 9: 43-67.

_____. (1978). Zur Phylogenese der terminalen Sproß-meristeme. Ber. Deutsch. Bot. Ges. 91: 699-716.

_____ (1982). Vergleichende Morphologie und Anatomie - Organismus und Zelle, ist eine Synthese möglich? Ber. Deutsch. Bot. Ges. 95: 45-56.

Leinfellner, W. (1969a). Über die Karpelle verschiedener Magnoliales. VIII. Überblick über alle Familien der Ordnung. Österr. Bot. Z. 117: 107-127.

_____. (1969b). Über peltate Karpelle, deren Schlauchteil außen vom Ventralspalt unvollkommen aufgeschlitzt ist. Österr. Bot. Z. 117: 276-283.

Leins, P. (1975). Die Beziehungen zwischen multistaminaten und einfachen Androeceen. Bot. Jb. Syst. 96: 231-237.

_____. (1979). Der Übergang vom zentrifugalen komplexen zum einfachen Androeceum. Ber. Deutsch. Bot. Ges. 132: 189-204.

Pauzé, F. und Sattler, R. (1979). La placentation axillaire chez Ochna atropurpurea. Can. J. Bot. 57: 100-107.

Stebbins, G. L. (1974). Flowering Plants, Evolution above the Species Level. Belknap Press of Harvard Univ. Press. Cambridge, Mass.

Troll, W. (1934). Über Bau und Nervatur der Karpelle von Ranunculus. Ber. Deutsch. Bot. Ges. 52: 214-220.

_____. Vergleichende Morphologie der höheren Pflanzen I. Bd. Teil 2. Bornträger, Berlin.

_____. (1957). Praktische Einführung in die Pflanzenmorphologie. VEB G. Fischer-Verlag, Jena.

Tucker, S. C. and Gifford, E. M. (1964). Carpel vascularization of Drimys lanceolata. Phytomorphology 14: 197-203.

Van Heel, W. A. (1981) A SEM-Investigation on the development of free carpels. Blumea 27: 499-522.

Vogel, St. (1978). Evolutionary shifts from reward to deception in pollen flowers. - In Richards, A. J. (Ed.) The Pollination of Flowers by Insects. Linnean Soc. Symposium Ser. No. 6: 86-96.

Weberling, F. (1981). Morphologie der Blüten und Blütenstände. Ulmer, Stuttgart.

v. Wettstein, R. (1935). Handbuch der systematischen Botanik. 4. Aufl. F. Deuticke, Leipzig und Wien.

Zimmermann, W. (1959). Die Phylogenie der Pflanzen. G. Fischer, Stuttgart.

ORIGIN OF SYMMETRY IN FLOWERS

Shirley C. Tucker

Department of Botany
Louisiana State University
Baton Rouge, Louisiana

Phylogenetically significant floral features include number, order of initiation and position of appendages, symmetry, aestivation, degree of fusion, and loss of parts. These features can be studied ontogenetically, and the ontogenies compared among related taxa to yield information on the relative time during ontogeny when the feature is determined or manifested. The family Leguminosae has three subfamilies which differ in floral symmetry, aestivation, and in location and degree of fusion of organs. It is proposed that these kinds of features arise early in ontogeny (during organogeny) and that as a consequence they are considerably more stable than those features which arise at mid-stage (during form change and enlargement) or at late stages of ontogeny (those which involve differentiation of tissues and cells). The mid-stage-determined features include many which separate related genera, while those determined late in ontogeny tend to characterize and separate related species. For any one feature, advanced taxa tend to express the feature precociously. For example, zygomorphy can be expressed even before organ initiation as an oval rather than circular shape of the floral apex. In other flowers, zygomorphy may be expressed first at sepal, petal, or stamen initiation, or it may be delayed until enlargement of petals or stamens. The order of organ initiation also is an expression of symmetry; in many legumes, members of each whorl are initiated sequentially from one side of the floral apex to the opposite side rather than simultaneously or helically. Elucidation of developmental differences between similar or related taxa can provide a basis for understanding how subfamilial and tribal differences could have evolved in the Leguminosae.

Contemporary Problems in
Plant Anatomy

ISBN 0-12-746620-7

I. INTRODUCTION

The determination of how differences in floral structure arise in related taxa is a basic biological problem underlying systematic distinctions. We take for granted that certain plant families have a characteristic type of floral symmetry; yet little work has been done to determine how and when during ontogeny these differences in symmetry are initiated. By comparing the early floral development of closely related taxa, it should be possible to show the developmental bases for diagnostic differences among such taxa. A second important problem is the question whether supra-generic distinctions are basically different from those separating related genera or related species. This question also can be addressed using comparative ontogenetic evidence from the taxa under consideration.

Fledgling botanists are imprinted early with the Besseyan system of phylogenetic arrangement for angiosperm families (Bessey, 1915). Henceforth such features as symmetry, fusion of floral parts, change from helical to whorled arrangement of parts, reduction in number of parts and loss of parts are regarded as a convenient framework within which families can be re-arranged as more information accrues. However, one can examine the framework from the standpoint of its evolution at the developmental level. How does the shift from radial to zygomorphic symmetry occur in ontogeny? And how does that shift affect all other floral characteristics such as order of appendage inception, relative sizes of parts, order of enlargement of parts, and tendencies toward morphological divergence among members of each whorl? The same considerations can be applied to the other aspects of the Besseyan framework.

II. RADIAL SYMMETRY IN MAGNOLIALES

Magnolialian flowers are prevailingly radial in symmetry. Organogeny proceeds acropetally along the ontogenetic helix by initiation of numerous free appendages around a large, domed floral apex, as in *Illicium floridanum* (Fig. 1-4). The 24-28 tepals and 30-39 stamens are consecutively initiated along a helix. Research on this flower (Robertson and Tucker, 1979) sought to determine the sequence of development of the 13 carpels, which appear to be whorled (Fig. 4) at maturity. Stages during or immediately after carpel initiation (Fig. 2) show that the 13 carpels stand at slightly different levels on the apical flanks and that the carpels differ slightly in size at this early stage; both of these features indicate a helical order of initiation rather than a whorl. The apparent whorled arrangement of carpels at anthesis therefore results from post-initiatory events.

Radially symmetrical flowers can have either whorled or helical arrangement, depending on the species, but the order of initiation of appendages can only be determined by ontogenetic study. Recent paper on floral ontogeny of *Monodora crispata* (Leins and Erbar, 1979), *Magnolia spp*. (Erbar and Leins, 1981), and *Ochna atropurpurea* (Pauzé and Sattler, 1978, 1979) exemplify the usual pattern in radially symmetrical flowers with helical, centripetal or acropetal development of appendages. A recent examination of floral development in *Silene coeli-rosa* (Lyndon, 1978a, 1978b) showed that the two whorls of five stamens actually arise as a continuous helix in which the members of a whorl are of quite different sizes during early development. Ross (1982)

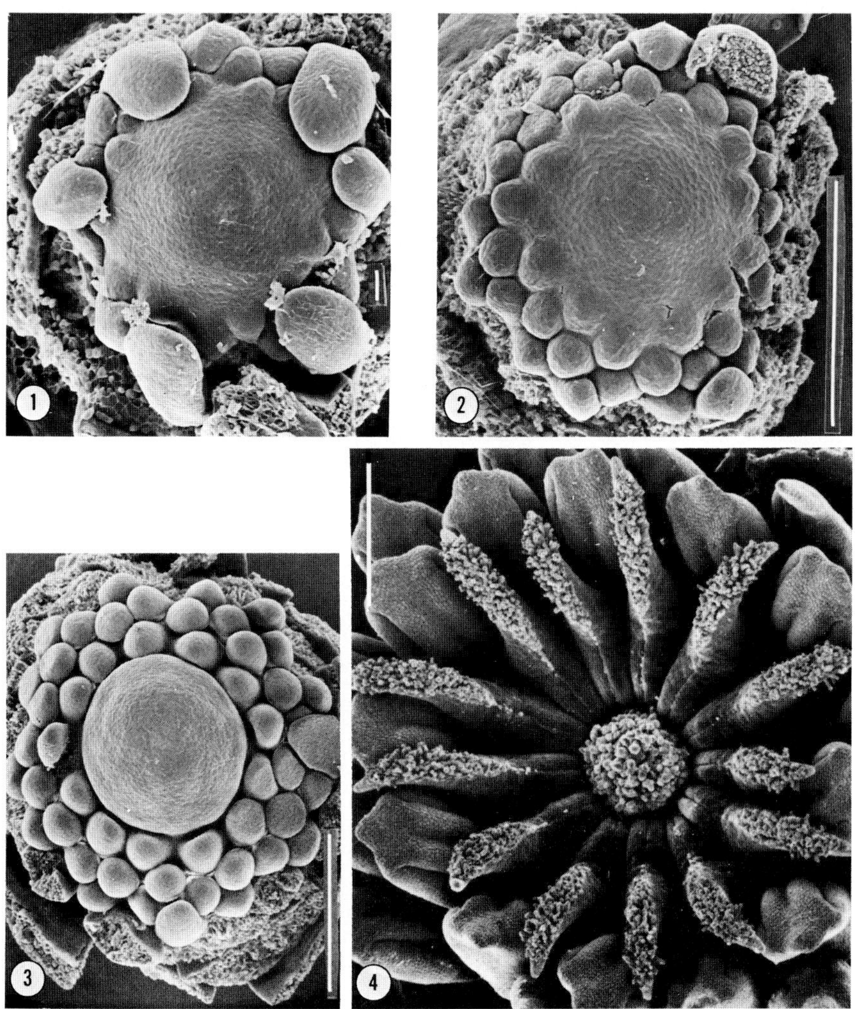

FIGURES 1-4. Flower development in radially symmetrical Illicium floridanum. All parts arise along a helix. Tepals removed in Fig. 2-4.
Fig. 1. Early stamen initiation; some tepals are also present. Bar = 50 μm.
Fig. 2. Initiation of the 13 carpels in a low helix is complete; a large apical residuum persists at center. Bar. = 500 μm.
Fig. 3. Older flower with 13 enlarged carpels, about 37 stamens, showing radial arrangement. Bar = 500 μm.
Fig. 4. Gynoecium and part of androecium in flower at anthesis. Bar = 500 μm.

compared the floral ontogeny in species of five genera of Cactaceae, all radially symmetrical. The carpels in several were shown to arise sequentially rather than as a simultaneous whorl.

Another variation on the theme in radially symmetrical flowers is centrifugal order in initiation, which has been documented in at least 32 families (Tucker, 1972). Examples have been described recently in several Primulales (Sattler, 1962, 1972), certain Hamamelidaceae (Endress, 1976), phytelephantoid palms (Uhl and Moore, 1977), *Limnocharis* (Sattler and Singh, 1977), and *Capparis* (Leins and Metzenauer, 1979). Sattler's atlas (1973) showed centrifugal sequence occurring in species of *Silene*, *Althaea*, *Hibbertia*, *Lythrum*, and *Lysimachia*.

III. TRANSITION TO ZYGOMORPHIC SYMMETRY

The shift from radial symmetry to zygomorphy can occur at many different stages in floral ontogeny, depending on the taxon. In *Coleus blumei* (illustrated in Troughton and Sampson, 1973) the floral apex is radially symmetrical in early stages and produces sepals and petals in whorls. However, only four stamens are initiated, with a gap at one expected stamen site. This omission of one primordium is the first manifestation of zygomorphy in *Coleus*. Later in enlargement and differentiation of the corolla, zygomorphy becomes pronounced. Consequently *Coleus* would illustrate a pattern in which zygomorphy is introduced relatively late in organogenesis. Good examples of flowers showing early manifestation of zygomorphy include *Begonia* (Merxmuller and Leins, 1971), *Couroupita* (Leins, 1972), *Downingia* (Kaplan, 1967, 1968a, 1968b), *Reseda* (Payer, 1857; Goebel, 1887; Leins and Sobick, 1977), *Pinguicula* and *Utricularia* (Buchenau, 1865). *Downingia*, *Pinguicula*, and *Utricularia* are unusual in showing bilateral symmetry from the earliest stages of floral development. Sattler (1973) showed the following five species of flowers in which zygomorphy is apparent early in ontogeny: *Juglans cinerea* (male and female flowers), *Populus tremuloides* (male flowers), *Peperomia caperata*, *Pisum sativum*, and *Habenaria clavellata*. He showed many others in which zygomorphy appears late in ontogeny.

IV. SYMMETRY AND INITIATION IN PIPERALES

Floral symmetry has been shown to be crucial in evaluating the presumed close relationship between Piperales and Magnoliales. Systematists for the past 80 years have consistently asserted that the Piperales are a derived group related to Magnoliales, based primarily on the structure of *Saururus cernuus* (Saururaceae), which is apocarpous, and whose six stamens and four carpels have been presumed to be radially symmetrical and helically initiated as in members of Magnoliales. Eichler (1878) for example described piperaceous flowers as basically trimerous and radial. Other features said to be shared by Piperales and Magnoliales include open stigma and style, long-decurrent stigma, monocolpate pollen, stipules, oil cells and vessel elements with scalariform perforation plates. Two of the most significant features (radial symmetry and helical initiation) have now been refuted for *Saururus* and other members of Piperales. Other developmental features such as common primordia have been revealed which strengthen the argument against close relationship to Magnoliales.

Saururus is the most primitive genus of the five in Saururaceae. It includes two species, *S. cernuus* in southeastern United States and *S. chinensis* in Asia. The flower of *S. cernuus* (Fig. 5) has four free carpels, six free stamens, and no perianth. Two other sauruaceous taxa, *Houttuynia cordata* and *Anemopsis californica*, differ from *Saururus* in having large petalloid bracts at the base of the spike. Connation of carpels and adnation of stamens to carpels occur in *Houttuynia cordata* (Fig. 6). The same features plus connate stamens and development of an inferior ovary occur in *Anemopsis californica* (Fig. 7). Piperaceae, the second family in Piperales, includes about ten genera; *Peperomia* and *Piper* are the largest with about 1000 species each. The floral structure is very simple in *Peperomia* (Fig. 8), since all species have two stamens and a unicarpellate gynoecium in each flower. The species of *Piper* (Fig. 9) vary in stamen number from 2-7, with 3,4, and 6 being characteristic of certain sections of the genus. The gynoecium of *Piper* consists of three or four fused carpels.

In order to evaluate assumed relationships between Piperales and Magnoliales, I compared floral ontogeny in *Saururus*, *Houttuynia*, *Anemopsis*, *Peperomia*, *Piper*, and other genera. Such studies necessarily involve inflorescence development as well. All of these taxa have elongate spikes containing numerous flowers. The pattern of the origin of the floral apex in *Saururus* and *Houttuynia* differs significantly from that in *Anemopsis* and all taxa of Piperaceae investigated. In *Saururus cernuus* (Fig. 10) the inflorescence apex produces "common" primordia which, later and at lower levels, bifurcate transversely to produce a floral apex above and a bract primordium below (at arrows in Fig. 10; Tucker, 1975, 1979). The type of floral initiation in *Houttuynia* is transitional in that the first floral apices on a spike originate, not from a common primordium like the majority of its flowers, but as delayed axillary buds in the axil of each of the large petalloid bracts (Tucker, 1981). In *Anemopsis californica* the inflorescence apex produces only bract primordia (B in Fig. 11); later and at lower levels, floral apices (at arrows) are initiated in the axils of the bracts (Tucker, in preparation). In all members of Piperaceae which have been investigated (Tucker, 1980, 1982a, 1982b), the inflorescence apex produces only bract primordia. The floral apices are initiated later in axillary position. The initiation of floral apices in *Saururus* and *Houttuynia* by bifurcation of common primordia had not been reported for other taxa when first reported (Tucker, 1975, 1979). Uhl reports origin of the floral apex from a common primordium in a palm, *Palandra* (Uhl and Dransfield, this volume).

Common primordia are far more frequent during initiation of floral appendages. The petal and stamen arise as a common structure in many Primulales (Pfeffer, 1872; Sattler,

FIGURES 5-9. Flowers of several taxa in Piperales. Fig. 5-7. Members of Saururaceae. Fig. 8,9. Members of Piperaceae.
Fig. 5. *Saururus cernuus*, apocarpous and with no fusion. The three stamens are labelled in order of initiation; the other member of each pair has been removed. G, gynoecium.
Fig. 6. *Houttuynia cordata*, showing fusion and reduced number of stamens.
Fig. 7. *Anemopsis californica*. One of three carpels and two of six stamens have been removed.
Fig. 8. *Peperomia metallica*. The two paired stamens are labelled "1"; the gynoecium, G.
Fig. 9. *Piper amalago*. B, bract; G, gynoecium; the stamens are numbered in order of succession, and illustrate bilateral symmetry. Bar = 500 μm.

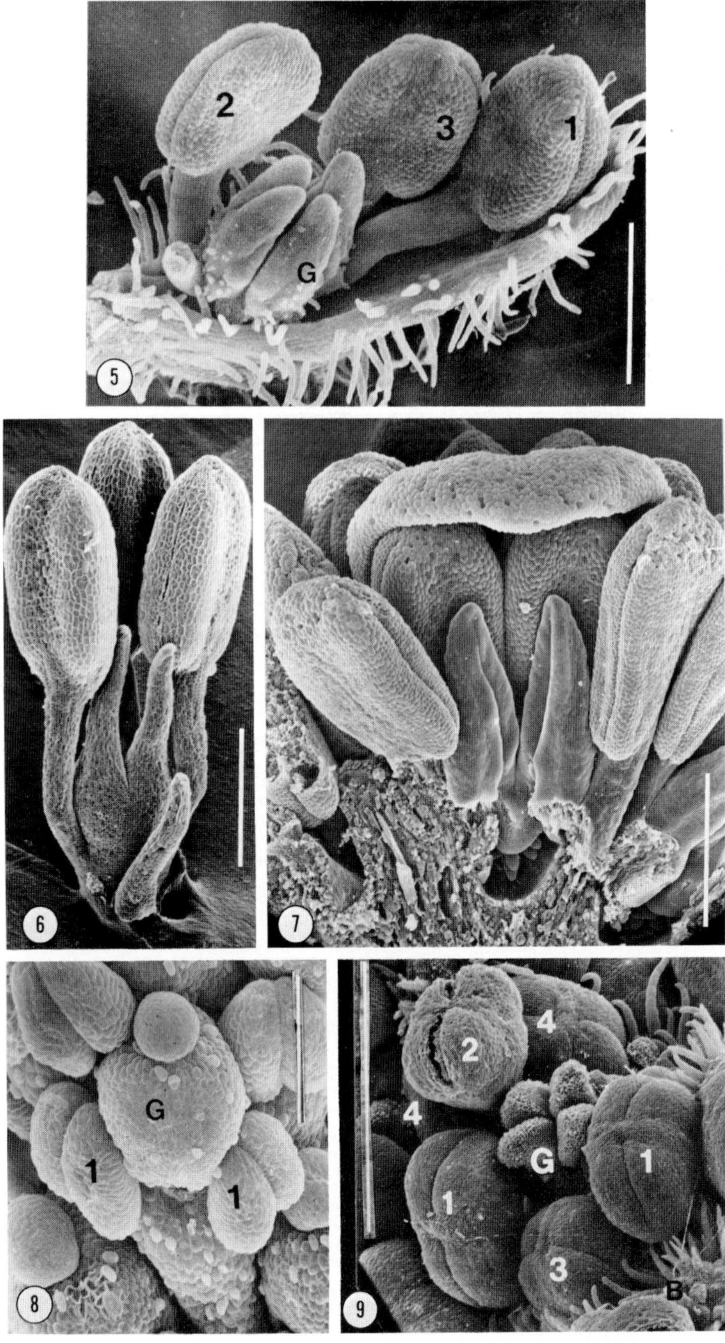
2
3
1
G
5
6
7
G
1
1
8
2
4
4
G
1
1
3
B
9

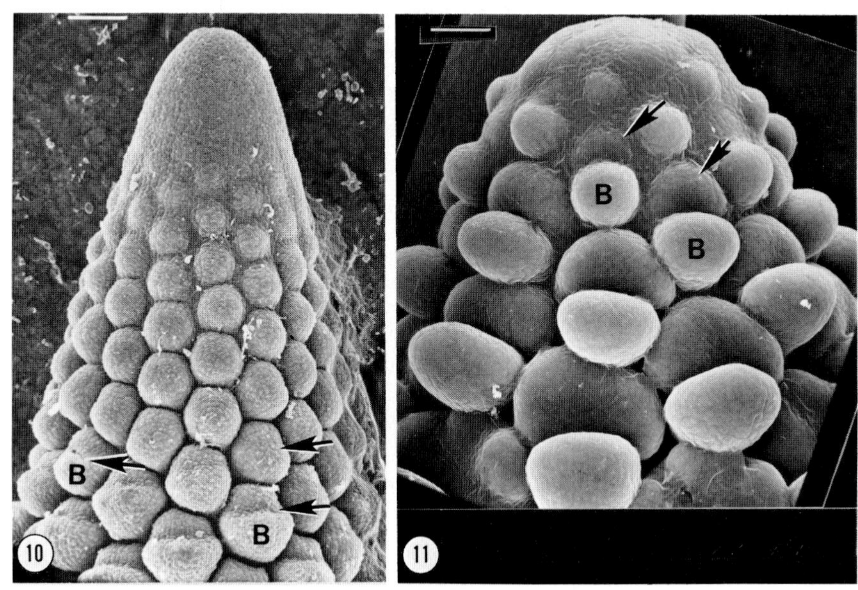

FIGURES 10, 11. Different patterns of floral apex initiation on the inflorescences of two genera of Saururaceae. Fig. 10. Saururus cernuus. Common primordia initiated which at lower levels bifurcate (at arrows) to produce a floral apex above and a bract below.
Fig. 11. Anemopsis californica, in which bracts are initiated by the apex. Later, at lower levels (at arrows) a floral apex is initiated in the axil of each bract. B, bract. Bar = 50 μm.

1962), Couroupita (Leins, 1972), and Iris decora (Pande and Singh, 1981) among others. Common primordia also may produce a number of stamens (Payer, 1857; van Heel, 1966; Leins, 1972) or a stamen and a nectary (Brett and Posluszny, 1982). Although Anemopsis californica lacks common primordia in floral inception, they play a role in stamen initiation. Each pair of stamens arises as a common primordium (Fig. 12) which bifurcates vertically to produce two stamens. Common primordia are reported mainly from relatively advanced plant families. They are not reported in any Magnolialian taxa.

A. Sequence of Appendage Initiation in Saururaceae

Two examples are illustrated here to show that zygomorphic symmetry is manifested from the earliest stages in flowers of Saururaceae. In *Anemopsis californica* (Fig. 12) each floral apex (at arrows) is tangentially broad before any appendages are initiated. At lower levels of the inflorescence one can observe origin of two large common stamen primordia (SP) laterally, followed by initiation of a median anterior stamen (S) or a stamen pair in the same position. Each common stamen primordium bifurcates vertically to form two stamens, as in the flower at lower right in Fig. 12. Subsequent carpel initiation (not shown here) follows a consistently zygomorphic symmetry during initiation events. In another saururaceous plant, *Saururus cernuus* (Fig. 14, 15) organogenesis consists of initiation of three pairs of stamens in succession, the first median sagittal, the latter two both lateral. The six stamens were first described as radial and trimerous by Eichler (1878) and others. The zygomorphic symmetry established early and maintained through stamen initiation is continued through initiation of two pairs of carpel primordia in decussate arrangement. Subsequent enlargement of stamens and carpels is unequal so that the mature arrangement appears helical. Ontogenetic analysis, however, shows that zygomorphic symmetry is established early and maintained in *Saururus* (Tucker, 1975, 1976, 1979), *Houttuynia* (Tucker, 1981), and *Anemopsis* (Tucker, in preparation).

Floral symmetry also is prevailingly zygomorphic in members of Peperomia from earliest stages (Fig. 13; Tucker, 1980) and in Piper (Tucker, 1982a, 1982b). The floral apex is tangentially broad from inception in all investigated, and floral organs are initiated in pairs or singly, with zygomorphic symmetry always apparent throughout (Fig. 16). The floral diagrams in Figure 16 compare floral structure in three Saururaceae (a-c) and five Piperaceae (d-i). Zygomorphic symmetry and paired or solitary initiation of parts are features which unify Saururaceae and Piperaceae in Piperales. The same features, plus the occurrence of the unusual common primordia, serve to distinguish Piperales from Magnoliales, in which none of these features occur, or occur only rarely.

FIGURES 12-15. Origin of bilateral or zygomorphic symmetry in flowers of Piperales.
Fig. 12. Inflorescence of Anemopsis californica (Saururaceae). Youngest floral primordia (at arrows) are broad and arcuate above the bracts (B), which appear pale compared to other organs. In slightly older flowers, lateral stamen pairs (SP) are initiated. Subsequently a fifth stamen or stamen pair arises in median anterior position as in the flower at lower right. Initiation shows zygomorphic symmetry at all stages.
Fig. 13. Inflorescence of Peperomia pellucida. Bracts are initiated around the apex; at lower levels a floral apex (at arrow) forms in the axil of each bract (B).
Fig. 14, 15. Young flowers of Saururus cernuus showing strong zygomorphic symmetry during stamen initiation (Fig. 14) and after the six stamens are formed (Fig. 15). S, stamen. Bar = 50 μm.

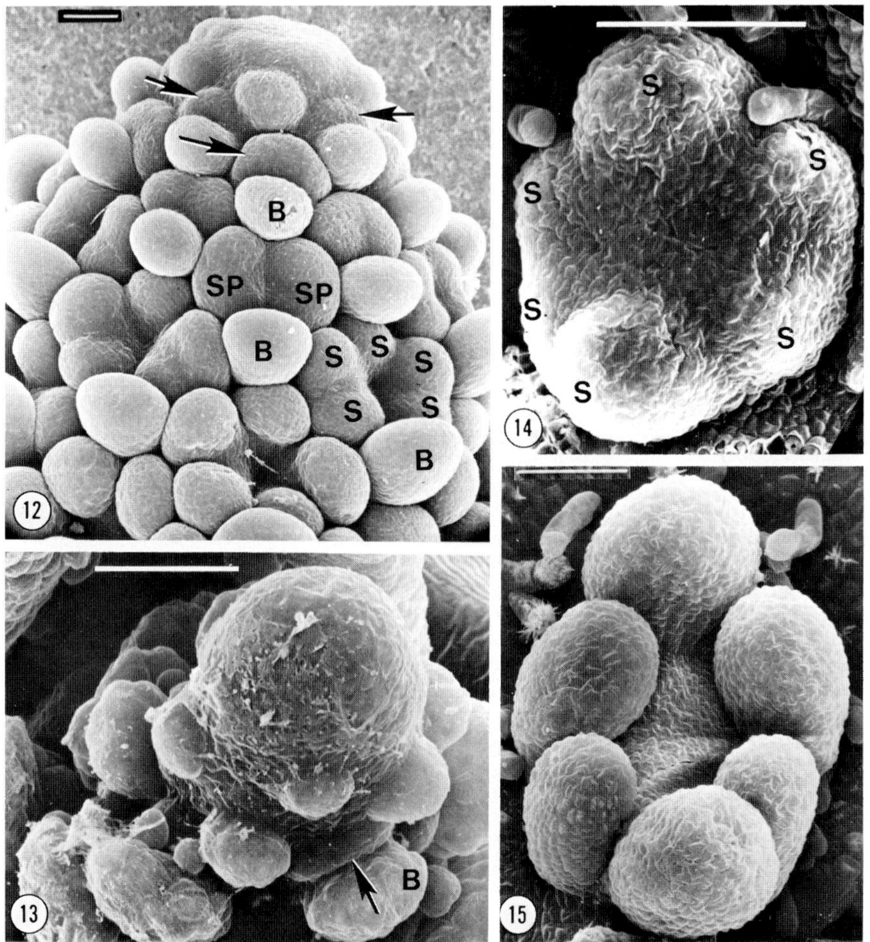

V. LEGUMINOSAE AS A MODEL

The importance of symmetry in ordinal or familial relationships involving Magnoliales and Piperales led to a search for a family in which floral ontogeny could be used to study shifts in such major floral features as symmetry, fusion, order of initiation of appendages, and increase or loss of parts. I selected the Leguminosae, in which the

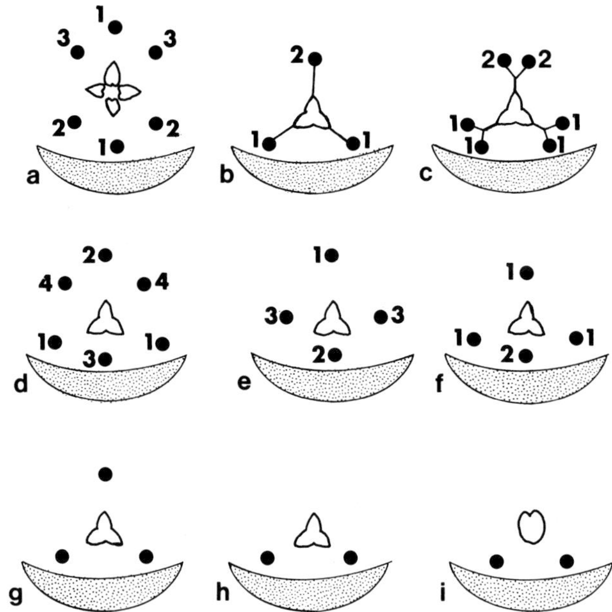

FIGURE 16. Types of floral symmetry in Piperales. The gynoecium consists of 1-4 parts, and the androecium of 2-6 parts. Each is subtended by a bract; there is no perianth. a. *Saururus cernuus*. b. *Houttuynia cordata*. c. *Anemopsis californica*. d. *Piper amalago* (*Enckea*). e. *Piper marginatum* (*Ottonia*). f. *Piper jamaicensis* (*Artanthe*). g. *Piper recurva* (*Artanthe*). h. *Pothomorphe umbellata*. i. *Peperomia rubella*. (Fig. e, g after Schmitz, 1872).

three subfamilies differ in 1) radial versus zygomorphic symmetry, 2) number of parts per flower, and 3) fusion, proliferation, or loss of parts. The three subfamilies also differ markedly in the type of petal aestivation, the developmental basis for which has never been investigated. We expect to be able to determine the developmental bases for differences in floral structure among the three subfamilies, based on study of selected taxa representing the various tribes in each subfamily. As Crisci and Stuessy (1980) state, "Because phylogenetic changes arise by modification of ontogeny, information on developmental patterns can be used to infer phylogeny." The Leguminosae is rich in morphological diversity (Dickison, 1981) particularly adaptations to pollinators. The wealth of up-to-date systematic background (Polhill and Raven, 1981) provides a basis for comparative ontogenetic studies.

The flowers of the three subfamilies are characterized as follows:

Caesalpinioideae - Taxa have radial to slightly zygomorphic symmetry, a floral tube or hypanthium, imbricate or valvate calyx, and imbricate petals with the uppermost petal inside the others (ascending cochlear pattern). Stamens are usually five to ten, and they may be free or fused (monadelphous). Many taxa have a gynophore, either short or elongate.

Mimosoideae - Taxa have radially symmetrical flowers, mostly five-merous (less commonly 3-,4-, or 6-merous). Sepals and petals are valvate in the bud (edges meeting but not overlapping). Stamens are numerous per flower in many taxa, and they are free (unfused) in many, or more or less fused in others.

Papilionoideae - Most taxa have strongly zygomorphic papilionaceous flowers, each having a posterior or adaxial "standard" petal, two lateral "wing" petals, and two anterior or abaxial "keel" petals. The stamens are usually ten, and either fused completely (monadelphous) or nine fused and one free (diadelphous). Sepal aestivation varies. Petals are imbricate with the uppermost petal (the standard) outside the rest (descending cochlear pattern). The basic papilionaceous flower is modified in hundreds of small ways among genera; petals for instance vary in shape, size, fusion, orientation, persistence, and color. In the androecium, diversity is expressed in heteromorphy, fusion, glands, dehiscence mechanisms, and pollen morphology. In the gynoecium there is similar range of diversity, particularly as expressed in fruit development.

A. Hypothesis

In the Leguminosae, I propose to test a hypothesis that the stable floral characteristics which separate subfamilies of legumes (or of supra-generic taxa in general) are usually manifested early in floral ontogeny during organogeny. The characteristics separating species are more labile and arise late in ontogeny. An intermediate group of characteristics is determined in mid-stages of ontogeny during enlargement of the floral bud. These features tend also to be stable and to characterize major groups of taxa. Stebbins (1974, p. 115) suggested that "increasing precocity of gene action" could be responsible for increasing degree of fusion in phylogenetically specialized flowers. We can extend this idea of increasing precocity from fusion to other aspects of floral specialization. The three categories of early, mid-stage, and late stage of floral ontogeny correspond roughly to organogeny, enlargement, and cell differentiation. Examples of the three categories of floral features are presented in Table 1. Placement of some features is tentative while data is accumulating, and re-arrangements of some features are likely. With these caveats, I will discuss two or three leguminous examples of features in the categories of early and mid-development.

TABLE 1. Diagnostic Floral Characteristics and Their Relative Time of Determination in Leguminosae

1. EARLY-DETERMINED FEATURES:
    - Floral symmetry
    - Order of appendage initiation
        - A. Vertical
        - B. Horizontal: simultaneous or successive
    - Number and type of whorls
    - Number of parts per whorl
    - Omission of some parts

2. MID-DEVELOPMENT-DETERMINED FEATURES:
    - Corolla aestivation
    - Organ abortion or arrest
    - Elongation of parts
    - Tube formation of calyx, corolla, androecium
    - Petal fusion
    - Gynophore
    - Differing size and shape of petals
    - Differing size and shape of stamens
    - Stamen dehiscence type

3. LATE-DETERMINED FEATURES
    - Minor petal shape changes
    - Petal sculpturing, hooks, pits, cuticular elaboration
    - Petal color
    - Filament elongation
    - Stigma form and hairs
    - Nectaries
    - Hairs
    - Fragrance and nectar
    - Carpel shape changes to produce fruit

B. Early Development Features

1. Pre-appendage symmetry

The shape of the floral apex before organ initiation seems to be correlated with symmetry at later stages in many plants. Radially symmetrical flowers such as *Mimosa pudica* in Mimosoideae (Fig. 17) have globose floral apices before appendage initiation begins. Floral apices of *Caesalpinia pulcherrima* in Caesalpinioideae (Fig. 18) are tangentially broad and radially narrow when first recognizable; the flowers show zygomorphic symmetry throughout organogenesis in this plant, as shown in older floral stages (Fig. 18). Although the floral apex is broadest tangentially at early stages, it expands radially before initiation of the first sepal on the anterior (abaxial) side next to the subtending bract.

2. Sequence of organogeny

Sequence must be considered in two different directions: vertical (acropetal versus basipetal) and transverse (whorled, spiral or helical, unidirectional, and other possibilities). In the radially symmetrical flowers of Leguminosae, one might suspect that vertical sequence would tend to be uniformly acropetal, while in more specialized and highly zygomorphic flowers there might tend to be irregularities or departures from strictly acropetal succession of floral organs. The majority of legume flowers reported in print have acropetal succession in organogeny. Exceptions occur among specialized polyandrous mimosoids (Gemmeke, 1982; van Heel, 1983) in which stamen succession is reported to progress laterally

FIGURES 17-18. Inflorescences to show origin of symmetry in floral primordia in two subfamilies of legumes.
Fig. 17. Mimosa pudica; some bracts removed. Youngest floral apices are globose.
Fig. 18. Caesalpinia pulcherrima with bracts and larger sepals removed. Youngest floral apices (at arrows) are tangentially broad and radially narrow. Bar = 50 μm.

from each of five original stamen sites. There also are some reports, not well documented, of centrifugal sequence among papilionoids. Some of the more refined techniques now available may help to resolve some of the uncertainties.

The order of succession within each whorl can be simultaneous, helical, or unidirectional; all three occur among various legumes.

3. Sepal Initiation

Sepals are commonly initiated in a 2/5 helix in many caesalpinioid taxa such as *Parkinsonia aculeata* (Fig. 34, 35) on a floral apex which is at least temporarily radially symmetrical. In other Caesalpinioideae such as *Caesalpinia pulcherrima* (Fig. 36, 37) and many Papilionoideae such as *Lupinus affinis* (Fig. 28-31), there is unidirectional sequence of organogenesis, from the anterior to the posterior (abaxial to adaxial) side of the floral apex. The first sepal is initiated anteriorly, followed by a lateral pair of sepals, and lastly by a pair of posterior (adaxial) sepals. In *Cadia purpurea* (Fig. 38), a relatively primitive papilionoid in the tribe Sophoreae, one sepal is larger than the others, but the initial distinction is later obscured by zonal growth of the entire calyx. A third type of sepal initiation is shown in *Calliandra portoricensis* (Fig. 39). Here all the sepals are initiated simultaneously and remain the same size and shape throughout development, since marginal growth ceases once they meet.

In the Mimosoideae there is tribal separation on the type of calyx aestivation (Elias, 1981; Lewis and Elias, 1981). All have valvate calyx (Fig. 39) and valvate corolla (Fig. 27) except the Parkieae and Mimozygantheae. The latter two groups are considered primitive in the subfamily because they share this feature with the other two subfamilies. Valvate aestivation in the remainder of the Mimosoideae is considered a derived condition.

One surprising discovery in our preliminary work on mimosoid taxa is that there may be occasional examples of zygomorphy. In *Mimosa pudica* the calyx arises precociously on the posterior side (Fig. 23-25). Sepals are "fused" into a ringlike calyx from the beginning, and the ring is first

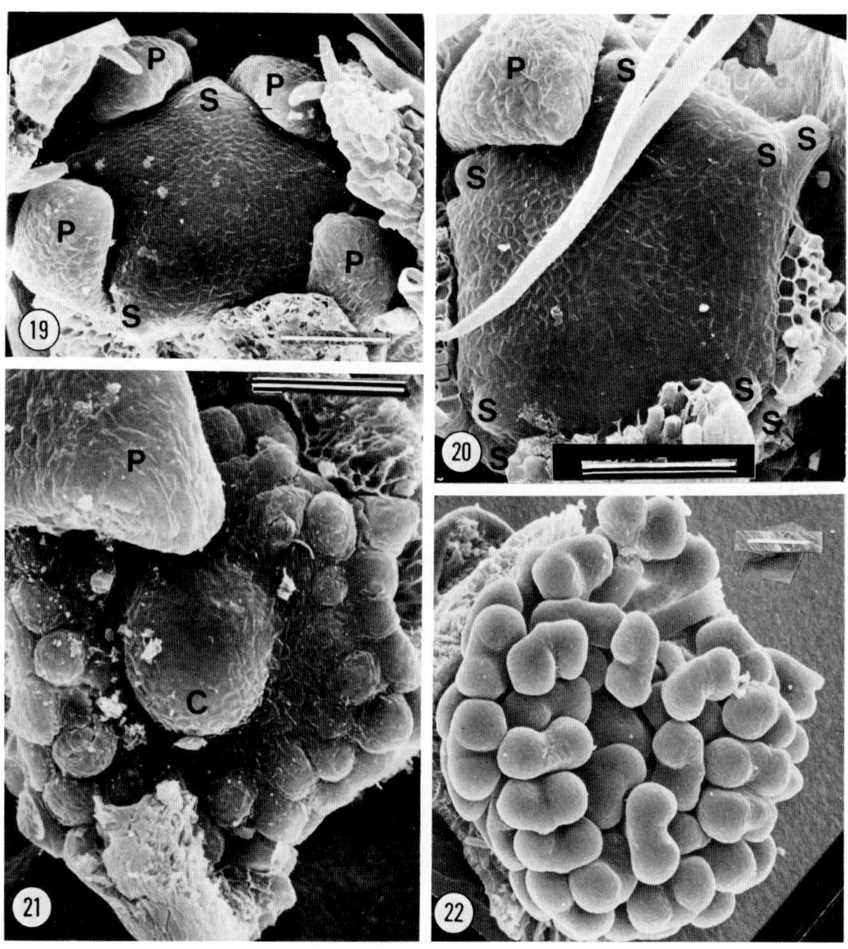

FIGURES 19-22. Radial symmetry in floral organogenesis, pro parte, in Calliandra portoricensis (Mimosoideae). Sepals have been removed in all.
Fig. 19. First five stamens arise at five points (S). P, petal.
Fig. 20. Second whorl of five stamens (both whorls labelled S) directly above the first.
Fig. 21. Numerous additional stamen primordia have formed in the intervals between the original five points. P, petal; C, carpel.
Fig. 22. Flower at anthesis; order of the numerous stamens cannot be detected at this stage. Bar = 50 μm.

evident posteriorly and is larger on that side, at least in early stages. Posterior precocity is especially unexpected, in view of the prevailingly anterior precocity in most zygomorphic legume flowers, to be discussed later. Later stages in calyx development in *Mimosa* (Fig. 26, 27) show the lobes elongating very unevenly among flowers in an inflorescence. However the corolla (Fig. 27) and stamens show completely radial symmetry and simultaneous development among members of each whorl.

4. *Petal and Stamen Initiation*

Whorled, simultaneous initiation is common for petals and the first stamen whorl in mimosoids such as *Calliandra portoricensis* (Fig. 19, 20). Subsequent stamen initiation, however, diverges from the patterns shown earlier in radial flowers (ie., *Illicium floridanum*). *Calliandra* is polyandrous; five stamens form at equidistant points, then members of a second whorl of five are initiated acropetally, directly above the first five (Fig. 20). Later additional stamen primordia are initiated laterally in both directions from each of the five original stamen sites (Fig. 21). Gemmeke (1982) and van Heel (1983) document this same pattern

FIGURES 23-27. Sepal development in *Mimosa pudica* showing some zygomorphic tendencies in a radial flower.
Fig. 23. Inflorescence with several flowers (bracts removed). The ringlike calyx in each is precociously developed posteriorly.
Fig. 24, 25. Individual flowers showing variation in calyx form.
Fig. 26. Older flowers with calyx lobes eccentrically and variably enlarged over the summit.
Fig. 27. Flower showing radial symmetry in the four-merous corolla. Bar = 50 μm.

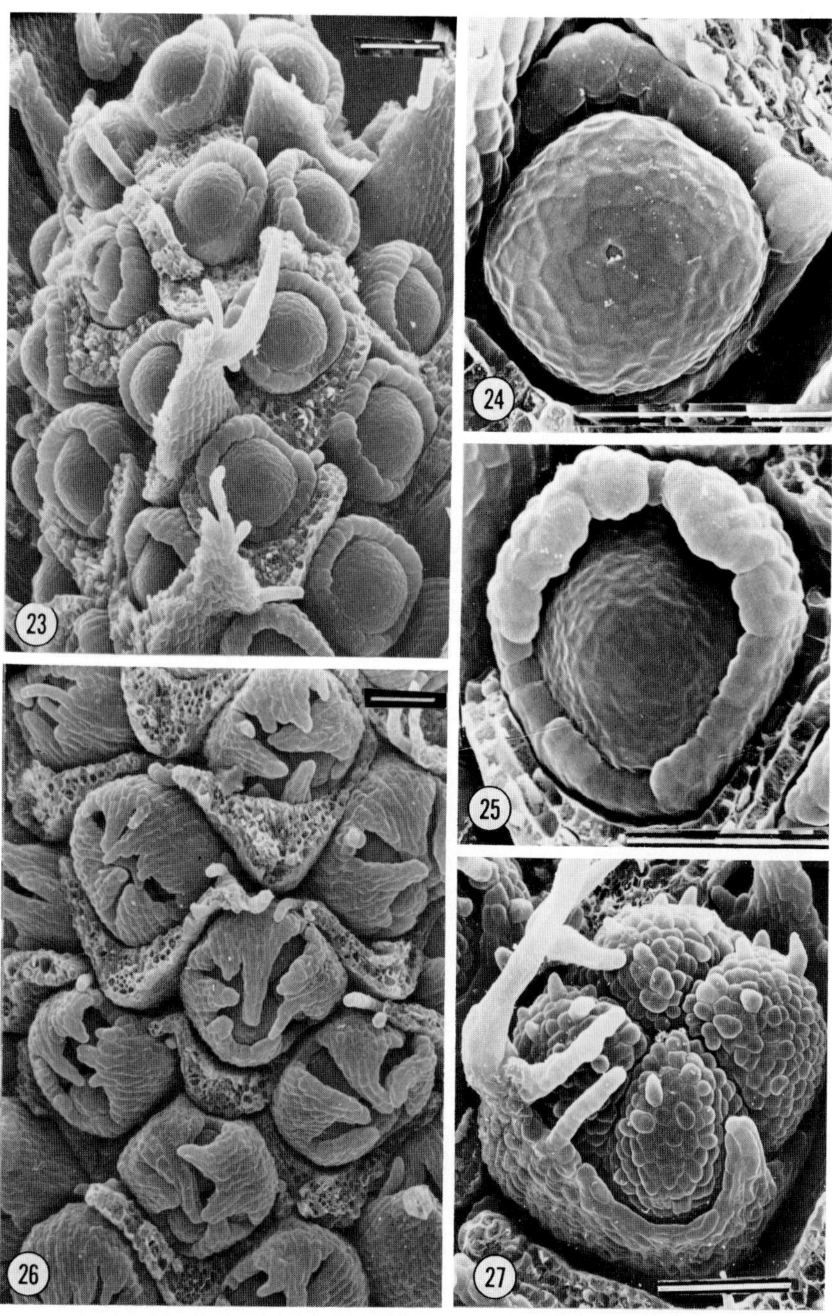
23
24
25
26
27

of stamen succession in other polyandrous mimosoids, except that they did not find a second whorl of five directly above the first. In open flowers (Fig. 22) it is impossible to determine the original order of initiation of stamens.

The unidirectional order of initiation in each whorl of petals and stamens is typical of numerous papilionoid and many caesalpinioid species. Examples are provided in *Lupinus affinis* (Fig. 28-31) and *Sophora japonica* (Fig. 32, 33), although only selected stages are shown for each species. Anterior (or abaxial) members of each whorl are initiated first, then the posterior (or adaxial) members. A precocious anterior sepal is visible in Fig. 28. Precocious anterior stamens are shown for the first stamen whorl in *Sophora japonica* (Fig. 33) and in the second stamen whorl in *Lupinus affinis* (Fig. 29, 30). Precocious organogeny beginning on the anterior (abaxial) side of the floral apex has been shown previously in 12 genera of papilionoids by several authors including Payer (1857). The members of a whorl are not

FIGURES 28-33. Floral organogenesis in *Lupinus affinis* (Fig. 28-31) and *Sophora japonica* (Fig. 32, 33) in which floral organs are initiated earlier on the anterior side.
Fig. 28. Floral primordium with bracteoles (Br) and anterior sepal ridge (Se).
Fig. 29. Flower with sepal ridge (Se), 5 petals (P), 5 stamens, (S), all larger on the anterior side of the flower. C, carpel.
Fig. 30. Slightly older flower similarly labelled, with two stamens of the second whorl, developed first on the anterior side of the flower (at arrows).
Fig. 31. Four stages of flowers, in all of which floral organs are precociously developed anteriorly.
Fig. 32. Floral primordia of *S. japonica*, tangentially broad (at arrows). Bracts removed.
Fig. 33. Flower of *S. japonica* with all but one sepal (Se) removed. All petals (P) and stamens (S) of the outer whorl are all present anteriorly; only petals are present posteriorly at this stage. Bar = 50 μm.

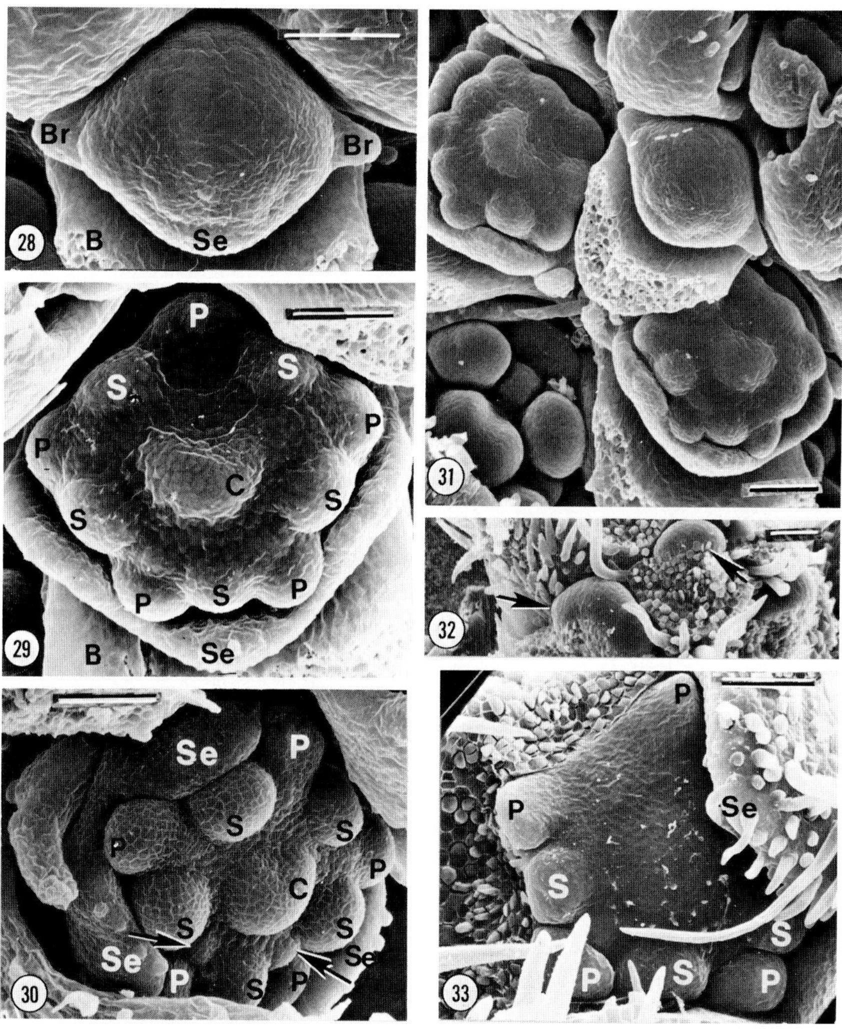

simultaneous; in fact the floral apex is producing petal primordia on one flank while it is producing stamen primordia on another. This type of apical activity, producing two kinds of organs at the same time, is unknown among vegetative apical meristems, as far as I am aware. It also casts some doubt on the view that a floral apex goes through a series of physiological stages at which it is competent to produce only one type of appendage (Cusick, 1956; Soetiarto and

Ball, 1969). It is difficult to see how this interpretation of stepwise floral control can be correlated with unidirectional sequence of organogeny in papilionoid legumes.

Curiously, in the flowers of Reseda (Resedaceae) and Pinguicula (Lentibulariaceae), organogeny proceeds from the posterior to the anterior side (adaxial to abaxial) of the apex (Goebel, 1882, 1887; Leins and Sobick, 1977; Buchenau, 1865), exactly the reverse order from that in legume flowers.

C. Mid-Development Features

1. Petal Aestivation

Although petal aestivation is a diagnostic difference among Caesalpinioideae, Mimosoideae, and Papilionoideae, there is almost no developmental information about how the different patterns of imbrication arise in ontogeny. Briefly, mimosoid petals (Fig. 27) are valvate (with margins in edge-to-edge contact) except in Tribe Parkiae and Mimozygantheae. In caesalpinioid flowers (Fig. 42) the lateral wing petals (W) enclose the standard petal (ST) in an ascending cochlear pattern. Papilionoid flowers (Fig. 43) have the standard petal enclosing the wing petals in a descending cochlear pattern.

FIGURES 34-39. Sepal initiation and origin of aestivation.
Fig. 34, 35. Helical sepal initiation and quincuncial aestivation in Parkinsonia aculeata. "1" is the first-initiated sepal; the rest are numbered in order of initiation.
Fig. 36, 37. Unidirectional sepal initiation in Caesalpinia pulcherrima. The first-initiated sepal (1) overlaps the succeeding two pairs (2, 2" and 3, 3") successively from the anterior toward the posterior side of the flower (2', 3' are not visible).
Fig. 38. Helically arranged calyx in Cadia purpurea. The first-initiated sepal is at the top.
Fig. 39. Valvate aestivation in Calliandra portoricensis. Bar = 50 μm.

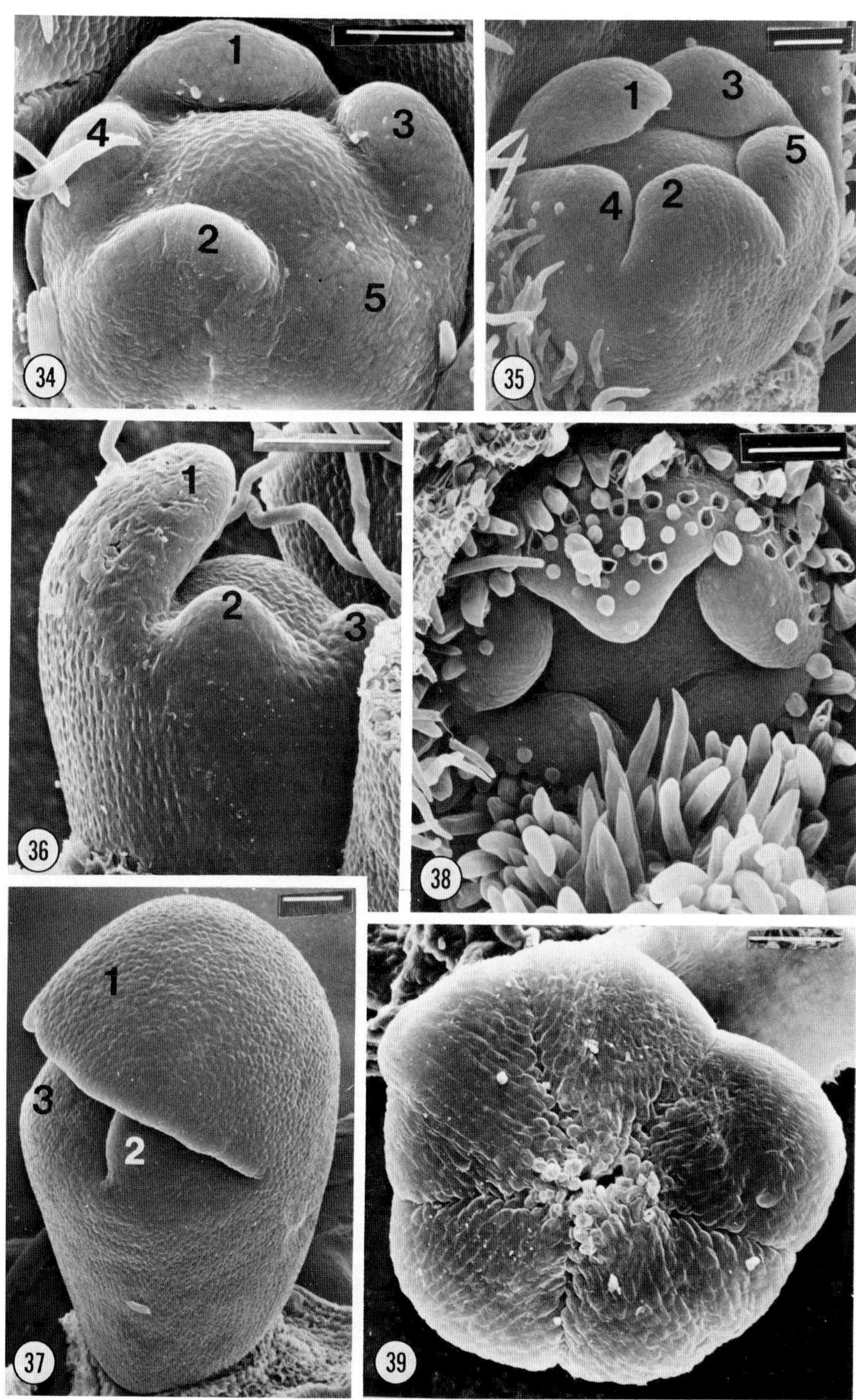
1
3
4
2
5
34
1
3
5
4
2
35
1
2
3
36
38
1
3
2
37
39

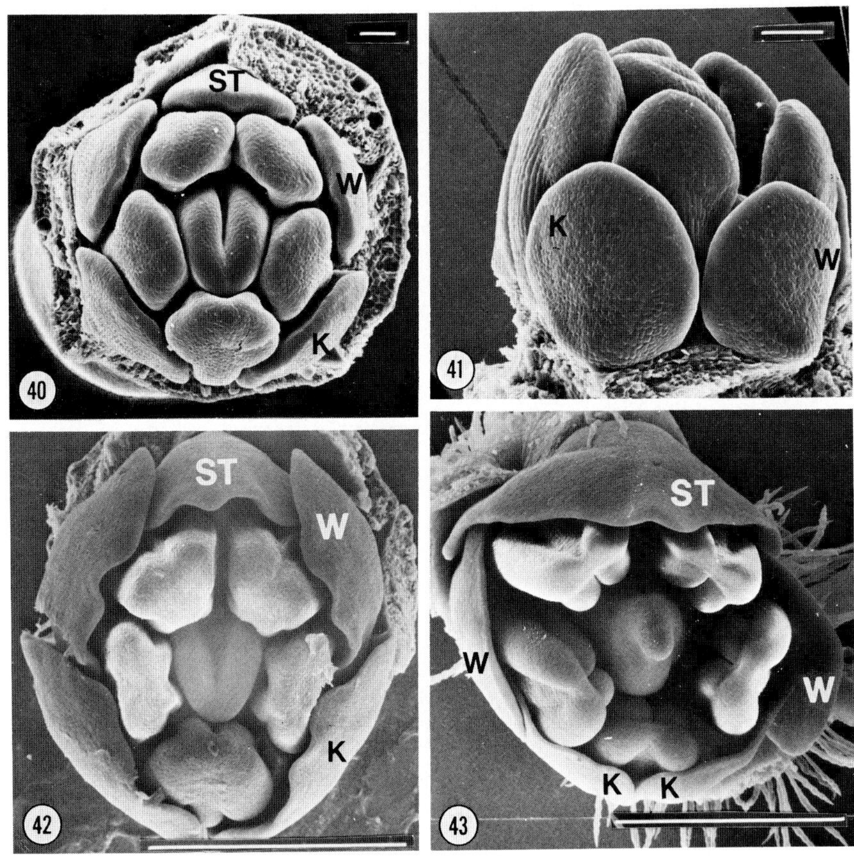

FIGURES 40-43. Corolla aestivation in caesalpinioid and papilionoid flowers.
Fig. 40-42. Caesalpinia pulcherrima with standard petal (ST) inside wing petals (W). K, keel petals.
Fig. 43. Lupinus affinis with standard (ST) outside the wing petals (W). Keel petals (K) are fused. Bar = 50 µm in Fig. 40, 41; bar = 500 µm in Fig. 42, 43.

The most puzzling problem is how the two imbricate patterns are controlled or determined. Originally I surmised that, since type of aestivation is such a stable feature, it could result from a) overlap of petal primordia bases at initiation, b) overlap of petal bases in the order of their initiation, c) oblique insertions, or d) lateral extension of separate petal bases soon after initiation. However, none of these has been found to occur among the legumes studied so far. Rather, aestivation is manifested in mid-development as the petals enlarge (Fig. 40, 41). The polar view of a young floral bud of Caesalpinia pulcherrima (Fig. 40) shows standard and wing petal margins nearly in contact, but with no indication in this view of why the wings will grow outside the standard margins. The keel petals are already growing marginally outside the wing petal margins. These SEM views suggest the idea that the transsectional shape of each type of petal is important in determining type of aestivation. The keel petals are relatively flat in these mid-stages, while the wings are arcuate on their lower margin. A side view of another flower of the same species (Fig. 41) shows keel and wing approaching each other. The keel will grow outside the wing margin. The area of overlap involves only the middle level of the two petals; the bases are not overlapping. With further growth in height and width of the petals, the overlap will become greater, but will not include the bases. The flower of papilionoid Lupinus affinis (Fig. 43) has wing margins incurved on the side where they are surrounded by the standard petal margins, and flattened margins of the sides which will grow outside the incurved keel margins. In both caesalpinioid (Fig. 42) and papilionoid (Fig. 43) types of aestivation, the wing petal shape appears to be a determining factor. The upper side of each wing is flattened or flared in a caesalpinioid flower, while it is incurved in a papilionoid flower. The lower side of each wing is incurved in a caesalpinioid flower, while it is flattened or flared in a papilionoid flower. Histological work is under way to test the assumption that differential marginal growth of the wings is responsible for differences in aestivation between the two subfamilies. It also is possible that different growth processes may produce similar aestivation in different taxa through convergence.

In one genus, Cadia, in the primitive tribe Sophoreae of Papilionoideae, petal aestivation varies greatly, from ascending cochlear as in Caesalpinioideae to descending cochlear as in Papilionoideae, plus aberrant patterns. Van der Maesen (1970) reported some data by Ross (Table 2) for Cadia purpurea; I collected some data on the same species from a plant at the U.S. Plant Introduction Station, Coral Gables, Fla. Besides the two types of aestivation mentioned earlier, C. purpurea has flowers with completely quincuncial aestivation, and others in which the standard has one "inside" margin and one "outside" margin. Ross reported 21 different patterns of petal arrangement; I found eight in a much smaller sample. Cadia, in other words, appears not to have the same "control" over aestivation that characterizes practically all legumes. Developmental studies were begun on C. purpurea, and we can report some preliminary findings. The flowers of C. purpurea (Fig. 44, 46) are compared with flowers of Sophora japonica (Fig. 45, 47) of a similar size to those in Cadia. The petals in Cadia purpurea are very much smaller than those in Sophora japonica; those of Cadia remain all the same size and shape at a stage when those of Sophora differ among themselves in size and shape. The margins of the standard in Sophora (Fig. 47) have overlapped those of the wing, while the petals in Cadia (Fig. 46) are not even close to each other. Overlapping occurs very late in Cadia, and appears to be the result of chance. Here also, histological documentation is under way.

TABLE 2. Variation in Petal Aestivation in Cadia purpurea

	Ross Unpubl., cited in van der Maesen, 1970	Tucker 1983
1. Papilionaceous (standard outside wings)	28	9
2. Caesalpinioid (standard inside wings)	48	1
3. Standard 1/2 outside, 1/2 inside	29	2
4. Wings both outside keel	22	4
5. Wings both inside keel	13	2
6. Wings 1/2 outside, 1/2 inside	13	6
7. Quincuncial imbricate	9	1
Total number of patterns seen	21	8
Total number of flowers examined	114	12

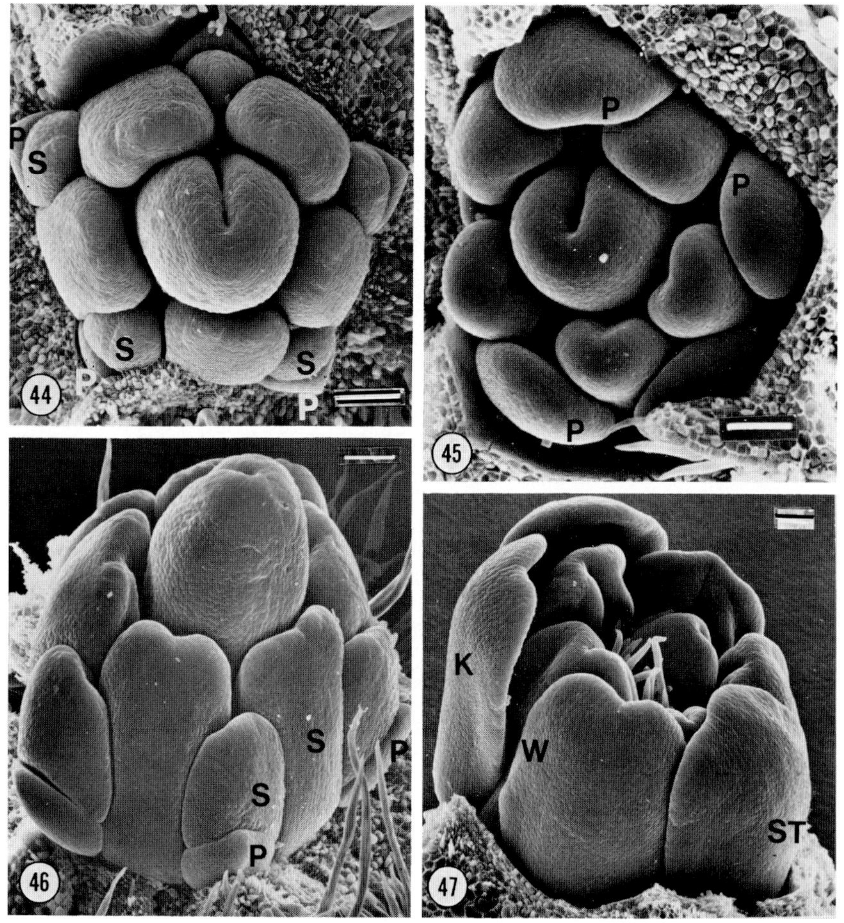

FIGURES 44-47. Comparison of petal enlargement in Cadia purpurea (Fig. 44, 46) and Sophora japonica (Fig. 45, 47). Sepals removed in all. Flowers in Fig. 44 and 45 are comparable in size but have greatly different-sized petals. Fig. 46. Petals are 85-100 μm high.
Fig. 48. Petals are 300-400 μm high, and keel petals at left are considerable larger than the wings and standard petals. Bar = 50 μm.

2. Fusion of Floral Organs

Examples of congenital connation (within a whorl) and adnation (between whorls) are common in legumes. Connation generally occurs by meristematic activity below the level of attachment of adjacent organs, thereby raising them together so that they appear to be branches of a single structure. Cusick (1966) called this type of growth "zonal growth", which is a preferable term to that of fusion, since the "fused" region has never consisted of discrete entities. Zonal growth is illustrated in the calyces of *Mimosa pudica* (Fig. 48) and *Lupinus havardii* (Fig. 49); it also is common in monadelphous and diadelphous androecia in legumes.

Postgenital fusion involves a true fusion between two organs already formed. Examples include the fusion of carpel margins in *Trifolium* and *Laburnocytisus* (Boeke, 1973; Boeke and van Vliet, 1979). The carpel in legumes arises as an open structure (Grégoire, 1924; Baum, 1948; van Heel, 1981, 1983). Walker (1975a, 1975b, 1978) examined postgenital fusion of carpel margins in *Catharanthus roseus* at the light microscope level and ultrastructural level. Fusion included appression of the margins, deposition of additional wall matrix by the epidermal cells to cement them together, thereby embedding some of the cuticle. Walker also reported some periclinal cell division in the marginal cells being appressed, and subsequent cell expansion which tends to obliterate the line of fusion.

In legumes the most common example of postgenital fusion outside the gynoecium is the fusion of the two keel petals to form a boat-shaped keel. In *Lupinus affinis* (Fig. 50-52) the adjacent margins of the keel petals become appressed distally (Fig. 52) in a conduplicate fashion so that the margins protrude slightly. More proximally (Fig. 51) the keel petals remain approximated but free. In some other papilionoids investigated such as *Hedysarum flavescens*, the keel margins are appressed edge-to-edge with no conduplicate ridge. Hence there are slight differences in the way in which postgenital fusion occurs in the same organs in different species.

3. *Loss of Floral Organs*

Appendages which fail to be initiated in the usual sites exemplify loss or omission of parts during organogeny. However, the same terms may be applied erroneously to situations of initiation followed by arrest of parts during mid-development. Careful ontogenetic studies are necessary to distinguish these two kinds of occurrences. Reduction in number of parts is an important feature of legume flowers which is especially important in *Amorpha* (Baillon, 1872; Eichler, 1878; Taubert, 1894) and in some species of *Swartzia* (Cowan, 1968, 1981). Ontogenetic studies of *Amorpha fruticosa* (Fig. 53) show that the young flower has not only the standard petal, but also rudiments of four other petals in the expected sites. Although the solitary petal is not the sole diagnostic feature of the genus *Amorpha*, it is still interesting that this feature depends on arrest, rather than loss of parts. Heslop-Harrison (1964) proposed a model to explain tendencies toward unisexuality in structurally hermaphrodite flowers. He

FIGURES 48-52. Fusion of floral parts.
Fig. 48, 49. Congenital fusion in calyx of *Mimosa pudica* (Fig. 48) and *Lupinus havardii* (Fig. 49).
Fig. 50-52. Postgenital fusion of keel petal margins in *Lupinus affinis*.
Fig. 50. Completely fused margins of keel petals (K).
Fig. 51. Distally fused, proximally open keel margins.
Fig. 52. Fused distal tips of two keel petals. Bar = 50 μm.

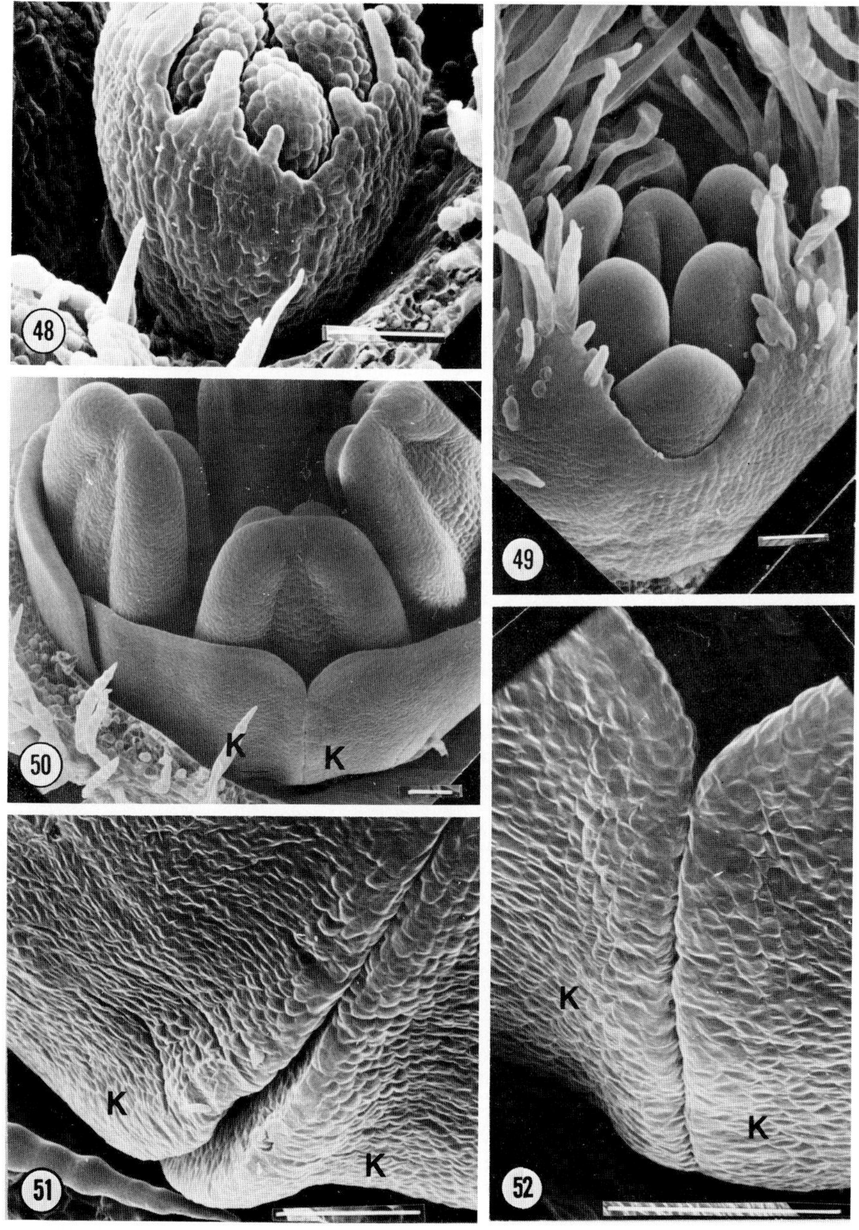
48
49
50
K
K
51
K
K
52
K
K

compared development to a relay system including three steps (initiation, early growth, and meiosis), at any one of which inactivation of one sex could be triggered. That analysis can be extended to perianth parts as well as reproductive organs. Once initiated, the *Amorpha* petal primordia undergo a series of critical steps at any of which cessation may occur. The exact nature of these stages remains to be investigated.

Bauhinia in the subfamily Caesalpinioideae is another leguminous genus in which loss of organs (petals and/or stamens) is common. One example is *B. divaricata* (Fig. 58) with a single stamen and two to four or five petals. Urban (1885) reported the species to be dimorphic: either andromonoecious (functionally male, as in flower in Fig. 58) or hermaphrodite (not illustrated). Early stages in ontogeny of *B. divaricata* (Fig. 54-57) show that several rudimentary organs are present around the base of the developed organs. Based on location of these rudiments, they appear to represent the remaining petals (to give a total of five), and four stamen primordia (to give a total of five).

FIGURES 53-58. Loss or arrest of floral organs.
Fig. 53. *Amorpha fruticosa*. Mature flowers have a single petal, the standard (ST); in this young flower two additional petal primordia are visible at arrows; two more are present but not visible in this view. Ten stamens and the carpel are also present.
Fig. 54-58. Floral development in *Bauhinia divaricata*.
Fig. 54. Young flower with large stamen (S) removed. Two large petals (P), three smaller petals (at arrows) and the carpel (C) are present.
Fig. 55, 56. The solitary stamen (S) and two large petals (P) are accompanied by 4 additional organ rudiments (at arrows in Fig. 55). C, carpel.
Fig. 57. Five organ rudiments from nearly-mature flower in Fig. 58.
Fig. 58. Large flower bud divested of calyx. Functional stamen is 2 mm high. Two long and two short petals are present, plus stamen rudiments (at arrow). Bar = 50 µm in Fig. 53-57; bar = 500 µm in Fig. 58.

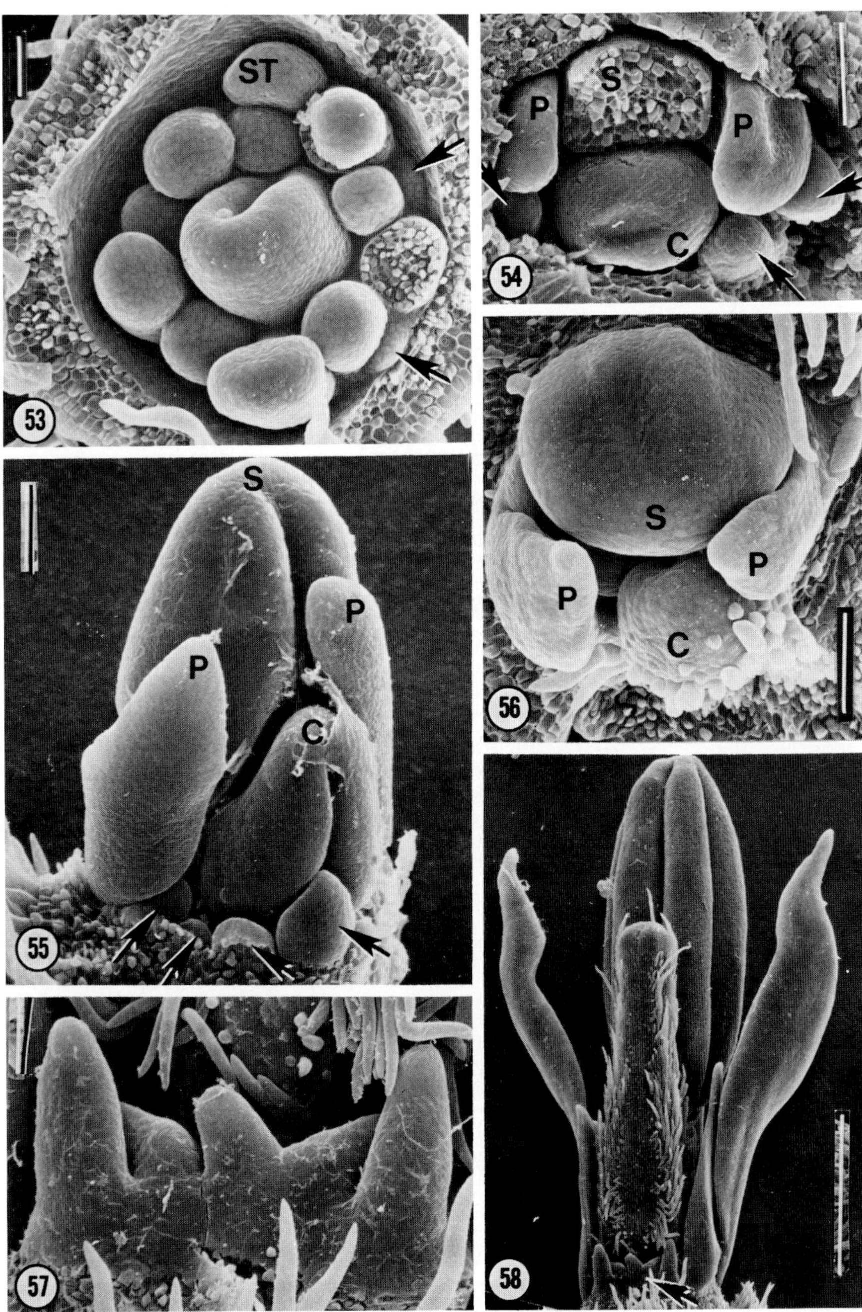
ST
53
S
P
P
C
54
S
P
P
C
55
S
P
P
C
56
57
58

The second or inner whorl of stamens is completely absent. Hence the marked reduction in apparent number of parts in *B. divaricata* is due to arrest of several petals and four stamens, and the loss of five additional stamens. Many other species of *Bauhinia* undergo conversion of stamen primordia into staminodia. We plan to study floral ontogeny in several other species of *Bauhinia* since the species vary in number of functional stamens: 1, 2, 3, 5, 7, 9, or 10 per flower (Urban, 1885; DeWit, 1956). Differences between species in *Bauhinia* seem to depend at least in part on different patterns of suppression, loss or conversion of floral appendages. This genus offers ample opportunities to analyze the developmental bases for distinctions between related species.

VI. SUMMARY

Differences and similarities of floral structure are the relatively stable framework of systematics. Determining the ontogenetic bases for diagnostic characteristics can distinguish cases of homology from those of convergence, and can clarify relationships among related taxa. Developmental evidence is presented to support lack of relationship between Piperales and Magnoliales. It is proposed that the morphological characteristics which distinguish major supra-generic levels of taxa (such as symmetry, order of initiation of parts, number and arrangement of parts, fusion, and loss of parts) are manifested early in ontogeny and are relatively stable. Some characteristics manifested during mid-development of flowers (during enlargement and shape changes) likewise tend to characterize tribes or genera in Leguminosae. Examples include connate calyces, androecial tubes, petal fusion, gynophores, elongation, and form determination in petals. Aestivation is an exception in that it distinguishes subfamilies of Leguminosae but is a characteristic which is manifested well after organogenesis, during elongation. Characteristics which separate related species tend to be manifested late in floral development, during the stage of cell differentiation. Examples in Leguminosae include petal color, minor shape changes (such as hooks or pits), petal sculpturing, nectaries, hairs, and stigma form. These features are more labile than those occurring early in floral ontogeny; they can change without major disruptions in subsequent development.

While exceptions undoubtedly abound to these general correlations between significance of morphologic characteristics and their time of manifestation in floral ontogeny, the proposal is made in the hope that others will test its basic points in various other plant families. Also it is hoped that it will stimulate ontogenetic studies in plant groups where details are poorly known.

ACKNOWLEDGMENTS

The research work described was supported by National Science Foundation grant DEB82-04132. The author is indebted to Kittie S. Derstine for her skilled assistance in scanning electron microscopy and photography. Russell Goddard, Michael T. Postek, and Seanna Rugenstein also contributed; their help is much appreciated. Plant material of the legumes was obtained at the Royal Botanic Garden, Kew, and at Fairchild Tropical Garden and the U.S. Plant Introduction Station, Miami; the author appreciated the opportunity to make collections and the help of the staff at these institutions.

REFERENCES

Baillon, H. (1872). "The Natural History of Plants", Vol. 2 (Engl. Transl. by M.M. Hartog), Reeve and Co., London.

Baum, H. (1948). Über die postgenitale Verwächsung in Karpellen. Oesterr. Bot. Zeit. 95, 86-94.

Bessey, C.E. (1915). Phylogenetic taxonomy of flowering plants. Ann. Mo. Bot. Gard. 2, 1-155.

Boeke, J.H. (1973). The postgenital fusion in the gynoecium of *Trifolium repens*: light and EM aspects. Acta Bot. Neerl. 22, 503-509.

Boeke, J.H. and van Vliet, G.J.C.M. (1979). Postgenital fusion in the gynoecium of the periclinal chimera *Laburnocytisus adamii* (Papilionaceae). Acta Bot. Neerl. 28, 159-167.

Brett, J.F. and Posluszny, U. (1982). Floral development in *Caulophyllum thalictroides* (Berberidaceae). Canad. J. Bot. 60, 2133-2141.

Buchenau, F. (1865). Morphologische Studien an deutschen Lentibularieen. Bot. Zeit. 23, 61-66, 69-71, 77-80, 85-91, 95-99.

Cowan, R.S. (1968). *Swartzia* (Leguminosae, Caesalpinioideae, Swartzieae). Flora Neotropica Monograph 1, 1-228.

______. (1981). Swartzieae. *In* "Advances in Legume Systematics" (Eds. R.M. Polhill and P.H. Raven), pp. 209-212. Royal Bot. Gard., Kew.

Crisci, J.V. and Stuessy, T.F. (1980). Determining primitive character states for phylogenetic consideration. Syst. Bot. 5, 112-135.

Cusick, F. (1956). Studies of floral morphogenesis. I. Median bisections of flower primordia in *Primula bulleyana*. Trans. Royal Soc. Edinb. 63, 153-166.

______. (1966). On phylogenetic and ontogenetic fusions. *In* "Trends in Plant Morphogenesis" (Ed. E. Cutter), pp. 170-183. Longmans, London.

Dickison, W.C. (1981). Evolutionary relationships of the Leguminosae. In "Advances in Legume Systematics" (Eds. R.M. Polhill and P.H. Raven), pp. 35-54. Royal Bot. Gard., Kew.

Eichler, A.W. (1878). "Blüthendiagramme". W. Engelmann, Leipzig.

Elias, T. (1981). Mimosoideae, In "Advances in Legume Systematics" (Eds. R.M. Polhill and P.H. Raven), pp. 143-151. Royal Bot. Gard., Kew.

Endress, P.K. (1976). Die Androeciumanlage bei polyandrischen Hamamelidaceen und ihre systematische Bedeutung. Bot. Jahrb. Syst. 97, 436-457.

Erbar, C. and Leins, P. (1981). Zur Spirale in Magnolien-Blüten. Beitr. Biol. Pflanz. 56, 225-241.

Gemmeke, V. (1982). Entwicklungsgeschichtliche Untersuchungen an Mimosaceen-Blüten. Bot. Jahrb. Syst. 103, 185-210.

Goebel, K. von. (1887). "Outline of Classification and Special Morphology of Plants", Engl. Transl. Clarendon Press, Oxford.

______. (1905). "Organography of Plants, Pt. 2. Special Organography", Engl. Transl. Oxford Univ. Press.

Grégoire, V. (1924). L'organogénèse de l'ovaire et la dehiscence du fruit. Bull. Soc. Roy. Bot. Belg. 56, 134-140.

Heel, W.A. van. (1966). Morphology of the androecium in Malvales. Blumea 13, 1-394.

______. (1981). A S.E.M.-investigation on the development of free carpels. Blumea 27, 499-522.

______. (1983). The ascidiform early development of free carpels, a S.E.M.-investigation. Blumea 28, 231-270.

Heslop-Harrison, J. (1964). Sex expression in flowering plants. In "Meristems and Differentiation", pp. 109-125. Brookhaven Symp. in Biol. 16.

Kaplan, D.R. (1967). Floral morphology, organogenesis, and interpretation of the inferior ovary in *Downingia bacigalupii*. Amer. J. Bot. 54, 1274-1290.

______. (1968a). Structure and development of the perianth in Downingia bacigalupii. Amer. J. Bot. 55: 406-420.

______. (1968b). Histogenesis of the androecium and gynoecium in Downingia bacigalupii. Amer. J. Bot. 55: 933-950.

Leins, P. (1972). Das zentrifugale Androecium von Couroupita guianensis (Lecythidaceae). Beitr. Biol. Pflanz. 48, 313-319.

Leins, P. and Erbar, C. (1979). Zur Entwicklung der Blüten von Monodora crispata (Annonaceae). Beitr. Biol. Pflanz. 55, 11-22.

Leins, P. and Metzenauer, G. (1979). Entwicklungsgeschichtliche Untersuchungen an Capparis-Blüten. Bot. Jahrb. Syst. 100, 542-554.

Leins, P. and Sobick, U. (1977). Die Blütenentwicklung von Reseda lutea. Bot. Jahrb. Syst. 98, 133-149.

Lewis, G.P. and Elias, T.S. (1981). Tribe Mimoseae. In "Advances in Legume Systematics" (Eds. R.M. Polhill and P.H. Raven), pp. 155-168. Roy. Bot. Gard., Kew.

Lyndon, R.F. (1978a). Flower development in Silene; morphology and sequence of initiation of primordia. Ann. Bot. 42, 1343-1348.

______. (1978b). Phyllotaxis and the initiation of primordia during flower development in Silene. Ann. Bot. 42, 1349-1360.

Maesen, L.J.G. van der. (1970). Primitiae Abricanae VIII. A revision of the genus Cadia Forskae (Caes.) and some remarks regarding Dicraeopetalum Harms (Pap.) and Platycelyphium (Harms) (Pap.). Acta Bot. Neerl. 19, 227-248.

Merxmuller, H. and Leins, P. (1971). Zur Entwicklungsgeschichte männlicher Begonienblüten. Flora 160, 333-339.

Pande, P.C. and Singh, V. (1981). Floral development of Iris decora (Iridaceae). Bot. Jour. Linn. Soc. 83, 41-56.

Pauzé, F. and Sattler, R. (1978). L'androcée centripete d'Ochna atropurpurea. Canad. J. Bot. 56, 2500-2511.

________________________. (1979). La placentation axillaire chez Ochna atropurpurea. Canad. J. Bot. 57, 100-107.

Payer, J.B. (1857). "Traité d'Organogénie Comparée de la Fleur". Paris.

Pfeffer, W. (1872). Zur Blütenentwicklung der Primulaceen und Ampelideen. Jahrb. f. wiss. Bot. 8, 194-215.

Polhill, R.M. and Raven, P.H. (Eds.) (1981). "Advances in Legume Systematics". Roy. Bot. Gard., Kew.

Robertson, R.E. and Tucker, S.C. (1979). Floral ontogeny of Illicium floridanum with emphasis on stamen and carpel development. Amer. J. Bot. 66, 605-617.

Ross, R. (1982). Initiation of stamens, carpels, and receptacle in the Cactaceae. Amer. J. Bot. 69: 369-379.

Sattler, R. (1962). Zur frühen Infloreszenz- und Blütenentwicklung der Primulales sensu lato mit besonderer Berücksichtigung der Stamen-Petalum-Entwicklung. Bot. Jahrb. 81, 358-396.

__________. (1972). Centrifugal primordium inception in floral development. Adv. Pl. Morph. 1972, 170-178.

__________. (1973). "Organogenesis of Flowers". Univ. Toronto Press.

Sattler, R., and Singh, V. (1977). Floral organogenesis of Limnocharis flava. Canad. J. Bot. 55, 1076-1086.

Schmitz, F. (1872). Die Blüthen-Entwicklung der Piperaceen. Bot. Abhandl., Hanstein, 2, 1-74.

Soetiarto, S.R. and Ball, E. (1969). Ontogenetical and experimental studies of the floral apex of Portulaca grandiflora. 2. Bisection of the meristem in successive stages. Canad. J. Bot. 47, 1067-1076.

Stebbins, G.L. (1974). "Flowering Plants". Belknap Press, Harvard Univ. Press.

Taubert, P. (1894). Leguminosae, In "Natürlichen Pflanzenfamilien", Teil III, Abt. III (Eds. A. Engler and K. Prantl), W. Engelmann, Leipzig.

Troughton , J.R. and Sampson, F.B. (1973). "Plants: a Scanning Electron Microscope Survey". J. Wiley and Sons Australasia, Sydney.

Tucker, S.C. (1972). The role of ontogenetic evidence in floral morphology. Prof. V. Puri Commem. Festschr. Adv. Pl. Morph. 1972, 359-369.

__________. (1975). Floral development in Saururus cernuus (Saururaceae). 1. Floral initiation and stamen development. Amer. J. Bot. 62, 993-1007.

__________. (1976). ______. 2. Carpel initiation and floral vasculature. Amer. J. Bot. 63, 289-301.

__________.(1979). Ontogeny of the inflorescence of Saururus cernuus (Saururaceae). Amer. J. Bot. 66, 227-236.

__________. (1980). Inflorescence and flower development in the Piperaceae. 1. Peperomia. Amer. J. Bot. 67, 686-702.

__________. (1981). Inflorescence and floral development in Houttuynia cordata (Saururaceae). Amer. J. Bot. 68, 1017-1032.

__________. (1982a). Inflorescence and flower development in the Piperaceae. II. Inflorescence development of Piper. Amer. J. Bot. 69, 743-752.

__________. (1982b). ______, III. Floral ontogeny of Piper. Amer. J. Bot. 69, 1389-1401.

Uhl, N.W. and Moore, H.E., Jr. (1977). Centrifugal stamen initiation in Phytelephantoid palms. Amer. J. Bot. 64, 1152-1161.

Urban, I. (1885). Morphologie der Gattung Bauhinia. Ber. Deut. Bot. Gesells. 3, 81-101.

Walker, D.B. (1975a). Postgenital carpel fusion in *Catharanthus roseus* (Apocynaceae). I. Light and scanning electron microscopic study of gynoecial ontogeny. *Amer. J. Bot.* 62, 457-467.

__________. (1975b). ______. III. Fine structure of the epidermis during and after fusion. *Protoplasma* 86, 43-63.

__________. (1978). Morphogenetic factors controlling differentiation and dedifferentiation of epidermal cells in the gynoecium of *Catharanthus roseus*. *Planta* 142, 181-186.

Wit, H.C.D. de (1957). A revision of Malaysian Bauhinieae. *Reinwardtia* 3, 381-539.

DEVELOPMENT OF THE INFLORESCENCE, ANDROECIUM, AND GYNOECIUM WITH REFERENCE TO PALMS

Natalie W. Uhl

L. H. Bailey Hortorium
Cornell University, Ithaca, N. Y.

John Dransfield

Royal Botanic Gardens, Kew, England

Preliminary information on the development of inflorescences, flowering units, and flowers are provided for two genera that are morphologically unusual in the Palmae. The study explains how the large terminal flowers in Eugeissona relate to the usual lateral flowers in palms, and describes the development of a new mode of polyandry, monopodial flower clusters, multiparted flowers, and gynoecia with ovules borne on floral axes. In Eugeissona, a lepidocaryoid palm, flowers are borne in pairs that are transferred from a lateral to a terminal position on each rachilla during very early ontogeny. The dyads are sympodial units, the second flower developing in the axil of a bracteole formed by the first floral apex. Stamens numbers in Eugeissona range from 20-70; the stamens arise in a whorl of three opposite the sepals, followed by a row of stamens on a common primordium opposite each petal. To accommodate

ISBN 0-12-746620-7

the higher numbers, the antepetalous ridge lengthens in the center; stamens develop from the ends of each ridge toward its center. In androecia of 70 stamens the last to arise develop centrifugally to the outside of and alternating with those of the primary ring. This pattern provides some insights on the significance of early stamen arrangement in other multistaminate palms. In polyandry, initial stamen arrangement relates to the expansion of the floral apex. Extra stamens develop toward where the apex is increasing or less frequently where residual meristem is present. In the second genus studied, *Palandra*, a phytelephantoid palm, staminate and pistillate inflorescences appear different at maturity but are developmentally homologous. Staminate flower clusters are primary inflorescence branches, each bearing two pairs of subopposite bracts and subtended flowers, and hence are monopodial. During early ontogeny, the apex of the branch aborts, and the fused pedicels of the four flowers elongate to form the stalk of the cluster. In pistillate inflorescences the primary branches develop similar pairs of subopposite bracts but a single flower is borne on the end of the primary branch. Pistillate flowers each bear a floral bracteole, usually followed by four spirally inserted sepals and four similar petals. Staminodes develop centrifugally after the gynoecium is initiated. The gynoecia of *Eugeissona* and *Palandra* are also unusual. Three carpels in *Eugeissona* and six in *Palandra* arise on the flanks of the floral apex. Subsequently the carpels and the floral apex enlarge together. The outer lateral walls of the carpels become continuous but their ventral sutures never close. One ovule "axillary" to each carpel develops on the floral axis in the center of the gynoecium. The floral apices enlarge along with the carpels and form part of the bases of both gynoecia. The significance of these findings in palms is discussed and related to developmental studies in other angiosperms.

I. INTRODUCTION

During the past decade observations of young stages of angiosperm reproductive systems have been greatly facilitated by two technical advances: first, by a new method for dissecting, handling, and observing primordia (Sattler 1968, 1973; Posluszny, Scott, and Sattler 1980); and second, by the improvement and increased availablity of the scanning electron microscope. Sattler's method is a relatively simple way of improving resolution and the SEM provides the essential depth of field for such observations. Much of the work done in the late 1960's and early 1970's may be profitably repeated now.

The potential of studies of development is largely unrealized. Developmental investigations are proving extremely valuable in determining homologies and in elucidating the mechanisms of morphological change. They may also reveal new characters and provide insights that can be used to indicate phylogenies. Sundberg (1982) has used sites of petal-stamen initiation to show phylogenetic lines in Primulales. Erbar and Leins (1981) have suggested derivation of Alismatales from Magnoliales because of presumed similarity in the order of petal and stamen development. Finally, the resulting information is beginning to influence concepts of inflorescence structure, of stamen form and arrangement, and of the shape of the angiosperm carpel.

In palms, studies of anatomy and development have been important in clarifying the structure of inflorescences, flower clusters, and flowers. Fisher and Dransfield (1977) showed that in rattan palms, the adnate inflorescence and its sterile form, the flagellum, found in some species of *Calamus*, represent axillary buds which become displaced and carried up on the leaf sheath above. Nine genera representing six major groups of palms have several inflorescences in a single leaf axil (multiple inflorescences). Fisher and Moore (1977) have shown that the multiple buds develop as separate apices on an originally single bud meristem. Other developmental studies have established that the triad is actually a short cincinnus (Uhl 1976), and that the flower cluster in Nannorrhops and the distinctive acervulus of chamaedoreoid palms are also cincinni (Uhl 1969, Uhl and Moore 1978). In the flower, different patterns of stamen development have been found in polyandric androecia of seven of the ten major groups of palms that include species with more than six stamens (Uhl and Moore 1980). Many questions remain. In this chapter we consider developmental patterns in two genera that are morphologically exceptional in the family.

In Eugeissona, a lepidocaryoid genus, the extremely large flowers appear terminal on the inflorescence branches whereas flowers elsewhere in palms are lateral. Eugeissona also has from 20-70 stamens and thus provides an opportunity to examine the development of polyandry in the lepidocaryoid major group of palms. In Palandra, one of three genera of phytelephantoid palms, several features call for developmental study. There is a greater degree of dimorphism between the staminate and pistillate inflorescences than in other palms. Flower clusters have been reported as monopodial but this has not been completely studied or documented. The number of bracts and perianth parts and the arrangement of stamens in the multipartite flowers are not easily determined at maturity.

Gynoecial structure is an important base in palm taxonomy (Uhl and Moore 1971). The gynoecia of both Eugeissona and Palandra are unusual and characteristic of their respective major groups. This study provides preliminary developmental information on inflorescences, flower clusters, and flowers of these two genera and considers the significance of such developmental studies in palms and in other angiosperms.

II. OBSERVATIONS

A. Eugeissona Griff.

The 21 genera of lepidocaryoid palms form an important and diverse major group whose members are easily distinguished by their scaly fruits (Fig. 3, Moore 1973). Eugeissona, a genus with two species in the Malay Peninsula and four in Borneo (see Dransfield 1970 for monograph), has been placed in a separate alliance in the group (Moore 1973). All species of Eugeissona are clustering palms with large pinnate leaves. Shoots occur in large clumps and may be acaulescent, have short erect stems, or stems elevated on stilt roots. Each shoot produces a large terminal inflorescence (Fig. 1) and is hapaxanthic, dying after flowering. The inflorescences are from 1-17 m long and bear primary branches which are equivalent to lateral inflorescences in pleonanthic palms (Tomlinson 1975). The primary branches are subtended by leaves with reduced blades or by tubular, pointed, sometimes spiny bracts and are branched to 2-4 orders. All branches end in a cupule of 11 to 13 tightly sheathing bracts which usually enclose a terminal pair of flowers. The flower pair consists of a staminate and a perfect flower. The staminate flower matures first and after anthesis is pushed out of the cupule of bracts by the enlarging perfect flower. Thus at any specific time individ-

FIGURES 1-4. Eugeissona. Habit. Fig. 1. E. utilis (Moore 9219), shoot, note large pinnate leaves and terminal inflorescence. Fig. 2. E. brachystachys, a primary branch of an inflorescence from Dransfield 621. x ca. 0.5. Fig. 3. Same, fruit with persistent petals. x ca. 0.4. Fig. 4. E. utilis, flowers at anthesis. Details: in, inflorescence; p, petal; st, stamens; r, cupule of rachilla bracts.

uals have all staminate or all perfect flowers which appear to be single (Figs. 2,4). The flowers represent the largest staminate and hermaphrodite flowers in the family. Only pistillate flowers of <u>Lodoicea</u> (borassoid) and of phytelephantoid palms are larger.

Two genera in the lepidocaryoid major group have more than six stamens. <u>Raphia</u> has some species with six and other species with seven to 20 stamens. <u>Eugeissona</u>, however, is always polyandric, stamen numbers ranging from 20 to 70 depending on the species.

1. <u>Organography</u>: <u>Staminate</u> <u>and</u> <u>Hermaphrodite</u> <u>Flowers</u>. Descriptions are of <u>E</u>. <u>minor</u> Becc. from <u>Dransfield</u> <u>777</u> unless otherwise noted. Staminate flowers (Fig. 5A,B) are ca. 5 cm long and borne on short flattened pedicels ca. 4 mm long. The three sepals are connate forming a tube 1.5 cm long with three short free lobes; the calyx is never exserted above the cupule of bracts. The three petals are 4.5 cm long, briefly united basally, valvate above, have spinose tips, and are very hard, woody and persistent. Twenty (-22) stamens are borne on an epipetalous ridge at the top of the petal tube. The stamens have short slender, terete filaments to 4 mm long and elongate anthers 2 cm in length. The pistillode when present is small and conical.

The hermaphrodite flower (Fig. 5C-G) closely resembles the staminate but is sessile and slightly larger. Two of the petals are flattened apically making the corolla asymmetrically pointed (Fig. 5C) and contrasting with the more symmetrical point of the staminate corolla (Fig. 5A). The stamens are like those of the staminate flower. They mature and shed their pollen long before the stigmas are receptive. The oblong gynoecium is composed of three carpels connate laterally but with their ventral sutures open in the center throughout the gynoecium; the floral axis is evident as a pointed projection in the basal ovarian region and bears three erect, anatropous, epitropous ovules, the micropyles facing the center of the gynoecium. The stigma consists of three short and rounded, fibrous lobes.

2. Organogeny: The Dyad. Complete young stages of inflorescences were not available but branches (rachillae) 1-2 cm long showed early stages of the flower pairs (Fig. 6). At this stage 2(-3) flower pairs, decreasing in age distally, are usually present near the end of each branch. These dyads are clearly borne in the axils of the upper cupular bracts. The apex of the rachilla is evident (Fig. 6,ra) as a low mound between the dyads. Apices of three staminate flowers (a) and of two hermaphrodite flowers (2a) are present in Figure 6. In later stages there is no evidence of the apex of the rachilla nor of more than one flower pair. The rachilla apex and additional flower pairs if present are obscured early in development by the expansion of one of the uppermost dyads which becomes terminal.

In the development of the flower pair, the first floral apex, a large dome, initiates a bract or prophyll in an adaxial position. This first prophyll, the dyad prophyll or bract, arises as two lateral adaxial bulges which rapidly become united and raised on a girdling base. Two stages of the dyad prophyll (db) are indicated in Figure 6. Subsequently the apex of the second flower of the dyad, the hermaphrodite flower, develops as a lateral bulge from the first apex (Fig. 6,2a). The second flower is clearly lateral to the first flower and borne in the axil of the dyad bract. Slightly later, on the apex of the hermaphrodite flower, a prophyll develops as two lateral wings (Fig. 7) and rapidly encloses the apex, continuing to enlarge around the perfect

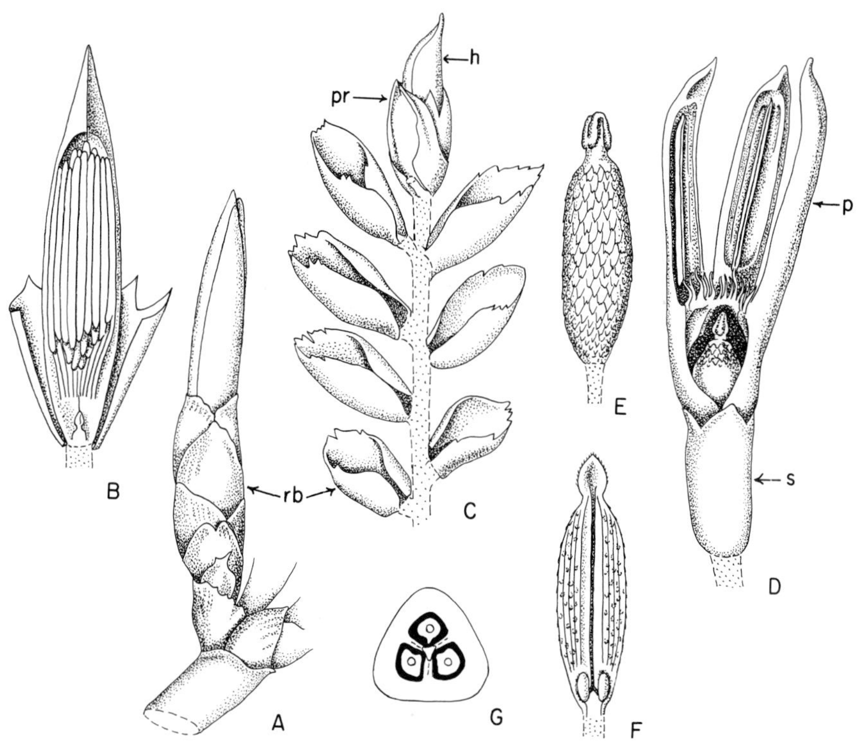

FIGURE 5. Eugeissona. Diagnostic drawings of flowers. A. Part of a branch with a staminate flower in bud. B. Longisection of a staminate flower. C. Branch showing upper bracts of rachilla and perfect flower. D. Perfect flower, petals pulled away to show gynoecium, all but two anthers shed. E. Gynoecium. F. Longisection of E, note basal ovules with projection of floral axis between them and central canal. G. Transection of E through ovarian region.

flower until split by the elongating bud after the staminate flower has been shed (Fig. 5C).

3. Organogeny: Staminate and Hermaphrodite Flowers. Development of the staminate and hermaphrodite flowers is identical until early stages of gynoecium formation. During sepal initiation the floral apex is a low dome. The calyx arises as a girdling primordium which soon produces three elongate ridges in spiral succession (Figs. 6,9). Subsequently the connate base and the free lobes enlarge as a unit (Fig. 6s). After sepal initiation the apex becomes flatter and the petals arise spirally as elongate bulges alternate with the sepals (Figs. 6,9,p). The apex becomes still flatter and widens opposite the petals before the first three stamens appear in close spiral succession opposite the sepals on the points of the apex between the petals (Fig. 8). The

FIGURES 6-9. Eugeissona. Organogeny. Fig. 6. Early stage of a branch tip with several flower pairs in different developmental stages. Fig. 7. Young dyad, upper staminate flower showing the first stamen primordia, one opposite each sepal, ridgelike common primordium opposite each petal, lower perfect flower with floral apex and 2 "wings" of prophyll only. Fig. 8. Staminate flower, similar stage, note definite antepetalous ridges. Fig. 9. First flowers of two dyads at different stages, upper flower with young petal primordia, lower flower showing the first antepetalous stamens on each side of the antesepalous stamens. Details: a, floral apex; 2a, second floral apex, that of the perfect flower; as, antesepalous stamen primordium; db, dyad bract; p, petal; pr, prophyll on perfect flower; rb, rachilla bract; s, sepal.

stamen primordia are more or less triangular and rounded distally. Following the initiation of the three antesepalous stamens, a ridge or common primordium develops opposite each petal; the next stamens to appear are on the ends of these ridges on either side of each antesepalous stamen (Fig. 9). Stamen primordia continue to arise in succession from each end toward the center of the antepetalous ridges. Six or seven stamens form on each ridge resulting in an androecium of a single triangular row of 21-24 stamens in E. minor (Figs. 9,10). In E. insignis Becc. (Dransfield 751) 52-60 stamens arise in the same manner; the additional stamens are

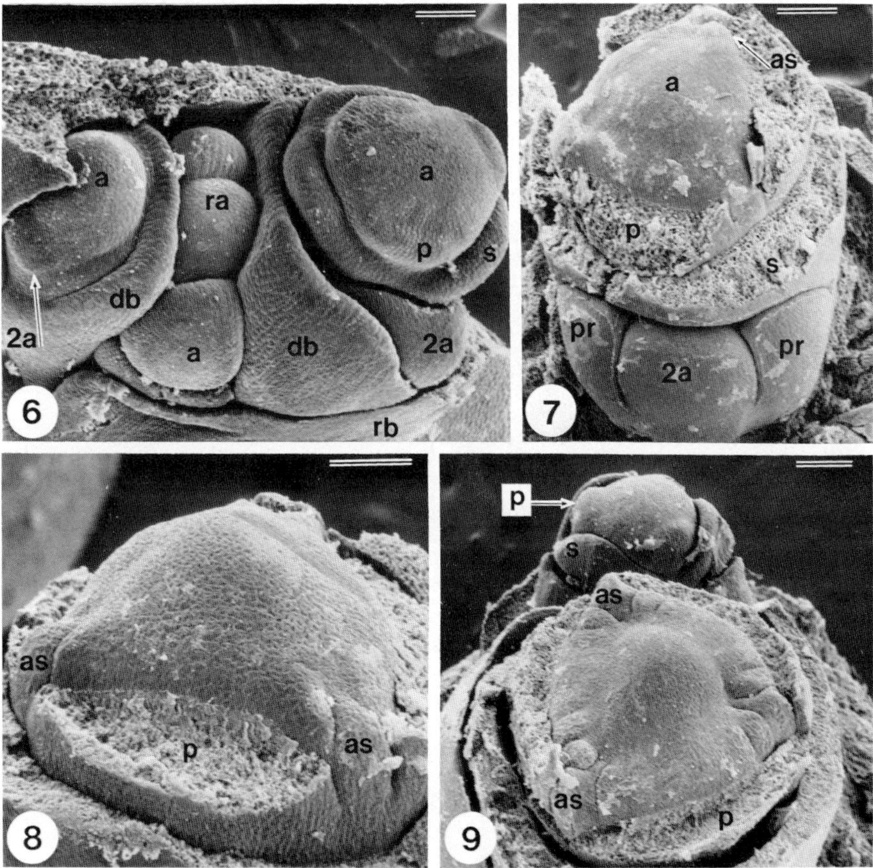

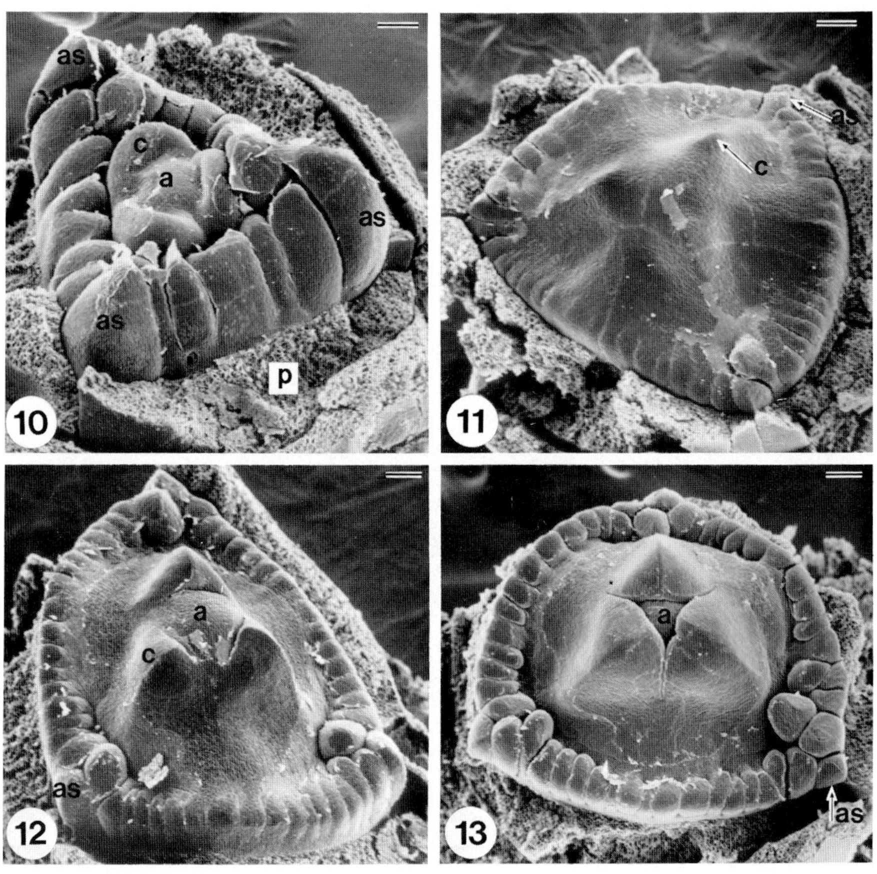

FIGURES 10-13. Eugeissona. Organogeny cont. Fig. 10. E. minor, stamen primordia and early stage of gynoecium. Fig. 11. E. tristis, note youngest (smaller) stamen primordia in centers of antepetalous ridges and first stage of carpel inception. Fig. 12. Same, later stage of gynoecium development. Fig. 13. Same, later stage, carpel primordia partly closed over floral apex. Details: a, floral apex; as, antesepalous stamen; c, carpel primordium.

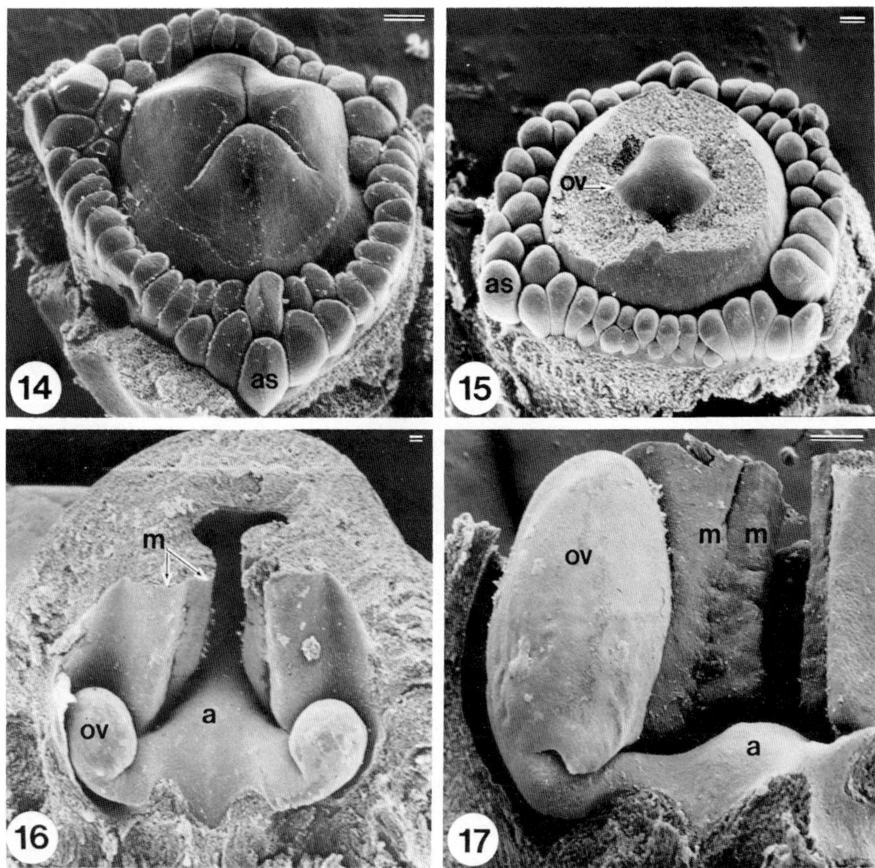

FIGURES 14-17. Eugeissona. Organogeny cont. Fig. 14. E. minor, stamen primordia larger, older ones in a double row, carpels united basally, closed over the floral apex. Fig. 15. E. utilis, free parts of carpels removed to reveal the floral apex with three ovule primordia, smaller stamens centrifugal to older ones on the antepetalous common primordia. Fig. 16. E. minor, base of gynoecium, ovules partly turned, carpels laterally connate, their margins abutting on the floral axis. Fig. 17. Same, later stage, ovule turned so that micropyle is toward the center of the gynoecium. Details: a, floral apex; as, antesepalous stamen; m, carpel margin; ov, ovule. Bars = 100 μm.

accommodated on wider antepetalous ridges (Figs. 11-14). In E. utilis Becc. (Dransfield 793) which has ca. 70 stamens, the highest number in the genus, several stamens develop centrifugally to the outside of and alternating with stamens of the first row on the antepetalous primordia (Fig. 15).

When the androecium is complete, three carpel primordia arise as triangular bulges on the flanks of the floral apex opposite the sepals. The carpel primordia rapidly become thick medianly and crescentic (Figs. 10-13) and expand marginally and distally to close over the apex (Fig. 14). During the early stages of carpel development, the floral apex also appears to increase in height and width (Figs. 11-13). In staminate flowers development of the gynoecium stops at about the time the carpels have closed over the apex and the small gynoecial primordium aborts or matures into the pistillode (Fig. 5B).

In the hermaphrodite flower, the ovules arise as three bulges on the edges of the floral apex opposite the carpels (Fig. 15). As the gynoecium develops further the carpels become connate laterally. The united walls of adjacent carpels project into the center of the gynoecium forming septa, which are incomplete since the ventral suture of each carpel remains open. The double nature of the septa is evident at their margins (Figs. 16,17). The width of the ventral sutural opening of the upper carpel is apparent between the two septa in Figure 16. The margins of adjacent carpels abut on the floral axis between the ovules (Figs. 16,17). During development the ovules turn through 180^0 and become anatropous with their micropyles facing the center of the gynoecium. A central canal is open throughout the long stylar region to the stigmas. The floral axis remains as a projection in the center of the ovarian region until late in fruit formation (Dransfield 1970).

FIGURES 18–20. Phytelephantiod palms. Fig. 18. Phytelephas sp. (Moore, Dransfield, and Giraldo 10224). Base of staminate tree with inflorescence near anthesis. Fig. 19. Ammandra decasperma O. F. Cook. Lower part of a pistillate tree with an inflorescence near anthesis, photo by J. Dransfield (Dransfield 4869). Fig. 20. Palandra aequatorialis. A. Fruit from one flower opened to show seeds. x 0.2. B. Infructescence composed of fruits from several flowers, arrow indicates limits of the fruit from one flower. x0.3. C. Flower cluster from staminate inflorescence. x 0.8. Details: bf, base of a staminate flower; pb, peduncular bract; s, seed.

B. *Palandra* O. F. Cook

Palandra is monotypic and one of three genera that comprise the phytelephantoid major group of palms. Materials of *P. aequatorialis* (Spruce) O. F. Cook studied were from the Rio Palenque Station, Ecuador (*Moore and Anderson 10210*). The three phytelephantoid genera are important in evolutionary studies because of their unusual morphology. Furthermore, the distribution of one of the genera, *Phytelephas*, overlaps certain Pleistocene refugia described for birds by Haffer (Moore 1973). The group is characterized by dioecism, dimorphic staminate and pistillate inflorescences, and multiparted flowers. Staminate and pistillate trees of *Palandra* have short erect trunks with persistent petioles, and large pinnate leaves with clustered pinnae. Figure 18 illustrates a similar habit in the related genus, *Phytelephas*. The axillary staminate and pistillate inflorescences appear very different (cf. Figs. 18–20). *Palandra* has over 900 stamens, the largest number in the family. The group has been considered one of the most highly specialized of the arecoid line (Moore 1973), but the genera are poorly known and their relationships within the family are not yet certain.

FIGURES 21–24. *Palandra aequatorialis*. Inflorescence structure. Fig. 21. Lower part of a staminate inflorescence. x 0.75. Fig. 22. Young stage, complete staminate inflorescence. x 1.3. Fig. 23. Young pistillate inflorescence. x 0.75. Fig. 24. Young stage of a pistillate inflorescence, note that floral primordia are obscured by elongate bracts. x 1.3. Details: br, primary bract; fc, cluster of four flowers; pb peduncular bract; pr, prophyll; sc, stalk of flower cluster.

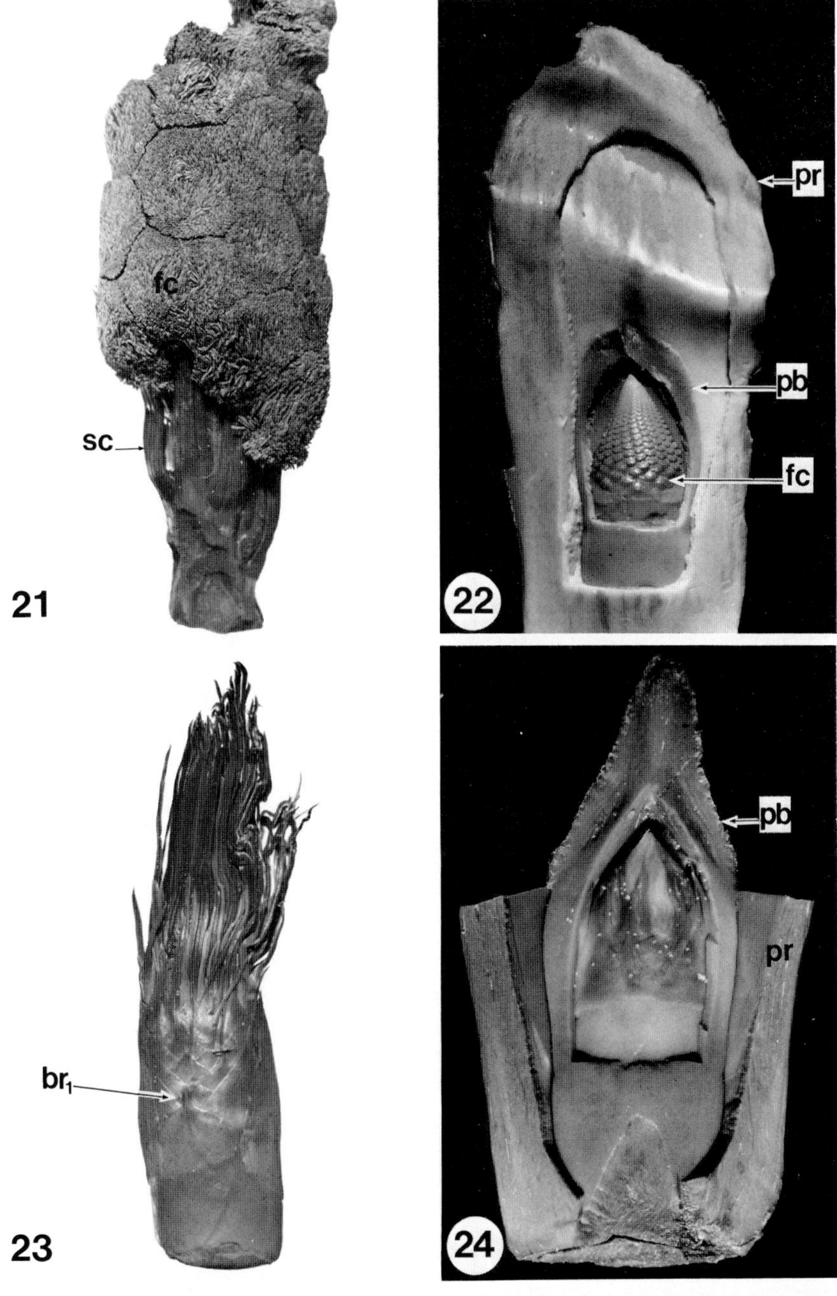
fc
sc
21
pr
pb
fc
22
br1
23
pb
pr
24

1. Organography: Staminate Inflorescence, Flower Cluster, and Flower. The staminate inflorescence is enclosed in bud by two bracts, a prophyll and one peduncular bract (Fig. 22). The prophyll remains more or less in the leaf axil but the peduncular bract is exserted as in *Phytelephas* (Fig. 18), splits abaxially as the inflorescence elongates, and remains above the inflorescence in flower. The inflorescence is a long spicate axis which bears spirally arranged, closely crowded flower clusters (Fig. 21). Each flower cluster consists of four flowers borne on a stalk formed by their united pedicels. At maturity the connate pedicels appear topped by a white mass of stamens (Fig. 21). Dissection of these units reveals four flat floral receptacles, each bearing between 900 and 1000 stamens surrounded by a low cupular perianth with 8-10 lobes (Fig. 20C,bf). Previous developmental studies have shown that the perianth is four-parted and that the stamens develop centrifugally. Pistillodes of one to three lobes are present in young stages of some flowers (Uhl and Moore 1977).

2. Organogeny: Staminate Inflorescence, Flower Cluster, and Flower. In early stages the staminate inflorescence is triangular, somewhat dorsiventrally flattened, and enclosed by two complete bracts, a prophyll and a peduncular bract (Fig. 22,pr,pb). Above these are a pair of large pointed bracts which tightly ensheath the lower lateral margins of the axis and a few shallow empty bracts below the parastichies of flower clusters. Figures 25-30 illustrate the development of the flower clusters and flowers. Primary bracts are produced in spiral order by the apex of the inflorescence (Fig. 25,br_1). At a later stage each of these bracts subtends the apex of a first order inflorescence branch (Fig. 26,ax_2). Subsequently the apex of each first

order branch produces a pair of subopposite elongate bulges. The bulges are common primordia, each of which divides to form a secondary bract (Figs. 26-28,br_2) and a floral apex. The first pair of bracts and their floral apices are followed in close succession by the inception of an anterior-posterior pair of floral apices and their subtending bracts (Fig. 27, F_{1-4}). Each floral apex increases in size and becomes dome-shaped during the initiation of an adaxial floral bracteole (Fig. 28,fb). Perianth development also begins at this time; a first sepal is evident on the right apex (Fig. 28). Inflorescence primordia are strongly compressed during initiation of these flowers and irregularities in shapes of floral apices, bracts, and perianth parts are common. Floral bracteoles are sometimes missing, and the numbers of sepals and petals per flower varies. The posterior floral apex is often compressed below the bract subtending the cluster and not visible as in the lower cluster in Figure 27. Most commonly four sepals are formed in spiral succession, followed by four petals (Uhl and Moore 1977 and Figs. 29, 30). However, two petals sometimes develop in one position. Sepals and petals are united basally at an early stage (Fig. 29). After the inception of the perianth, the floral apex becomes flat and expands marginally. Stamens are produced in centrifugal succession (Fig. 30). The apex of the primary branch (ax_2) is visible during the initiation of sepals and petals but is then obscured as the floral apices expand. The pedicels develop later as a four-lobed structure which is intercalated as the flowers enlarge.

3. Organography: Pistillate Inflorescence, Flower Cluster, and Flower. The pistillate inflorescence appears very different as shown for the related genus Ammandra (Fig. 19). The inflorescence is spicate but much shorter than the

staminate. It is also enclosed in bud by a prophyll and one peduncular bract (Fig. 24). When the flowers are at anthesis a curly white mass of bracts, perianth parts, and stigmas are exserted from an opening near the tip of the peduncular bract (Fig. 19). These are bracts and perianth parts of about 18 pistillate flowers which are borne near the end of the inflorescence axis (Fig. 23). The flowers are so closely appressed that it is difficult to determine the limits of a single flower at maturity; Figure 31A represents two flowers. Dissections show that each flower is subtended by a rather long pointed bract (Fig. 31F), and bears about 12-14 bracts and bractlike perianth parts (Fig. 31H plus J). Development

FIGURES 25-30. *Palandra aequatorialis*. Development of staminate flower clusters and flowers. Fig. 25. Apex of the main axis of the inflorescence with spirally arranged primary bracts in various stages of development. Fig. 26. Part of an older staminate inflorescence, primary bracts with subtended primary branch apices and early stages of secondary bracts. Fig. 27. Later stage, primary branch with four floral apices. Fig. 28. Later stage, floral apices enlarged, primordia of floral bracteoles and of first sepals evident. Fig. 29. Floral apex expanded just before stamen initiation, perianth intact. Fig. 30. Stamen initiation, note sepals and petals in 4-parted whorls and meristematic edges of apex. Details: ax_1, apex of inflorescence; ax_2, apex of primary branch; br_1, primary bract borne on main axis of inflorescence and subtending a primary branch; br_2, secondary bract borne on the primary branch and subtending a flower; f_{1-4}, staminate flowers in order of their development; fa, apex of a staminate flower; fb, bracteole or prophyll produced by the floral apex; p, petal; s, sepal.

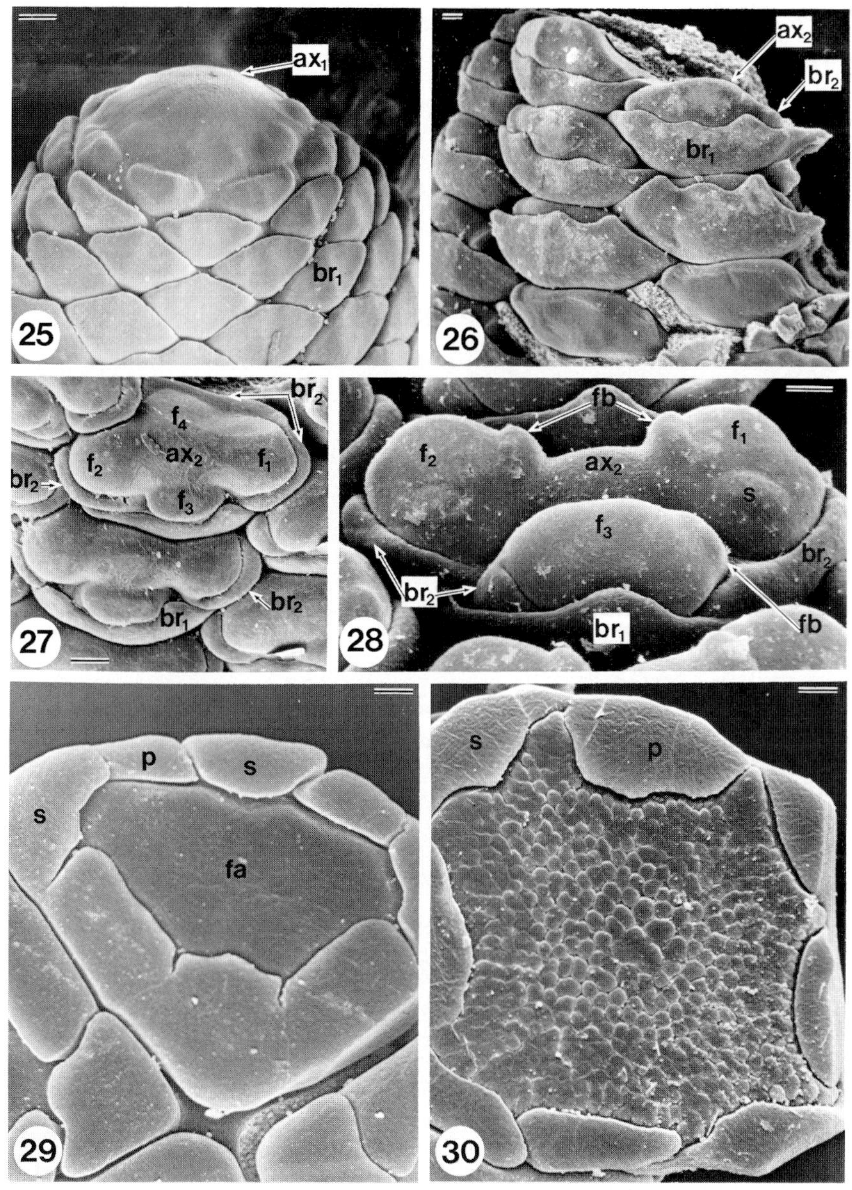
ax1
br1
25
ax2
br2
br1
26
br2
f4
f2
ax2
f1
br2
f3
br1
br2
27
fb
f2
f1
ax2
s
f3
br2
br2
fb
br1
28
p
s
s
fa
29
s
p
30

has clarified the arrangement of these organs as explained below. About 35 staminodes with terete filaments of various lengths and elongate anthers closely surround the central gynoecium which has a rounded, shallowly six-lobed ovarian part, a long terete style, and six long filiform stigmas. The gynoecium is composed of six carpels, each bearing one ovule (Fig. 31G,I). The carpels are partially united peripherally. Their ventral sutures are open and their free conduplicate lateral sides and margins can be seen in a section of the style (Fig. 31C). The stigmas are flat and conduplicately folded (Fig. 31D), and their inner surfaces are lined with stigmatoid tissue.

4. Organogeny: The Pistillate Inflorescence, Flower Cluster, and Flower. The pistillate inflorescence, like the staminate, is enclosed in bud by a prophyll and a complete peduncular bract (Fig. 24). Above these as in the staminate inflorescence there is a pair of empty lateral peduncular bracts which are well developed. A large number of the lower primary bracts are empty but enlarge to cover the base of the inflorescence axis (Fig. 23); in the staminate inflorescence these bracts remain small and subtend the flower clusters.

Development of the flowering units is illustrated in Figures 32-41. At an early stage (Fig. 32), the apex of the inflorescence is a low dome which produces spirally arranged primary bracts (Fig. 32br_1); the lower primary bracts remain empty, only about the distal 18 produce primary branches (Figs. 32,33,ax_2). The first structures produced by the apex of each primary branch are a pair of subopposite lateral bracts homologous to those subtending the first two flowers in the staminate inflorescence; as in the staminate inflorescence a second anterior-posterior pair of bracts arises next (Figs. 33,34br_2). Subsequently the apex of each primary

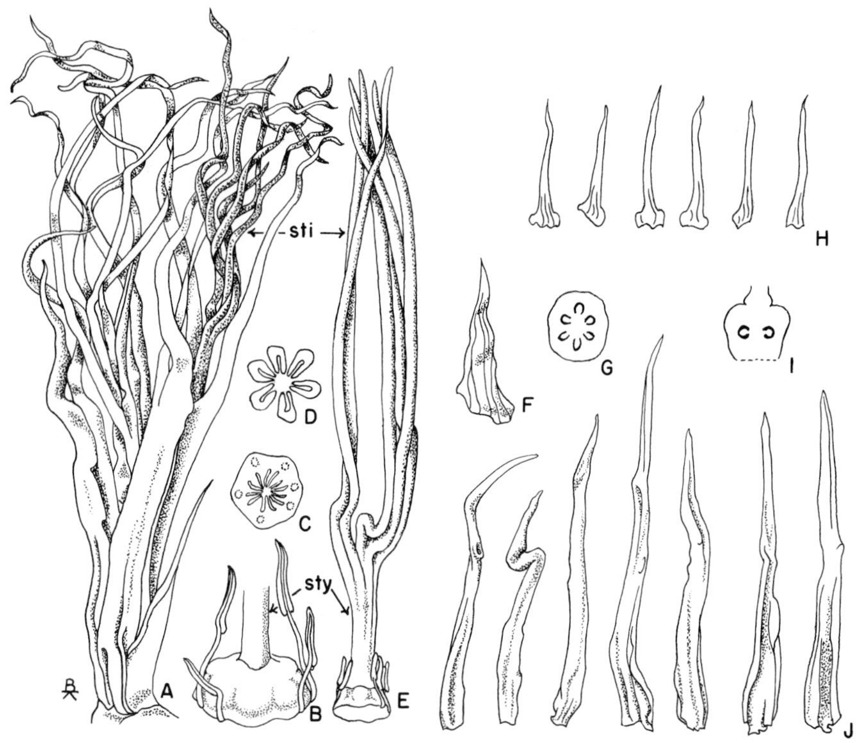

FIGURE 31. Palandra aequatorialis. Diagnostic drawings, pistillate flower. A. Two flowers from the pistillate head. x ca. 0.7. B. Ovarian part of the gynoecium with several attached staminodes. C. Transection of the cylindrical style. D. Transection of the stigmas. E. Complete gynoecium from a young flower ca. one-half the size of a mature flower. F. Bract that subtends the flower. G. Transection gynoecium. I. Longisection ovarian part of gynoecium. H, J. Appendages of the flower, interpreted as four bracts, one floral bracteole, four sepals, four petals. Details: sti, stigmas; sty, style.

branch becomes the apex of a pistillate flower. The floral apex is at first a broad rather flat dome (Fig. 34) which forms a floral bracteole and a whorl of four sepals in spiral order followed by four similar petals. The four primary bracts, the floral bracteole, and the perianth parts elongate and cover the floral apex. In later stages these organs appear similar and have all been considered perianth parts. Even in early stages they are irregular in shape (Figs. 33-35), but it seems clear that they are homologous to similar structures in the staminate flower cluster and flower. After perianth inception, the floral apex increases in height and six carpels arise as circular bulges (Fig. 36) on its flanks. The carpels expand distally and laterally and become conduplicate (Fig. 37). During early stages of carpel expansion,

FIGURES 32-37. *Palandra aequatorialis*. Development of the pistillate flower. Fig. 32. Young pistillate head showing apices of the primary branches, several of the subtending primary bracts removed to reveal secondary bracts in various developmental stages. Fig. 33. Slightly later stage showing young anterior secondary bract. Fig. 34. Enlargement of a part of Fig. 33 - the apex of each primary branch will become a floral apex. Fig. 35. Later stage, bracts and perianth segments pulled back to show the increased height of the apex. Fig. 36. Early carpel formation. Fig. 37. Later stage carpel development. Details: ax_1, apex of the main axis of the inflorescence; ax_2, apex of a primary branch; br_1, primary bract on main axis; br_2, secondary bracts borne by primary branches; c, carpel; f, floral apex (primary branch); fb, floral bracteole; p, petal; s, sepal. Bars equal 100 micrometers, except in Figure 36 where the bar equals 1000 micrometers.

the floral apex continues to increase in height (cf. Figs. 36 and 37). As they enlarge the carpels become more strongly conduplicate, extend over the apex (Fig. 38), and increase rapidly in length. Their bases become connate laterally around the floral axis and their free conduplicate edges slant upward above the floral apex. In a flower 2 mm long, the carpels (Fig. 40c) are nearly twice the length of the perianth parts. In Figure 41 five of the six carpels have

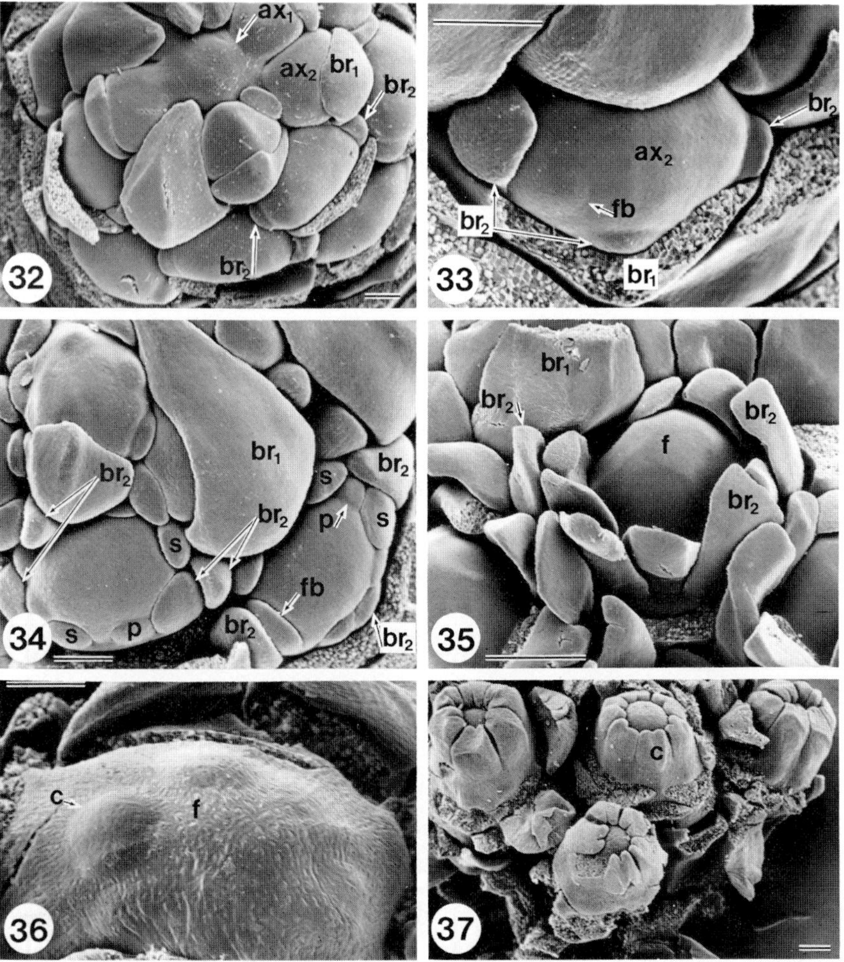

been removed revealing the free margins (m) of the sixth carpel; note that the carpel walls slant toward the center above the periphery of the floral apex (Fig. 41). An ovule arises from the floral apex in an "axillary" position to each carpel. Later, zonal growth of the united lateral sides of the carpels elevates the ovarian wall and the cylindrical style; the long conduplicate parts elongate further and become the stigmas. Parts of the sides of the carpels and their margins remain free in the center of the gynoecium as noted above. When the flower is ca. 2 mm long, the first staminodes become evident as small circular bumps on a ring-shaped common primordium at the base of the gynoecium. (Fig. 41). The staminodes develop in oblique centrifugal rows (Fig. 39). This represents the first known instance where floral organs do not arise in acropetal succession in palms.

III. DISCUSSION

A. Inflorescence

1. In Palms. Since the inflorescences of *Eugeissona* and *Palandra* are unusual in several respects, some introductory comments on inflorescence structure in the Palmae are necessary. Inflorescences in palms range from the huge terminal panicle of *Corypha* to spicate axes a few centimeters long in *Chamaedorea* and *Reinhardtia*. Although many of them appear complex, the inflorescences have been shown to represent variations on a structurally simple theme (Moore and Uhl 1982). The axillary inflorescence is a monopodial, indeterminate branching system. Its variations are remarkable and not completely understood. The diagrammatic drawing (Fig.

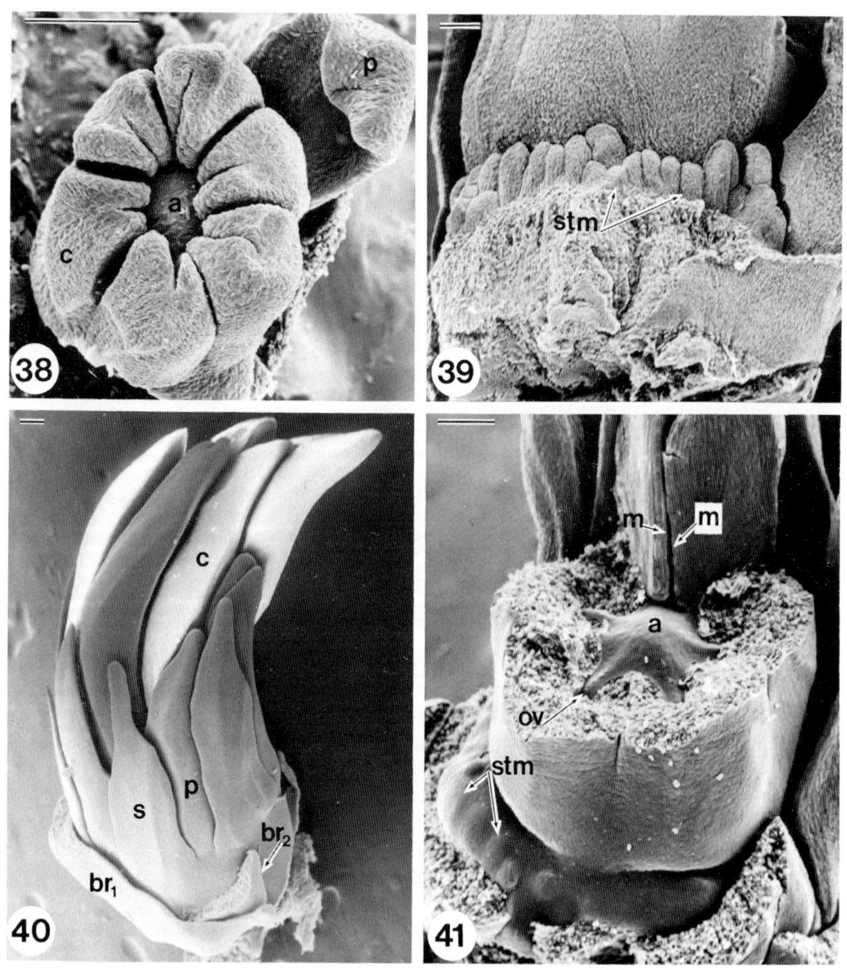

FIGURES 38-41. Palandra aequatorialis. Pistillate flower cont. Fig. 38. Later stage carpels. Fig. 39. Young staminodes, note centrifugal decrease in size. Fig. 40. Young pistillate flower. Fig. 41. Perianth and all but one carpel removed to show the floral axis with ovule primordia, note staminodial primordia on basal ring. Details: a, floral apex; br_1, primary bract; br_2, part of a secondary bract; c, carpel; m, carpel margin; ov, ovule primordium; p, petal; s, sepal; stm, staminodes. Bars equal 100 micrometers.

42) represents the basic pattern as we currently understand it. The first branch, the main axis, ax_1, is borne in the axil of a leaf or less frequently a modified leaf or bract (Tomlinson 1975). It is convenient to divide the main axis into the peduncle, the basal unbranched part, and the rachis, the distal branched part. The peduncle bears a prophyll (Fig. 42pr) and none to several empty peduncular bracts (Fig. 42pbr); subsequent bracts (rbr), rachis or primary bracts, each subtend a primary branch. The characteristics of the prophyll, peduncular bracts, rachis bracts, and the overall branching pattern are useful diagnostic characters. The branching pattern of the main axis may be repeated in the lateral branches and may result in as many as five or six orders of branching (Uhl 1972). If primary branches are short or lacking, the main axis is essentially unbranched and the inflorescence spicate as in some species of *Chamaedorea*.

FIGURE 42. Schematic diagram representing the structure of the main inflorescence branches in palms. Details: ax^0, the palm stem; ax^{1-5}, successive branches of the inflorescence; f, flower or flower cluster; fb, the bract subtending the flower or flower cluster (these are the shaded bracts in Figure 43); fpr, prophyll borne on the floral stalk; lf, leaf sheath; pr, prophyll; pbr, empty peduncular bract; rbr (and all bracts of similar shape), rachis or primary bract subtending an inflorescence branch. Spaces indicate that a branch may be lengthened or shortened; the number of orders of branching may be reduced to only ax^1. The initial prophyll is always present but varies in size and shape; other bracts may be enlarged, reduced, or missing. Adapted by B. Everhart from an original by H. E. Moore, Jr. and M. Nakayama.

All branches including the main axis usually end in a flower-bearing part, which it is useful to designate as a "rachilla" (Moore and Uhl 1982 and references therein). Rachillae are distinct units from early in ontogeny. The parts of branches bearing the primordia of all flowers and flower clusters are present early in the bud. The basal parts of branches are intercalated later as the inflorescence enlarges (Uhl 1976 and unpublished).

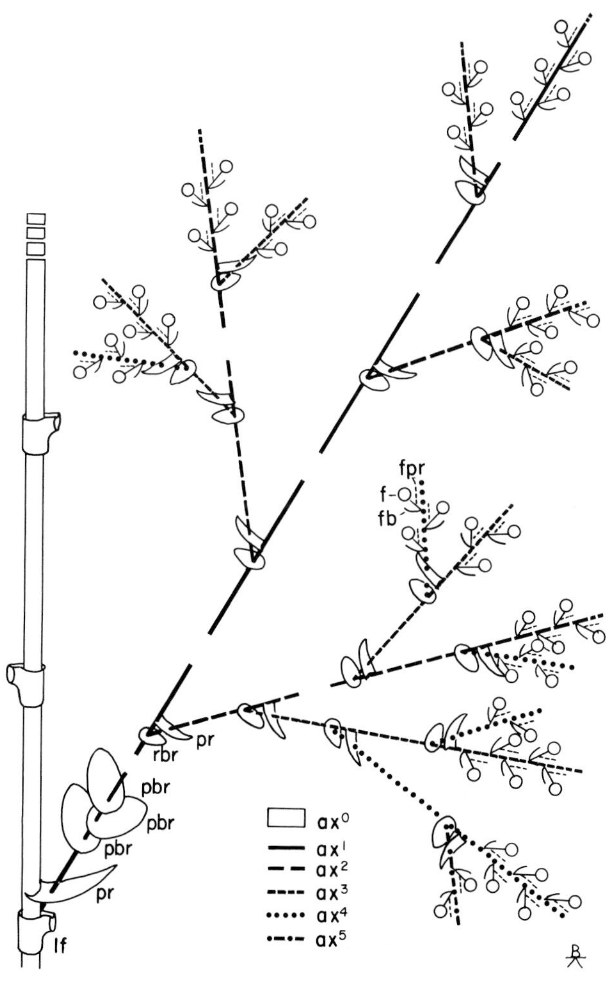

The structure of the flowering units usually differs from that of major inflorescence branches. The basic flowering unit is a single flower which is borne in the axil of a bract and bears a bract in an adaxial position, the floral bracteole or prophyll (Figs. 42,43). If the floral bracteole subtends a second flower, and the bracteole of the second flower a third flower, etc., a cincinnus, which is a sympodial unit, results. Various forms of cincinni characterize different groups and genera of palms (Fig. 43, Moore and Uhl 1982).

a. *Eugeissona*. As noted above most palm inflorescences are indeterminate. The apices of all branches simply peter out; usually no terminal flower bud is produced. The most obvious exception to this occurs in *Eugeissona* where all branches of a massive, much branched inflorescence appear to end in large terminal flowers. Dissections and developmental studies (Dransfield 1970 and above) have shown that not one but a pair of flowers occurs on the end of each branch. This study indicates that the pair of flowers is transferred from a lateral axillary position to a terminal position during ontogeny. The cupular bracts below the flower pair represent the rachilla bracts which in other palms subtend solitary flowers or flower clusters; the flowering branches of *Eugeissona* are thus homologous to rachillae elsewhere in palms. As far as we know this is the only instance where flowers are transferred from a lateral to a terminal position during ontogeny. It must be considered a derived condition unique to *Eugeissona*, and strengthens the isolated position of the genus in an alliance of its own within the lepidocaryoid major group. A few other terminal flowers occur in the family and will be discussed below.

The development of the flower cluster in *Eugeissona* is also of interest. A pair of flowers, enclosed by a well developed prophyll (the dyad bract) with the second flower

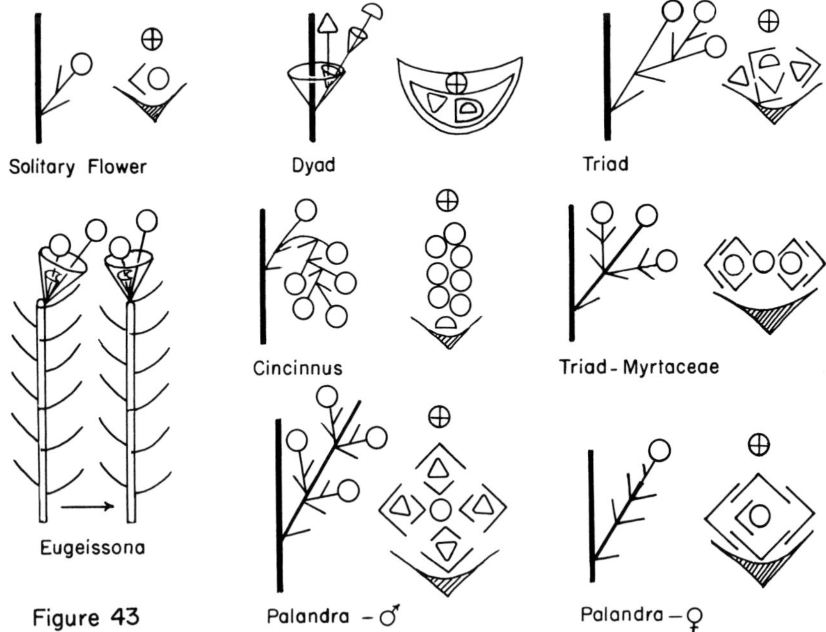

FIGURE 43. Diagrams of flower clusters.

enclosed by a second prophyll situated at right angles to the first, is the basic flowering unit in the lepidocaryoid group of palms (Dransfield 1970 and unpublished). In Eugeissona the second floral apex clearly arises from the apex of the first flower and the second prophyll develops from the apex of the second flower. This indicates that the second flower is lateral to the first and that the unit is a short cincinnus (Fig. 43, dyad). Various modifications of this unit occur in the lepidocaryoid major group and will be considered elsewhere.

b. Palandra. This study confirms that staminate and pistillate flowers in this genus are borne in monopodial units. Staminate flowers are in clusters of four consisting of two subopposite pairs of flowers borne on a very short primary branch of the inflorescence. After the second pair of flowers is formed, the apex of the primary branch stops growing and is obscured by allometric growth of the pedicels of the four flowers. In pistillate inflorescences, a single flower develops directly on the apex of each short primary branch.

Staminate and pistillate inflorescences are homologous although they appear very different at maturity. The prophyll and the peduncular bract are similar in young stages in the two inflorescences (Figs. 22,24). Splitting of the peduncular bract differs somewhat; the peduncular bract opens abaxially in staminate inflorescences and more or less terminally in pistillate inflorescences. A pair of incomplete lateral peduncular bracts are present in both. Subsequent primary bracts are fewer but larger in pistillate inflorescences and cover the developing flowers.

In staminate inflorescences the four flowers per cluster are borne on subopposite secondary branches. In pistillate inflorescences single flowers are borne directly on the primary branches, but before the branch apex becomes the floral apex, each primary branch produces two subopposite pairs of lateral bracts homologous to those that subtend the staminate flowers.

In both inflorescences flowering units are borne in parastichies. Many more flowers are produced by the staminate inflorescence. In staminate inflorescences 14 to 16 rows of ca. 25 flower clusters each cover the spicate main axis; thus, there are ca. 400 clusters of four flowers each, or 1600 flowers. Each flower bears over 900 stamens. The pistillate head bears only about 18 single flowers.

Staminate and pistillate flowers are similar in that floral apices are alike in shape and the perianth in both flowers consists of two four-parted whorls. Androecial development is centrifugal in both flowers. Thus there is developmental correspondence in branching patterns, in bracts, in structure of flower clusters and of flowers. Differences between the flowers result largely from allometric growth patterns. Staminate flowers have long pedicels while pistillate flowers are sessile. Perianth parts differ in shape. Sepals and petals of staminate flowers remain small and become united basally while the perianth parts of the pistillate flower become extremely elongate.

c. Flower size and position. The flowers of *Eugeissona* are the largest staminate and hermaphrodite flowers in palms. The pistillate flower of *Palandra* is also one of the largest. It represents the longest pistillate flower in the family; the female flower of *Lodoicea* is much larger in diameter. Elsewhere large pistillate flowers are terminal on short branches in several genera of borassoid palms, including *Lodoicea*, *Borassus*, and *Borassodendron*, and in some lepidocaryoid palms. Briggs and Johnson (1979) have noted that some inflorescence modifications are associated with the support of large flowers and fruits. They mention reduction in branching and in flower number, and shortening and thickening of axes. Reduction in flower number seems evident in *Eugeissona* and in pistillate inflorescences of *Palandra*. Reduction in branching and thickening of axes characterizes phytelephantoid inflorescences. In *Eugeissona* fruits are large (Fig. 3) and in *Palandra* fruits are very large and heavy (Fig. 20). Some of the differences in inflorescence structure in *Eugeissona* and *Palandra* may relate to the support of large flowers and fruits.

d. Relationships of phytelephantoid palms. This developmental information suggests that phytelephantoid palms may not be as closely related to cocosoid, arecoid, and geonomoid palms as their present placement indicates. In the three groups in question the inflorescence bears a prophyll and one complete peduncular bract. Phytelephantoid palms also have a pair of rather well developed lateral peduncular bracts perhaps indicating closer relationship to ceroxyloid or other major groups where more peduncular bracts are present. The monopodial nature of the flower clusters strengthens this conclusion. Flowers occur in triads in cocosoid, arecoid, and geonomoid palms while ceroxyloid palms have solitary flowers.

2. _In Other Monocotyledons and in Dicotyledons_. Developmental studies seem certain to contribute significantly toward a better understanding of inflorescence structure in angiosperms. As in _Eugeissona_ and _Palandra_ basic patterns and their modifications are often evident during ontogeny. There is at present no standard terminology nor any unified assessment of inflorescence structure. In a paper on the adaptive significance of inflorescence architecture, Wyatt (1982) has attributed some of this lack of knowledge to attempts to establish an overall system. To a large extent Rickett's (1944, 1955) classification has been followed in North America and the detailed analyses of Troll (1964, 1969) in Europe. Troll's work is unfinished and was based largely on annual dicotyledonous herbs with relatively few references to woody plants and monocotyledons. Briggs and Johnson (1979) in a masterful survey of inflorescences in Myrtaceae found Troll's system not always applicable and have presented a critique of his system and a new glossary. Some modification of their terms are indicated by Kuijt (1981) in an evolutionary survey of the Loranthaceae, and still others are needed by us for palms. Until basic patterns and their modifications are established for many families, a standard terminology seems unattainable.

Developmental studies are being used increasingly to clarify enigmatic structure. Wisniewski and Bogle (1982) have described the development of the inflorescence and flower of Liquidambar and found a monotelic inflorescence with basitonic modifications. Extensive work by Endress (1978) has elucidated other patterns in Hamamelidales. Leins and Gemmeke (1979) have shown that the head of Echinops exaltatus (Compositae) is simple and have described various aspects of the flowers.

In the monocotyledons, Charlton (1973) has found a common plan based on a main axis with bracts (and branches if present) in false whorls of three. Studies of development have been particularly useful in understanding the reduced inflorescences and flowers of aquatic monocotyledons. Posluszny and Tomlinson (1977) have found fairly regular sympodia in Zannichelliaceae. Problems remain as categories of inflorescence and flower merge in this group (Tomlinson 1982). The increased feasibility of developmental studies should lead to the needed evolutionary surveys in many families.

B. The Androecium

1. In Palms. The basic androecium of palms is clearly six stamens arranged in a whorl of three opposite the sepals and a second trimerous whorl opposite the petals. Reduction to the three antesepalous (AS) stamens occurs in seven genera representing five major groups while more than six stamens are found in androecia of 70 genera distributed in ten of the 15 major groups (Moore and Uhl 1982). Previous developmental studies (Uhl and Moore 1980) have shown different patterns of stamen initiation in seven of the major groups exhibiting

polyandry. This paper describes another pattern for *Eugeissona*, representing an eighth major group. Multistaminate members of coryphoid and cocosoid palms have yet to be studied.

The largest stamen numbers in the palms (900-1000) occur in *Palandra* and develop centrifugally (Uhl and Moore 1977). In all other genera studied, stamens arise in antesepalous (AS) and antepetalous (AP) arrangements. During the initiation of sepals and petals floral apices of palms are dome-shaped. Prior to stamen initiation in polyandrous genera, the floral apex expands in different and remarkable ways and stamen primordia arise in definite order and characteristic arrangements as shown in Figure 44.

The following configurations have been observed. The diagram of *Lodoicea*, a borassoid palm, shows three flowers from the cincinnus which has ca. 70 flowers and is enclosed in a pit formed by tightly appressed and fused bracts. The basic pattern here consists of alternate whorls of three stamens, one opposite each sepal, and whorls of 1-4 stamens opposite each petal. Lower AP whorls where the apex is wider have more stamens than the higher AP whorls. The arrangement of the stamens reflects the slight asymmetry of the flowers which in turn relates to pressures imposed by the inflorescence axis and the bracts that form the pit (see Uhl and Moore 1980). Note that only two stamens (#5) occur in the outer narrower AP sector of the flower.

In two other major groups developmental patterns are similar, but flowers are symmetrical and stamen numbers usually larger than in *Lodoicea*. Stamen numbers up to 200 develop according to the pattern shown for *Caryota mitis* in other species of caryotoid palms. The floral apex expands opposite each petal and in height, only three stamens occurring in each successively higher AS whorl. This arrangement

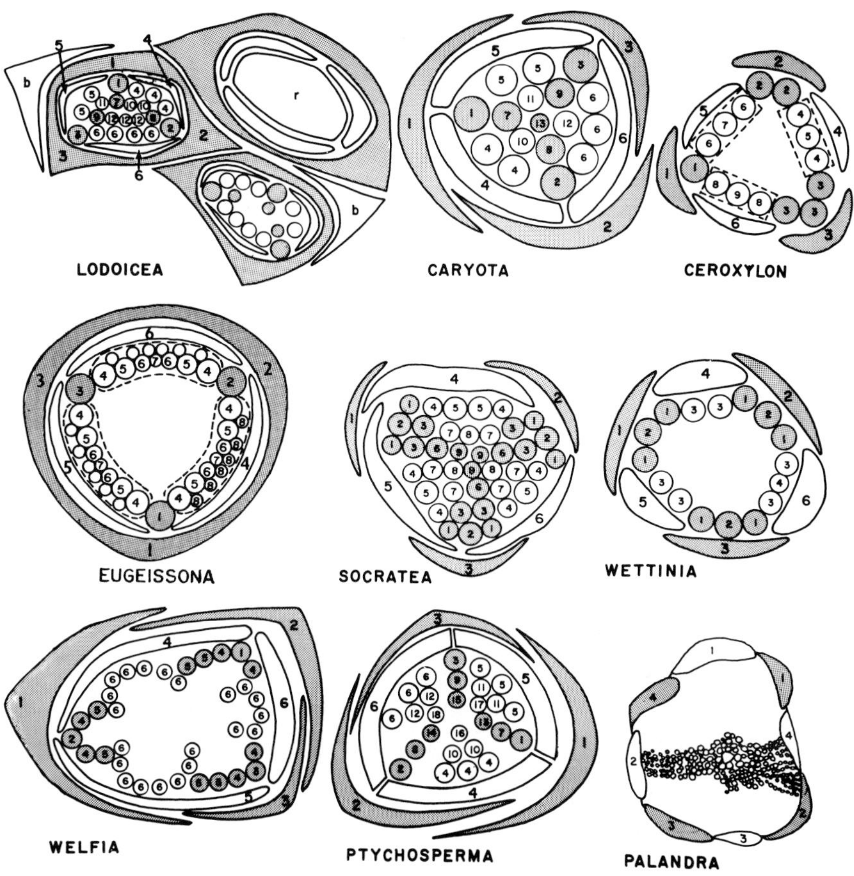

FIGURE 44. Diagrammatic drawings showing the order of inception of perianth members and stamens in palm genera. Shading: sepals and antesepalous stamens shaded, petals and antepetalous stamens clear. Dotted lines in Ceroxylon and Eugeissona indicate common primordia. Major groups as follows: Lodoicea, borassoid; Caryota, caryotoid; Ceroxylon, ceroxyloid; Eugeissona, lepidocaryoid; Socratea, Wettinia, iriarteoid; Welfia, geonomoid; Ptychosperma, arecoid; Palandra, phytelephantoid.

of stamens is similar to that of Ptychosperma of the arecoid major group where equally high stamen numbers occur.

In Ceroxylon, the only multistaminate genus in ceroxyloid palms, stamen numbers increase first on common primordia opposite the petals. Species with nine to twelve stamens have two or three stamens opposite each petal and one opposite each sepal. If numbers are higher, several stamens arise in each AS position.

In iriarteoid palms (see Fig. 44, Socratea and Wettinia), the floral apex expands opposite the sepals and in Socratea distally also. Here the order of initiation is puzzling; two lateral primordia appear first in each AS sector followed by a central primordium. Study of Eugeissona has provided some clues to this sequence as discussed below.

In Welfia distal expansion of the apex appears blocked as the flowers are enclosed in a shallow pit. The apex becomes flat and wider opposite the petals and spreads out in lobes opposite the sepals, with stamens arising in AS and AP arcs.

The present work on Eugeissona has revealed another pattern. As noted above, the floral apex becomes flatter and wider opposite each petal before the first stamens arise. The first three stamens develop opposite the sepals and subsequent stamens in a row on a common primordium opposite each petal. In each AP row the first stamens arise next to the AS primordia. In species with higher numbers, more stamens arise in the center of each AP row. In E. utilis, with ca. 70 stamens, the highest number in the genus, extra stamens arise in each AP row and finally the later primordia are intercalated centrifugally, outside of and alternating with the initial row. The androecium of Eugeissona provides some insights on the significance of the order of stamen initiation. As stamen number increases, new stamens are intercalated toward the center of the AP ridges by increase in meristematic tissue in the center of each ridge. It appears that stamens develop in succession toward where the apex is expanding or toward where some residual meristem is present. The sequence in Socratea and Wettinia parallels that in Eugeissona except that the meristem is expanding in the center of an AS rather than an AP sector. The first stamens arise on the edges of the AS sector and subsequent ones between the first pair and then in alternate whorls higher on the apex. Peripheral expansion of the apex is evident in Palandra where rows of smaller primordia develop toward the meristematic edge of the apex (Figs. 30,44). The centrifugal inception of the last stamen primordia in E. utilis is of interest. It would seem that no further expansion in width of the AP sectors is possible but that residual meristem is present on the outer side of the common AP primordia. The situation seems parallel to that in Datisca cannabina where on the whole floral members arise in centripetal sequence but smaller primordia are inserted between and outside the larger stamens (Leins and Bonnery-Brachtendorf 1977). This sort of intermediate centrifugal development has also been reported in Melaleuca nesophila, Aegle marmelos and some Malvales as noted by Pauzé and Sattler (1978). These authors suggest that centripetal and centrifugal developmental patterns are not mutually exclusive.

Some insight as to the significance of common primordia is also provided by Eugeissona. In both Ceroxylon and Eugeissona, distinct common primordia develop opposite the petals. Stamens of both genera are epipetalous and it seems possible that the common primordia are associated with epipetally. In zonal growth during later ontogeny the ridge and its stamens become joined and elevated with the adaxial parts of the petals.

In summary for palms:

a. After sepal and petal initiation in polyandric taxa the apical meristem of the flower expands in different and remarkable ways in different major groups of palms.

b. Except in phytelephantoid palms where stamen development is centrifugal, a basic trimery is evident, stamens arising in AS and AP sectors.

c. Centrifugal development represents a different form of apical expansion which may be intermediate as in Eugeissona.

d. Early stamen arrangement relates to the expansion of the apex; stamens develop in the direction that the apex or residual meristematic areas in the flower are expanding.

e. The different patterns of apical expansion and stamen arrangement seem to indicate that polyandry has developed several times in the family, presumbably in each of the major groups where it occurs. There are similarities in stamen development in polyandrous caryotoid and arecoid palms, perhaps due to similar growth patterns in staminate flowers of the two groups.

f. The different forms of the apex and consequent stamen arrangements appear related to pressures imposed on the expanding apex by perianth parts and bracts (Uhl and Moore 1980). This leads to the conclusion that bracts and perianth parts were specialized before polyandry developed in each major group.

2. In Other Monocotyledons and in Dicotyledons. Extensive developmental studies have been done for polymerous flowers of Alismatales by Sattler and colleagues and by Kaul in North America and by Leins and co-workers in Europe. A review of this work by Sattler and Singh (1978) showed that the primary pattern is trimerous in that pleiomerous androe-

cia and gynoecia develop when secondary primordia form on common primordia of a trimerous arrangement. Stamen (or staminodial) pairs develop from the expansion of AP regions which in extreme cases represent common petal-stamen primordia. However, the authors interpret the common primordia as resulting from expansion of the floral apex in alternisepalous positions. Space as well as fields for inception of the pairs of stamens is thus provided. Centrifugal inception occurs in Limnocharis flava. The authors believe polyandry is probably secondary. Their findings parallel those in palms except that no common petal-stamen primordia have been found in palms. Sattler and Singh did not find any developmental similarities between Magnoliidae and Alismatales. More recently, Erbar and Leins (1981) have suggested that certain parts of parastichies in petal-stamen initiation are similar and that therefore Alismatales may be derived from Magnoliidae. Elsewhere in monocotyledons, developmental studies are needed for Cyclanthaceae and Pandanaceae where large numbers of stamens occur.

In the dicotyledons developmental investigations of polyandrous androecia have revealed many patterns which cannot be completely assessed here. Stamens may arise in spiral order or in whorls directly on the apex, on a primary or secondary ring-wall in different configurations, or on five or ten sectors opposite petals, sepals, or both. Inception may be centripetal or centrifugal. The importance of centrifugal development to separate large groups of dicotyledons is being seriously questioned as more and more instances of both developmental modes in single families are reported (Pauzé and Sattler 1978). Leins (1975, 1979) has summarized dicotyledonous patterns and suggested possible phylogenies. Reduction series from many spirally arranged stamens to fewer in whorled arrangements have been suggested. However, polyandry is also interpreted as derived in some groups. Changes in size and shape of the floral apex have not been especially noted, but Ross (1982) has found differences in receptacle shape resulting from cessation of apical growth and the activation of an intercalary meristem in Cactaceae.

C. The Gynoecium

1. In Palms. The basic gynoecium in palms consists of three carpels. The carpels are free in thirteen genera of the coryphoid major group and in Phoenix and Nypa, both monotypic in their major groups. Evolutionary trends in the gynoecium (Uhl and Moore 1971) include both reduction and increase in carpel number, various degrees and types of fusion of carpels, and the development of septal nectaries. The carpel varies in form. It may be stipitate and its ovarian region, style, and stigma differ widely in shape. Increase in width and in histological specialization of carpel walls especially in stylar regions occurs in different major groups. Other trends are changes in position and size of the locule and in the position, size, and form of the single ovule that occurs in each carpel.

Primitive carpels in the family are of wide interest since palms are one of the first families to be definitely recognizable in the fossil record (Moore and Uhl 1982). Two unspecialized forms are recognized: the follicular carpel as illustrated for Trachycarpus fortunei (Fig. 45), and the tubular or ascidiform carpel of Nypa fruticans (Uhl 1972). The ventral suture is open for most of the length of the carpel of Trachycarpus, but there is a short tubular basal part. The vasculature consists of a dorsal bundle, a pair of lateral, and a pair of ventral bundles. The carpel in Nypa has a longer basal tubular part and a flaring conduplicate stigmatic opening. The vasculature is different from that of other palm carpels in that there is no recognizable dorsal bundle and bundles branch dichotomously.

Floral axes or receptacles are remarkably plastic in palm flowers. Elongation of the receptacle may occur below all floral organs or between any two whorls of organs (Moore and Uhl 1982). Such elongation serves to exsert anthers and stigmas or to elevate staminate flowers. The position of the carpels varies with respect to the end of the floral axis. Carpels may be terminal as in Trachycarpus and Thrinax or more lateral in position with the tip of the floral axis evident as a central projection (Phoenix and others).

The gynoecia of Eugeissona and Palandra are of special interest for several reasons. Both are representative of their respective major group. The gynoecia of the 21 genera of lepidocaryoid palms are composed of three carpels united laterally with locules described as incomplete (ventral carpellary sutures are open) and bearing three basal anatropous ovules with their micropyles facing the center of the gynoecium. Distinctive reflexed scales cover gynoecial surfaces. The only differences known among the genera are relatively minor variations in the form of the scales and in the shape of the styles and stigmas.

The pistillate flowers of the three genera of phytelephantoid palms are very much alike. Gynoecia of 6-10 carpels have rounded ovarian parts, elongate terete styles, and as many conduplicate stigmas as locules. The ovule is inserted adaxially near the base of the locule in each carpel. This study has shown that carpel number in Palandra is six in one whorl; no evidence of two whorls was seen.

In Eugeissona and Palandra the carpels are initiated and develop on the flanks of the floral axis (Figs. 10-14,36,

37). The carpels become crescentic, expand laterally, and become connate by their outer lateral surfaces. The ventral sutures of the carpels do not close and their margins abut on the floral apex in the base of the gynoecium. The floral axes in both genera enlarge with the carpels and form part of the base of the gynoecium. The shape of the carpel differs somewhat in the two genera. Carpels in *Eugeissona* are arcuate (Figs. 14,17), and those of *Palandra* more strictly conduplicate from early in ontogeny (Fig. 38). In *Eugeissona* the fusion of the lateral walls of the carpels extends to their margins but in *Palandra* only the outer lateral walls of the carpels are continuous. The inner walls of adjacent carpels are free within the ovary and throughout the style.

Ovules are borne in various positions on the adaxial surfaces (in the locules) of palm carpels. The ovule may be attached at the base, laterally, or near the top of the locule. The gynoecia of *Eugeissona* and *Palandra* are unusual in that the ovule primordia develop directly on the floral axis in an "axillary" position to the carpel. The two genera differ somewhat in the structure of the ovarian region. In *Eugeissona* the floral axis remains as a projection in the center of the ovary. The erect ovules are attached to the floral axis by curved funiculi (Fig. 17). In *Palandra* the apex is wider and flatter and the ovules are borne on six points of the apex which extend "into" the carpels. Histologically the apex can be recognized by larger cells and the presence of tannins as a central six-lobed area in the base of the gynoecium.

Cauline ovules borne on floral axes characterize a number of taxa rather widely scattered throughout the angiosperms (Sattler and Perlin 1982). This structural pattern has influenced theories as to the nature of the flower (see Macdonald and Sattler 1973 for review). In many cases the floral apex becomes the nucellus of the ovule (Tucker 1980) which is subsequently enclosed by the carpels. The evolutionary significance of the gynoecia of Eugeissona and Palandra within palms is not yet clear. They appear to represent particular growth patterns involving lateral inception of the carpels on the floral axis and some increase in the diameter and length of the floral axis as the carpels enlarge. As noted above, the floral axis forms part of the base of the gynoecium at maturity. The gynoecia of Eugeissona and Palandra are alike in that carpels are fused laterally, ventral sutures are open, and ovules are borne on the floral axes. Some similarities are seen in gynoecia of some genera of arecoid palms (Uhl unpublished). Further studies of anatomy and development of these and similar gynoecia are needed to resolve their relationships.

Very few other developmental studies of palm gynoecia have been done. Carpels in Nannorrhops ritchiana (Uhl 1969) are crescentic at inception, ovules are submarginal, and fusion of the carpels begins in the styles. In Ptychosperma (Uhl 1976) carpels are also crescentic and free in earliest stages and become continuous later. Mature carpels are tubular or ascidiform in both genera. This study has shown that carpels in Palandra are initially circular but rapidly become conduplicate and remain conduplicate in the mature gynoecium. The only SEM study of separate carpels (DeMason et al. 1982) has indicated that the carpels of Phoenix dactylifera are nearly circular in early stages and then become cup-shaped. The ovule is borne adaxially (Fig. 45).

Most palm carpels resemble that of Trachycarpus in having a tubular base and a plicate distal part. Carpels of Phoenix are ascidiform at maturity (Uhl and Moore 1971). More developmental information on Nypa fruticans and other apocarpous palms is essential. Nypa represents the oldest known palm and is also one of the first recognizable angiosperms in the fossil record. Pollen is known from the Senonian (Moore and Uhl 1982). The carpel is unusual in shape, vasculature, and growth patterns (Fig. 45,K,l). We need to ascertain how developmental patterns of the ascidiform carpel of *Nypa* compare with those of *Trachycarpus* and other follicular carpels, and with the strictly conduplicate carpels of *Palandra*.

2. *In Other Monocotyledons and in Dicotyledons*. Questions regarding gynoecial structure and evolution in palms apply in angiosperms as a whole. One question which developmental studies may help to resolve is the basic shape of the angiosperm carpel. It has long been recognized that some carpels are conduplicately folded with open ventral sutures and others are tubular basally or throughout (Eames 1961). How these two forms relate to each other remains controversial. For the past quarter century two theories have been in vogue. The conduplicate hypothesis (Bailey and Swamy 1951) describes the basic carpel as stipitate with a thin conduplicately folded lamina and three traces, a dorsal and two ventral bundles (Fig. 45A,B). Ovules are borne in two rows. According to this theory a carpel with a tubular base represents a derived form.

The peltate theory strongly endorsed by Leinfellner in the late 1960's (Leinfellner 1966) and widely used by botanists especially in Europe, compares the carpel with a peltate leaf, stating that the unspecialized carpel has a tubular base and a ventral cross zone ("Querzone"), a meristematic lip which forms the adaxial part of the carpel and which may bear the ovules (Fig. 45D,E). The question of whether carpel ontogeny parallels that of a peltate leaf is not completely resolved (Franck 1976). Many carpels are

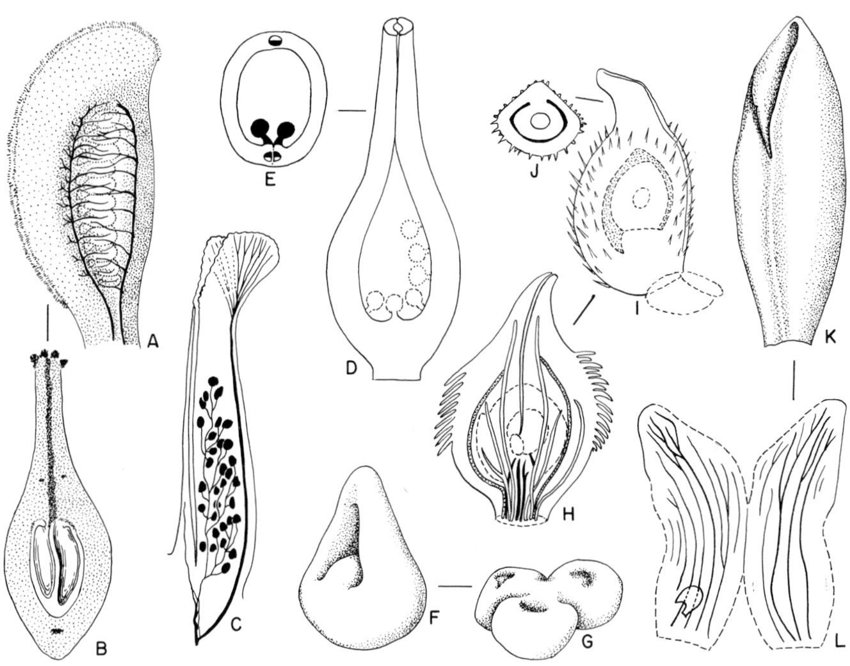

FIGURE 45. Carpel structure. A, B. Conduplicate carpel, Drimys piperita Hook f. (Bailey and Swamy 1951). C. Conduplicate carpel of Hydrocleis nymphoides (Kaul 1976). D, E. Hypothetical representation of a peltate carpel. F. Young stages of carpels of Phoenix dactylifera (De Mason et al. 1982). H, I. Ascidiform carpel, Trachycarpus fortunei (Uhl and Moore 1971). K, L. Ascidiform or tubular carpel, Nypa fruticans (Uhl 1972). (A-C, F-L redrawn with permission.)

indeed tubular or ascidiate at the base as illustrated for Trachycarpus and Nypa (Fig. 45). However, a cross zone is not always present.

Carpels are described as conduplicate, completely ascidiform, peltate, and variously intermediate. In Monimiaceae and Laurales, the ovary is wholly ascidiate (Endress 1980). Eupomatiaceae, Himantandraceae, Magnoliaceae, and Annonaceae have more or less epeltate or weakly peltate carpels (Endress 1977). The carpel of Illicium is conduplicate (Robertson and Tucker 1979). By many workers "peltate" is restricted to forms where the ovule or ovules are adaxial and a cross zone can be recognized as in Figure 45F,G.

It should be noted that a somewhat different conduplicate carpel is considered a primitive form in some monocotyledons (Kaul 1976) as illustrated for Hydrocleis nymphoides (Fig. 45C). Ventral bundles are lacking and ovules are scattered over the ventral surface. The significance of differences in conduplicate carpels also needs to be explored.

Intermediate forms with tubular bases and conduplicate distal parts, as found in many palms (Uhl and Moore 1971), exist in all possible degrees between conduplicate and ascidiform carpels. An SEM investigation of single carpels (Van Heel 1981) has shown that some carpels develop an oblique cup-shaped rim early in ontogeny while others remain open adaxially and conduplicate. Study of anatomy to locate cellular configurations and cell divisions combined with SEM observations of form (see Tucker 1980) is needed to indicate structural and growth patterns and lead to a better assessment of carpel form and evolution.

An understanding of many other parameters of gynoecial structure will certainly result from such study. The gynoecium is another area where the magnifications and depth of field provided by the SEM seem certain to reveal structures we have not previously been able to discern.

ACKNOWLEDGMENTS

The drawings, Figures 5, 20A-C, 31, 42-45, were done by Bente Everhart. Albert Bednarick assisted in the dissection and scanning and prepared all photographic plates. We are grateful to Anthony B. Anderson for collections of _Palandra_. This work was supported in part by National Science Foundation Grant DEB-8109374.

REFERENCES

Bailey, I. W. and Swamy, B. G. L. (1951). The conduplicate carpel of dicotyledons and its initial trends of specialization. _Amer_. _J_. _Bot_. 38, 373-379.

Briggs, B. G. and Johnson, L. A. S. (1979). Evolution in the Myrtaceae - evidence from inflorescence structure. _Proc_. _Linn_. _Soc_. _New_ _S_. _Wales_ 102, 157-256.

Charlton, W. A. (1973). Studies in the Alismataceae. II. Inflorescences of Alismataceae. _Can_. _J_. _Bot_. 51, 775-789.

De Mason, D. A., Stolte, K. W. and Tisserat, B. (1982). Floral development in _Phoenix_ _dactylifera_. _Can_. _J_. _Bot_. 60, 1437-1446.

Dransfield, J. (1970). Studies in the Malayan palms _Eugeissona_ and _Johannesteijsmannia_. Ph.D. Thesis. Cambridge Univ., Cambridge.

Eames, A. J. (1961). Morphology of the angiosperms. McGraw-Hill, New York.

Endress, P. K. (1977). Über Blütenbau und Verwandtschaft der Eupomatia und Himantandraceae (Magnoliales). _Ber_. _Deutsch_. _Bot_. _Ges_. _Bd_. 90, 83-103.

__________. (1978). Blütenontogenese, Blütenabgrenzung und systematische Stellung der perianthlosen Hamamelidoideae. Bot. Jahrb. Syst. 100, 249-317.

__________. (1980). Floral structure and relationships of Hortonia. Pl. Syst. Evol. 133, 199-221.

Erbar, C. Von und Leins, P. (1981). Zur spirale in Magnolien-Blüten. Beitr. Biol. Pflanzen 56, 225-241.

Fisher, J. B. and Dransfield, J. (1977). Comparative morphology and development of inflorescence adnation in rattan palms. Bot. J. Linn. Soc. 75, 119-140.

__________ and Moore, H. E., Jr. (1977). Multiple inflorescences in palms (Arecaceae): their development and significance. Bot. Jahrb. Syst. 98, 573-611.

Franck, D. H. (1976). The morphological interpretation of epiascidiate leaves - an historical perspective. Bot. Rev. 42, 345-388.

Heel, W. A. Van. (1981). A S.E.M.-investigation on the development of free carpels. Blumea 27, 499-522.

Kaul, R. B. 1976. Conduplicate and specialized carpels in the Alismatales. Amer. J. Bot. 63, 175-182.

Kuijt, J. (1981). Inflorescence morphology of Loranthaceae - an evolutionary synthesis. Blumea 27, 1-73.

Leinfellner, W. (1966). Über die Karpelle verschiedener Magnoliales. I. Oesterr. Bot. Z. 113, 383-389.

Leins, P. (1975). Die Beziehungen zwischen multistaminaten und einfachen Androeceen. Bot. Jahrb. Syst. 96, 231-237.

__________. (1979). Der Übergang vom zenetrifugalen Komplexen zum einfachen Androeceum. Ber. Deutsch. Bot. Ges. Bd. 92, 717-719.

Leins, P. und Bonnery-Brachtendorf, R. (1977). Entwicklungsgeschichtliche Untersuchungen an Blüten von Datisca cannabina (Datiscaceae). Beitr. Biol. Pflanzen 53, 143-155.

__________ und Gemmeke, V. (1979). Infloreszenz- und Blütenentwicklung bei der Kugeldistel Echinops exaltatus (Asteraceae). Pl. Syst. Evol. 132, 189-204.

Macdonald, A. D. and Sattler, R. (1973). Floral development of Myrica gale and the controversy over floral concepts. Can. J. Bot. 51, 1965-1975.

Moore, H. E., Jr. (1973). The major groups of palms and their distribution. Gentes Herb. 11, 27-141.

Moore, H. E., Jr. and Uhl, N. W. (1982). Major trends of evolution in palms. Bot. Rev. 48, 1-69.

Pauzé, F. et Sattler, R. (1978). L'androcée centripète d'Ochna atropurpurea. Can. J. Bot. 56, 2500-2511.

Posluszny, U. and Tomlinson, P. B. (1977). Morphology and development of floral shoots and organs in certain Zannichelliaceae. Bot. J. Linn. Soc. 75, 21-46.

__________, Scott, M. G., and Sattler, R. (1980). Revisions in the technique of epi-illumination light microscopy for the study of floral and vegetative apices. Can. J. Bot. 58, 2491-2495.

Rickett, H. W. (1944). The classification of inflorescences. Bot. Rev. 10, 187-231.

__________. (1955). Materials for a dictionary of botanical terms. III. Inflorescences. Bull. Torrey Bot. Club 82, 419-445.

Robertson, R. E. and Tucker, S. C. (1979). Floral ontogeny of Illicium floridanum with emphasis on stamen and carpel development. Amer. J. Bot. 66, 605-617.

Ross, R. (1982). Initiation of stamens, carpels, and receptacle in the Cactaceae. Amer. J. Bot. 69, 369-379.

Sattler, R. (1968). A technique for the study of floral development. Can. J. Bot. 46, 720-722.

__________. (1973). Organogenesis of flowers. A photographic text-atlas. University of Toronto Press, Toronto.

Sattler, R. and Perlin, L. (1982). Floral development of Bougainvillea spectabilis Willd., Boerhaavia diffusa L. and Mirabilis jalapa L. (Nyctaginaceae). Bot. J. Linn. Soc. 84, 161-182.

__________ and Singh, V. (1978). Floral organogenesis of Echinodorus amazonicus Rataj and floral construction of the Alismatales. Bot. J. Linn. Soc. 77, 141-156.

Sundberg, M. D. (1982). Petal-stamen initiation in the genus Cyclamen (Primulaceae). Amer. J. Bot. 69, 1707-1709.

Tomlinson, P. B. (1975). Flowering in Metroxylon (the sago palm). Principes 15, 49-62.

__________. (1982). Anatomy of the monocotyledons. VII. Helobiae (Alismatidae). (Ed. C. R. Metcalfe). Clarendon Press, Oxford.

Troll, W. (1964). Die infloreszenzen: Typologie und Stellung im Aufbau des Vegetationskörpers; Erster Band. Fischer, Jena.

__________. (1969). Die Infloreszenzen: Typologie und Stellung im Aufbau des Vegetationskörpers: Zweiter Band, erster Teil. Fischer, Stuttgart.

Tucker, S. C. 1980. Inflorescence and flower development in the Piperaceae. I. Peperomia. Amer. J. Bot. 67, 686-702.

Uhl, N. W. (1969). Anatomy and ontogeny of the cincinni and flowers in Nannorrhops ritchiana (Palmae). J. Arnold Arb. 50, 411-431.

__________. (1972). Inflorescence and flower structure in Nypa fruticans (Palmae). Amer. J. Bot. 59, 729-743.

__________. (1976). Developmental studies in Ptychosperma (Palmae). I. The inflorescence and the flower cluster. II. The staminate and pistillate flowers. Amer. J. Bot. 63, 82-109.

Uhl, N. W. and Moore, H. E., Jr. (1971). The palm gynoecium. Amer. J. Bot. 58, 945-992.

__________ and __________. (1977). Centrifugal stamen initiation in phytelephantoid palms. Amer. J. Bot. 64, 1152-1161.

__________ and __________. (1978). The structure of the acervulus, the flower cluster of chamaedoreoid palms. Amer. J. Bot. 65, 197-204.

__________ and __________. (1980). Androecial development in six polyandrous genera representing five major groups of palms. Ann. Bot. 45, 57-75.

Wisniewski, M. and Bogle, A. L. (1982). The ontogeny of the inflorescence and flower of Liquidambar styraciflua L. (Hamamelidaceae). Amer. J. Bot. 69, 1612-1624.

Wyatt, R. (1982). Inflorescence architecture: how flower number, arrangement, and phenology affect pollination and fruit-set. Amer. J. Bot. 69, 585-594.

CLEISTOGAMY: A COMPARATIVE STUDY OF INTRASPECIFIC FLORAL VARIATION

Elizabeth M. Lord

Department of Botany and Plant Sciences
University of California
Riverside, California

This paper describes the results of developmental studies on two annual and one perennial cleistogamous species in which dimorphic, hermaphroditic flowers are produced on an individual. The cleistogamous (CL) flowers remain closed and self in the bud, while the chasmogamous (CH) flowers open and may outcross.

Though variations exist in the manner of floral expression in these dimorphic species, there are structural and developmental features with respect to the CL flower which are common to them all. Divergence in development of the CL flower occurs early in ontogeny; the more divergent the mature form from the ancestral CH flower, the earlier the onset of divergence occurs. The CL flower always shows precocious differentiation of its organs with a consequent reduction in some or all of the floral parts. By combining information about shape changes in ontogeny with actual rates of growth, a more adequate representation of the ontogeny and phylogeny of the CL flower has been obtained than was proposed in the past. Environmental controls on induction of the two floral forms may be mediated by hormones such as gibberellins and abscisic acid. Initial studies on pollination indicate that significant differences exist between the two flowers with respect to, at least, form and behavior of the pollen. A synthetic approach to the study of floral development is proposed combining data on structure and function to explain form in an evolutionary context.

Contemporary Problems in
Plant Anatomy

ISBN 0-12-746620-7

INTRODUCTION

Darwin (1897), in his book The Different Forms of Flowers on Plants of the Same Species, devoted the first six chapters to the phenomenon of heterostyly, the seventh to that of dioecy and gynodioecy, and the eighth to cleistogamy.

Study of both the population biology and pollen-stigma interaction in the self-incompatible heterostylous species, has since provided us with much information about the intricacies of floral biology (Vuilleumier, 1967; Ganders, 1979; Shivanna, 1982; Richards and Barrett, 1983). Species with unisexual flowers have also been of particular interest to population and evolutionary biologists (Bawa, 1980; Lloyd, 1965; Thomson and Barrett, 1981), plant physiologists studying hormonal controls on sexuality (Nitsch et al., 1952; Heslop-Harrison, 1959) and morphologists (Uhl, 1966; Sattler, 1973; Lazarte and Palser, 1979; DeMason et al., 1982). With the exception of early descriptive (Goebel, 1904, 1905; Ritzerow, 1908) and ecological studies (Bergdolt, 1932; Uphof, 1938), until recently the third phenomenon of cleistogamy has been neglected. In cleistogamous species, two types of hermaphroditic self-compatible flowers are borne by an individual. The closed or cleistogamous (CL) flowers set seed by self-fertilization, while the open or chasmogamous (CH) flowers achieve anthesis and may outcross. The advent of population biology has brought a renewed interest in how breeding systems affect population structure. At the same time, structural botanists are attempting a more functional approach to the study of floral biology. In this context, cleistogamy is being reexamined.

Reasons that this floral polymorphism has been less popular than the others may be: (1) that the variation is exhibited by an individual rather than by separate plants in the population, and (2) that this phenomenon shows much phenotypic plasticity. In a recent theoretical treatment of the biology of sex, Charnov (1982) restricts his examples of special sexual types in the higher plants to heterostyly and unisexual flowers. However, about 70% of the angiosperms are hermaphroditic (Yampolsky and Yampolsky, 1922) and most are self-compatible (Fryxell, 1957); thus, investigation of a system like cleistogamy, which occurs in at least 56 families, should provide information more generally applicable for the flowering plants as a whole (Lord, 1981).

This paper summarizes the data gathered on two annuals, *Lamium amplexicaule* (Labiatae) and *Collomia grandiflora* (Polemoniaceae), as well as a perennial, *Viola odorata* (Violaceae), in an attempt to provide an explanation of form and function for cleistogamous species in an evolutionary context.

In a system as labile as cleistogamy, where a strong environmental component to variation exists, knowledge of those factors responsible for floral expression is necessary. Once we have a defined population of primordia with predictable developmental fates, comparative ontogenetic study may reveal the processes by which form divergences take place. Such structural data is the basis for experimental manipulations to determine internal controls on floral organ growth. Finally, mature floral function can be studied by following the processes of pollination and fertilization in these dimorphic flowers with contrasting breeding systems. The CL flower is derived from the CH evolutionarily by reduction and modifications of the floral parts which ensure self-pollination. Knowledge of events that lead to morphological change during ontogeny allows us to devise more adequate models to explain how the CL form arose ontogenetically and perhaps phylogenetically. From this data on cleistogamous species, broader generalizations can be made since similar modifications in form characterize both CL flowers and those of inbreeding species throughout the angiosperms (Arroyo, 1973; Frankel and Galun, 1977; Lloyd, 1965; Moore and Lewis, 1965; Ornduff, 1969; Rollins, 1963; Stebbins, 1957).

A. Floral Variation

The proportion of CL and CH flowers produced on a plant is controlled by both genetic and environmental factors, but when the organism is grown under natural conditions, both floral types occur at some point in the life cycle (Lord, 1981). There are many such cases where reproductive systems demonstrate a versatility controlled by environmental factors (Heslop-Harrison, 1982; Nitsch, 1965).

In the perennial *Viola* species, seasonal changes in photoperiod and temperature determine floral type, with only one floral form being produced at a time (Chouard, 1948; Evans, 1956; Mayers, 1983). The annuals, *Lamium* and *Collomia*, are less plastic in their response to environmental triggers. They exhibit an ontogenetic shift in floral form typically from CL to CH and back to CL before senescence. This occurs as an acropetal sequence in the raceme of *Lamium* and a basipetal sequence in the cymose inflorescence of *Collomia* (Fig. 1a-d). This is a type of heteroblasty, which can be arrested at the CL stage under poor growth conditions (nutrient or water stress, and poor light intensity) (Minter and Lord, 1983). In *Lamium*, the ratio of CH/CL flowers is

Fig. 1. (a) Mature CL and CH corollas of Lamium amplexicaule. X 3.0. (b) Flowering shoots of Lamium grown under long days (L) and short days (R). X 1.1. (c) Flowering shoot of Collomia grandiflora. X 0.75. (d) Mature CL and CH corollas of C. grandiflora. X 1.5.

increased under long day lengths (Table I, Fig. 1b), but in neither of the two annuals do completely CH-flowering plants occur in the population (Correns, 1930; Wilken, 1982).

These inflorescence patterns, reported for other cleistogamous species as well (Uphof, 1938), demonstrate the existence of ontogenetic constraints on floral plasticity. With the shift in floral type, in both the annuals and perennials, intermediate forms occur. In *Lamium*, they exist in small numbers and are found in the CL population (Lord, 1980b, c).

B. Developmental Studies

Flower positional information as well as the capacity to predictably induce the two floral forms with environmental cues, facilitates comparative ontogenetic studies.

It is agreed that CL flowers are derived from the CH form and are characterized at least by a reduction in the androecium and lack of corolla expansion at anthesis (Darwin, 1897; Goebel, 1904; Lord, 1981; Ritzerow, 1908; Uphof, 1938). These are the common denominators for all CL species, but further reductions and form modifications may occur in the other floral organs as well. Among the three organisms we have studied, there is a gradient of organ reduction and shape alteration in the CL flowers. *Lamium* shows the least modification with reduction in cell number and size in the anthers and corolla of the CL flower. The calyx is unmodified, as is the gynoecium except for an aborted nectary. The anthers mature precociously, as do the stigma and style (Lord, 1979, 1980c, 1982b). In *Collomia*, similar reductions occur, but here, in addition, the anthers are modified in their shape as well as size with two front locules aborted

TABLE I. Effect of Photoperiod on Open (CH) Flower Production in Lamium amplexicaule L.

Photoperiod	Temperature (°C)	Light intensity ($\mu E\ m^{-2}\ s^{-1}$)	CH flowers (%)	Sample number
Expt 1				
SD 10 h D	21	140	1.0 ± 0.83	40
14 h N	10			
LD 16 h D	21	140	15.2 ± 9.53	30
8 h N	10			
Expt 2				
SD 10 h D	25	581.2	18.75 ± 6.7	30
14 h N	13			
LD 16 h D	25	410.25	48.65 ± 12.9	19
8 h N	13			

(E. M. Lord. 1982a. Ann. Bot. 49:261-263.)

(Minter and Lord, in press). In Viola, the mature CL flower shows modification in both size and shape of all organs. It is the most highly divergent species of the three in terms of structural features (Mayers, 1983).

Comparative developmental study has established that, in the annuals, both CL and CH flower primordia are indistinguishable at inception (Fig. 2a-g) (Lord, 1982b; Minter and Lord, in press). Divergence occurs during anther differentiation in both species. In contrast, Viola with

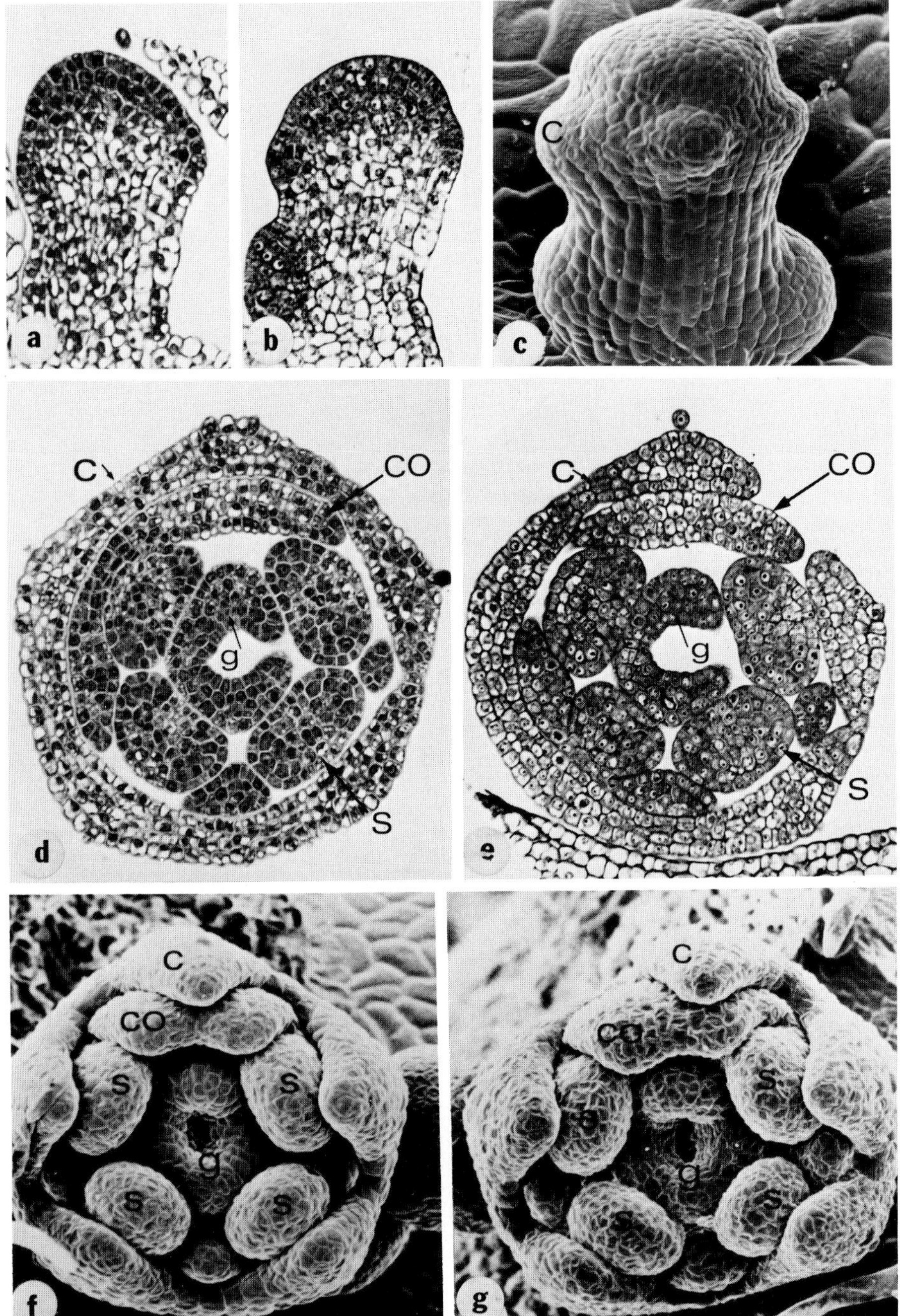
a
b
C
c
C
CO
g
S
d
C
CO
g
S
e
c
CO
s
s
g
s
s
f
c
CO
s
g
s
g

the most highly divergent mature forms, shows primordial distinction from inception; the primordium of the CL flower is smaller than that of the CH, and this difference is maintained throughout development. In fact, the shoot apex producing the CL flower is smaller than the CH apex and though the phyllotactic pattern is the same, the plastochron is shorter, the phyllotactic index reduced, and a faster rate of flower and leaf development is seen at the CL apex. Though all organs in the CL flower of *Viola* are reduced in size, the result is not merely a dwarf form of the CH flower; rather, a unique CL form is produced (Mayers, 1983).

Since we have a comparative system in the annuals whereby initially identical floral primordia become gradually canalized along paths leading to divergent mature forms, we can study the processes, in terms of cell division and elongation, as well as tissue differentiation patterns that are responsible for the ultimate form changes. In particular, since correlative growth is so evident in the tight succession of organs in the flower (Wardlaw, 1965) this system can perhaps give us clues as to the regulatory processes responsible for such a close coordination of events as occurs in floral differentiation.

Fig. 2. *Lamium amplexicaule*. (a) Longitudinal section of CH floral primordium. X 232. (b) Longitudinal section of CL floral primordium. X 232. (c) SEM view of CH primordium with calyx being initiated. X 300. (d) Cross section of CH floral bud with all organs initiated. X 88. (e) Cross section of CL floral bud. X 88. (f) SEM face view of CH floral bud [corresponds to (d)]. X 255. SEM face view of CL floral bud [corresponds to (e)]. X 255.

(E. M. Lord. 1982b. Bot. Gaz. 143:63-77.)

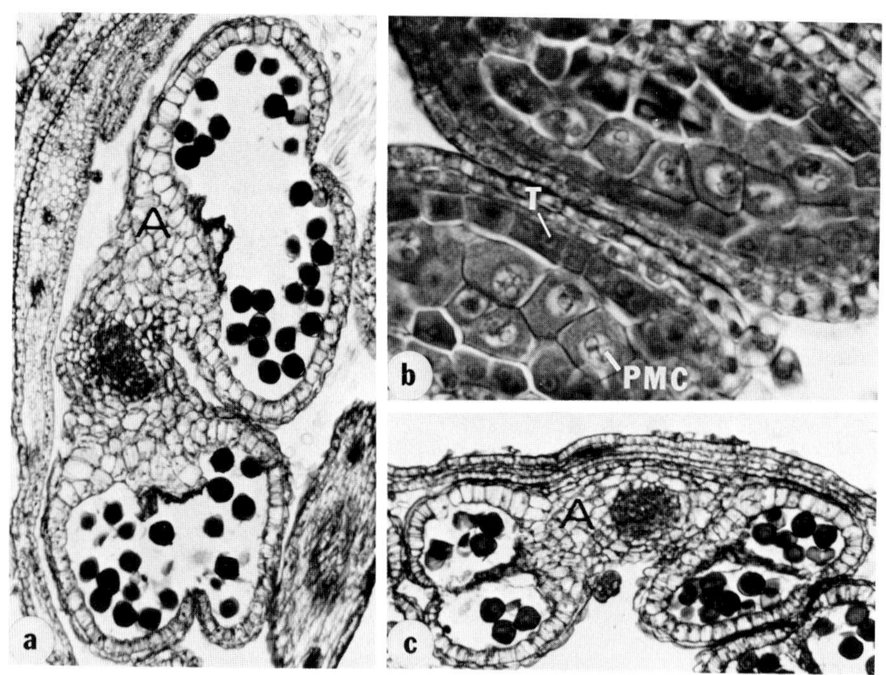

Fig. 3. Lamium amplexicaule. (a) Cross section of mature CH anther. X 170. (b) Cross section through portions of two anther sacs at pollen mother cell stage. X 600. (c) Cross section of mature CL anther. X 170.
(E. M. Lord. 1980c. Amer. J. Bot. 67(10):1430-1441.)

In both annuals, divergence in structure occurs first during anther differentiation. Few sporogenous cells are produced in the CL anther due either to fewer cells being initiated or from a shorter duration of division in this tissue resulting, after meiosis, in fewer pollen grains in a smaller anther. In Lamium, the result is a dwarf form of the CH anther (Fig. 3a-c) (Lord, 1980c); in Collomia, the CL anther is further modified by abortion of two of the four locules leading to a unique CL anther form (Minter and Lord, in press). Calyx and ovary development appear unmodified in these CL flowers.

In both species, this initial anther divergence seems to trigger subsequent events such as corolla divergence (Fig. 4a-g). A dissociation occurs between the corolla apex which surrounds the epipetalous stamens and its base which encloses the gynoecium (Fig. 5). Lack of cell expansion alone explains the form of the CL corolla base at maturity, but a reduction in cell division as well as expansion is responsible for the mature form of the corolla apex (Fig. 6a-h).

A number of studies have reported precocious development of the CL flowers in cleistogamous species (Lord, 1981). In the three species studied, the CL flower is indeed faster in its development from initiation to pollination and hence to seed set. Since calyx growth is the same in both floral types in the annuals, calyx dimensions can be used as a developmental index to age other organs and events in the flower, a technique pointed out by Erickson (1948) in his study of anther differentiation in Lilium. In Viola, the solitary axillary flowers are tightly enclosed in stipules of the subtending leaf and early stages of development cannot be observed directly. In addition, the calyx as well as all other organs are modified in the CL flower so absolute measures of rate must be used. This was accomplished by utilizing the plastochron index in combination with chronological aging (Mayers, 1983).

By incorporating the element of time into our studies, we have shown that changes in absolute and relative growth

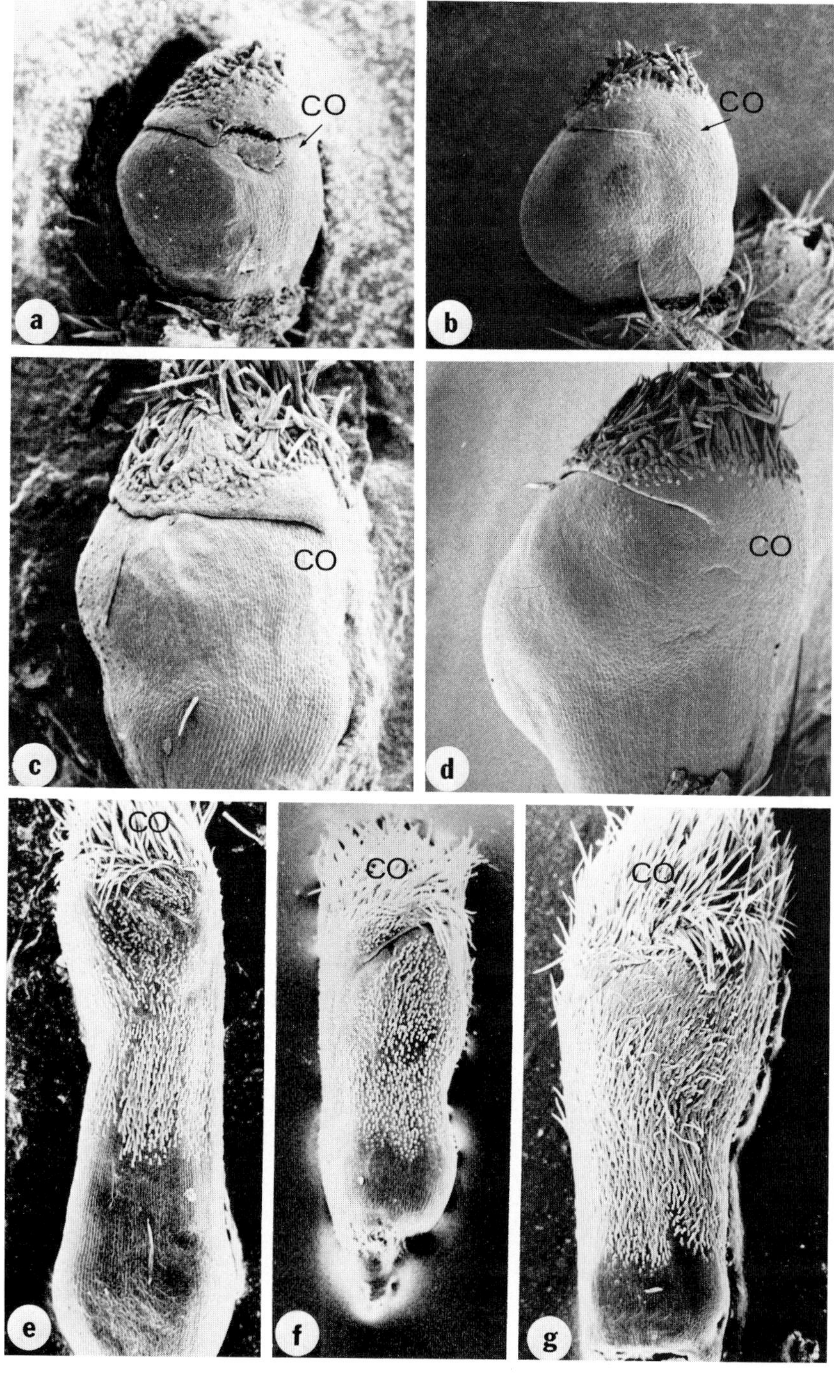
CO
CO
CO
CO
CO
CO
CO
a
b
c
d
e
f
g

rates of CL organs contribute to the modifications seen in mature form and function in these flowers.

In Lamium, the duration of development up to meiosis in the anthers appears to be the same for both flowers, but subsequent differentiation accelerates in the CL flower to result in a reduced mature corolla and anthers with fewer pollen grains. The anther differentiates precociously as does the stigma and stylar component of the gynoecium (Lord, 1979, 1982b).

In Collomia, the fact that meiosis in the CL anthers occurs measurably earlier argues for the hypothesis that a shorter duration of sporogenous division activity results in their precocious differentiation (Minter and Lord, in press). The rate of growth in anther length after meiosis is the same for both floral types in Collomia, but the curve is merely shifted in the CL anther due to the earlier onset of differentiation.

A developmental coordination appears to exist between stamens and corolla in the epipetalous species, Lamium and Collomia. Since all CL species show a reduction in both androecium and corolla, there may be a common denominator in CL floral development with an initial reduction in anther tissue causing a subsequent reduction in corolla growth rates and lack of anthesis (Lord, 1981). In Collomia particularly, differentiation events in the anther seem to

Fig. 4. SEM side views of floral buds of Lamium amplexicaule with calyx removed to show corolla shape changes in development. (a,b), (c,d), (f,g) of comparable ages. (a) CL. X 58. (b) CH. X 58. (c) CL. X 58. (d) CH. X 58. (e) CL. X 58. (f) CL. X 58. (g) CH. X 58. (Modified from E. M. Lord. 1982b. Bot. Gaz. 143:63-72.)

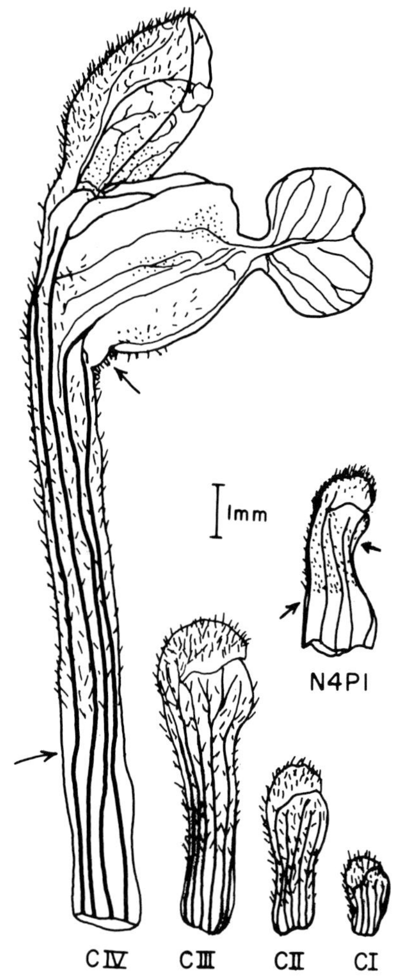

Fig. 5. Mature CL corolla (N4P1) and stages (CI to CIII) in the development of the mature CH corolla (CIV) of *Lamium amplexicaule*. Arrows indicate comparable areas, ventral and dorsal regions, in the two corolla types which were investigated anatomically.

(E. M. Lord. 1980c. Amer. J. Bot. 67:1430-1441.)

trigger other developmental events in the floral organs. Changes in relative growth rates in dimensions of both the CH corolla and calyces of both flowers occur following microspore mitosis in the anthers (Minter and Lord, in press).

When a constant ratio is maintained between growth rates of the different dimensions of an organ, its growth is said to be allometric. Allometric studies serve to pinpoint stages in organ growth when divergence in shape is occurring (Fig. 7a, b). In *Lamium*, dissociation between parts of a single organ, the corolla, occurs resulting in a new shape being achieved, one not found in the ontogeny of the CH corolla. The androecium is attached to the upper corolla region and the modification of its form may be simply a correlative effect of anther reduction (Lord, 1980c). In *Collomia*, the earlier onset of meiosis in the CL anther explains its reduced mature size, but it is not merely a dwarf form of the CH anther as appears to be the case in *Lamium*. The absence of ventral locule differentiation in the CL anthers of *Collomia* results in a qualitative divergence in their form (Minter and Lord, in press). Further study of the patterns of cell division, expansion, and tissue differentiation throughout these dimorphic anthers is necessary to determine the processes which lead to their form divergence.

Comparative systems like cleistogamy, where divergent forms occur within a single species, provide "experiments" in nature which can be used to identify some of the processes responsible for form variation. As yet, we have an incomplete picture of how floral organs develop, no doubt, because of their three-dimensional complexity as compared to the vegetative organs. Evidence suggests that these determinate organs are homologous to leaves and show similar patterns

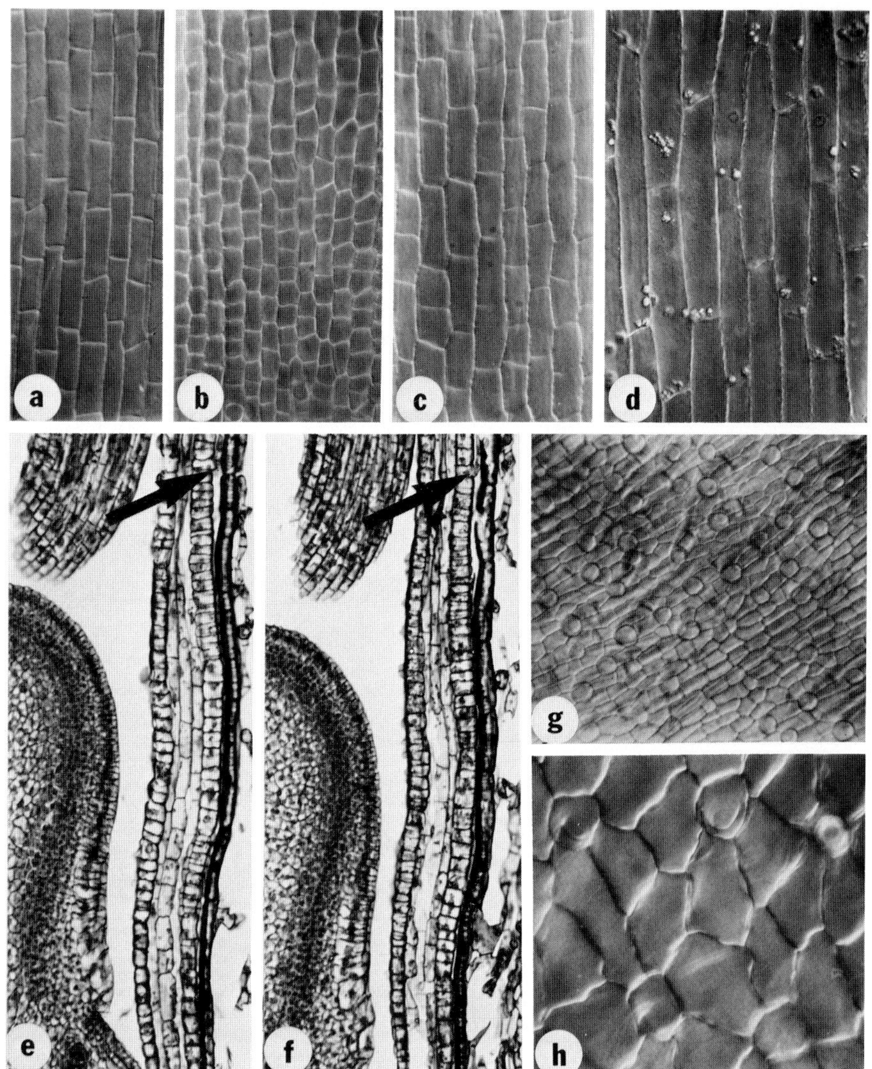
a
b
c
d
e
f
g
h

of growth (Boke, 1949; Tepfer, 1953; Wardlaw, 1965), but little work has been done on the histogenesis of floral organs after their initiation. Of the two most modified organs in cleistogamous species, the corolla and stamens, the corolla has been more carefully studied (Boke, 1948, 1949; Daniel and Sattler, 1978; Kaplan, 1968; Tepfer, 1953). We need detailed quantitative studies of shape, size, and age changes in floral organ primordia during ontogeny to reveal the processes by which a collective unit as complex as the flower is achieved.

Such work has been done on the angiosperm leaf, using the element of growth rate, detecting both surface as well as internal differentiation patterns and visualizing growth in three dimensions (Avery, 1933; Erickson, 1976; Kaplan, 1970b, 1973; Larson and Isebrands, 1971; Maksymowych, 1973; Williams, 1975).

Fig. 6. Lamium amplexicaule. (a) Epidermal cells of mature CL corolla-dorsal region. X 150. (b-d) Epidermal cells from progressively older corollas of CH flower-dorsal region. X 150. (e) Dorsal longisections through corolla of CL flower showing cell numbers from hair line to base of corolla. X 150. (f) CH corolla. X 150. (g) Epidermal cells of mature CL corolla-ventral region. (h) Epidermal cells of mature CH corolla-ventral region.

(E. M. Lord. 1980c. Amer. J. Bot. 67:1430-1441.)

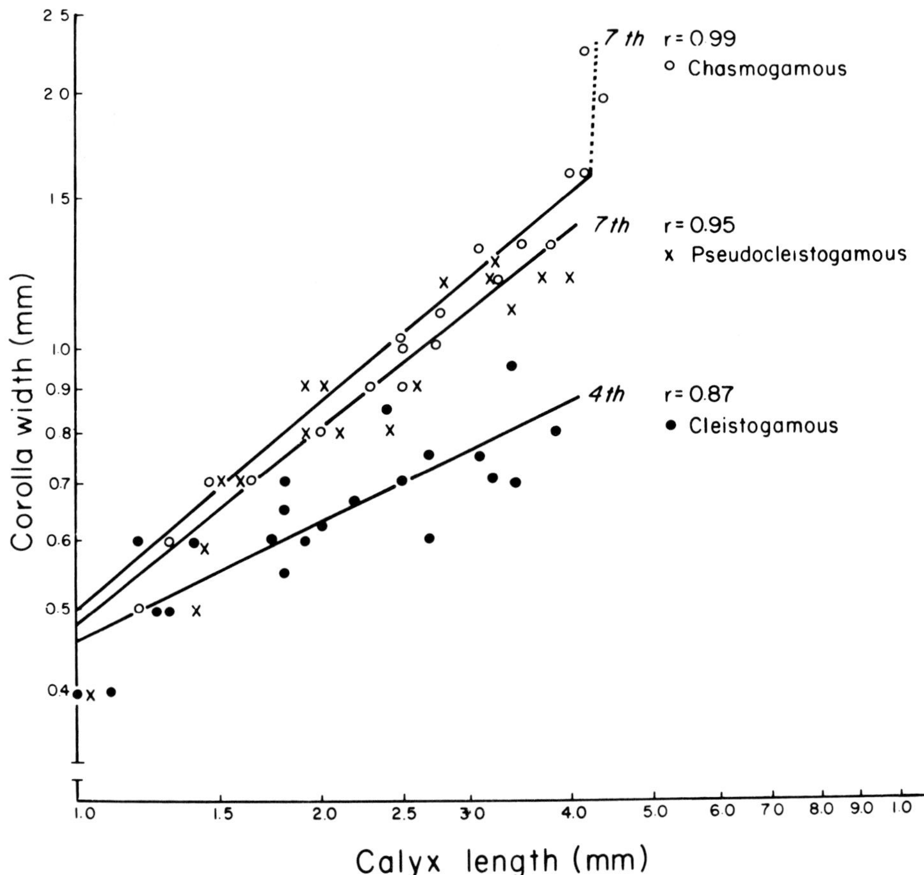

Fig. 7. Comparative allometric growth plots of corolla and calyx dimensions in CL and CH flowers of Lamium amplexicaule. (a) Calyx length vs. corolla width of CL flower from node 4, CH and pseudocleistogamous flower from node 7. (b) Calyx length vs. corolla length in CL and CH flowers.

(E. M. Lord. 1979. Bot. Gaz. 140:39–50 and 1982b. Bot. Gaz. 143:63–72.)

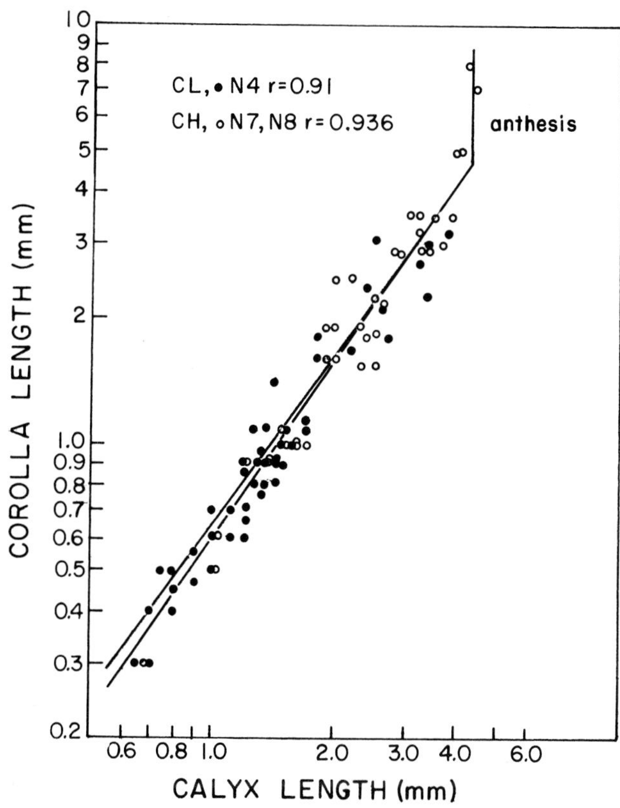

FIGURE 7 (b).

C. Hormonal Controls

We have used the developmental information gained from study of the two annual species in attempts to characterize some of the physiological factors responsible for both correlative effects between organs in each flower and for the switch from one floral type to the other in the inflorescence.

Few such studies have been done on cleistogamous species. Lee et al. (1978) suggested that a burst of ethylene production in the CL flower of *Salpiglossis* at pollination was responsible for arrested corolla expansion. Indeed, exogenous ethylene retards corolla development in both floral types, but the divergence in stamen development occurs at a much earlier stage before ethylene levels appear different.

In *Lamium*, we proposed a photoperiodic control on the amount of CH flower production whereby a build-up of

TABLE II. Organ Culture of CL (N4) and CH (N7,8) Floral Buds of *Lamium amplexicaule* L.

Expt no.	Node	Growth medium	Buds developed / buds cultured	Open corollas
1	7,8	K -	32/40	1
		KGA_3	31/38	31
	4	K -	16/20	0
		KGA_3	17/25	17
2	7,8	K -	6/10	0
		KGA_3	6/10	6
	4	K -	8/10	0
		KGA_3	10/10	10
3	7,8	K -	27/31	0
		KGA_3	18/29	16

(E. M. Lord and A. M. Mayers. 1982. Ann. Bot. 50:301-307.)

gibberellins in plants growing under long-day conditions triggers corolla expansion in the potentially open flowers at the upper nodes during spring and summer (Lord, 1980a, 1982a). When GA_3 is applied to plants growing under conditions that normally induce predominately CL flowers, entirely CH flowers are produced. In addition, a gibberellin inhibitor can be shown to prevent CH flower production under inductive conditions. We postulated that the underlying developmental phenomenon of increasing anther size from lower to upper node flowers had local control of corolla expansion via gibberellins produced in the anthers. Pollen is a known source of gibberellin (Letham et al., 1978), and since gibberellins in anthers have been implicated in the expansion of floral parts in a number of species, it is possible that the smaller anther size of the CL flowers results in gibberellin levels too low to permit anthesis. Murakami (personal communication), using the rice seedling bioassay, found gibberellin activity in this species, though negative results were obtained when we used the dwarf corn and barley half-seed bioassays (compliments of B. O. Phinney).

Using the established developmental markers in corolla and anthers, we studied the effect of GA_3 on floral expression _in vivo_ and _in vitro_ (Lord and Mayers, 1982). When morphologically and histologically undifferentiated primordia of the two flower types (Fig. 2d-g) were placed on a defined growth medium (MS salt mix, sucrose, vitamins, inositol, 0.1 ppm kinetin), they developed into CL forms unless GA_3 was added to the medium, in which case anthesis occurred (Table II). The results demonstrate that gibberellin is responsible for anthesis and organ elongation in the CH flowers; in its absence, flowers are CL in culture. But, GA_3 fails to induce a complete CH flower with respect to cell number in the corolla and anther. Apparently, controls on early developmental processes responsible for divergence in form in the two flowers are more complex than the action of a single hormone.

In *Collomia grandiflora*, GA_3 also induces anthesis in flower primordia fated to be CL (Minter and Lord, 1983). Here the hormone-induced CH corolla is more normal, with both cell number increase and expansion occurring to form an intermediate between CL and CH flowers. Again though, GA_3 fails to induce a CH-type anther. Since water stress induces entirely CL flowers in a number of cleistogamous species (Uphof, 1938) as well as *Collomia* and *Lamium*, it may be that an increase in the plants' ABA levels, often correlated with water stress (Quarrie and Jones, 1977), triggers CL flower production. Indeed, ABA treatment to plants under good watering conditions induced CL flowers throughout the plant, just as GA_3 had induced entirely open flowers (Minter and Lord, 1983). Since ABA and gibberellins are known to interact in various growth phenomena (Letham et al., 1978), we postulated that they play a role together in regulating the ontogenetic shift in floral form in the inflorescence.

In contrast to *Lamium*, in which long days induce an increase in the ratio of CH/CL flowers, *Viola* is induced under this photoperiod to produce only CL flowers. This implies a very different mechanism of control on floral expression in these two species.

Goebel (1900) stated that "size and construction of one organ is frequently determined by those of another. These reciprocal influences are termed correlations." He espoused the idea that competition for nutrients was responsible for the reduced CL floral form (Goebel, 1904). Indeed, poor nutrient status induces predominantly CL flower production in many species (Uphof, 1938), but not in *Viola* where the photoperiod is an overriding factor (Mayers, 1983). There is enough data to demonstrate that hormones balance sex expression in the angiosperms (Letham et al., 1978) and little doubt that they play a role in the correlative features of floral organ development (Greyson and Raman, 1975; Greyson and Tepfer, 1966; Heslop-Harrison, 1959; Hicks, 1975, 1980; McHughen, 1980).

Our preliminary experimental studies on two annual cleistogamous species suggest that at least gibberellins and abscissic acid are involved in mediating the observable environmental effects on floral expression. Other, as yet, unknown factors are involved in the induction of the CH flower. It is necessary to establish that endogenous levels of these hormones fluctuate during development and to characterize the type of gibberellin involved before further morphogenetic study is warranted. Ontogenetic gradients of floral form analogous to those seen in cleistogamous species exist as well in monoecious species such as the Cucurbitaceae (Atsmon and Galun, 1960; Nitsch et al., 1952).

In the comparative study of unisexual flower formation, workers established that the balance of various hormones determine the floral type produced (Atsmon and Galun, 1960; Galun, 1961; Galun and Atsmon, 1960; Heslop-Harrison, 1959). In these studies, the authors claim that floral buds all pass through a common bisexual phase early in development and then, depending on their position in the plant, are induced to be either male or female flowers. Species which have been studied morphogenetically have not been examined carefully with the ontogenetic studies necessary to test this assumption of common primordial type.

In *Lamium* and *Collomia*, primordia of both floral forms appear indistinguishable at inception and during early development, and then, depending on their position in the plant and the external environmental conditions, they differentiate into CL or CH floral forms. In both cucumber and *Cannabis sativa*, male flower formation is induced by gibberellins (Letham et al., 1978). A distinguishing feature of the CH flower in cleistogamous species is its increased androecial size, or increased 'maleness,' compared to the CL flower. If increases in endogenous gibberellins are responsible for induction of the CH flower in *Lamium* and *Collomia*, then they provide more evidence for considering gibberellins to be involved in male expression. In cleistogamous species, the shift in sexuality is a quantitative one from less (CL) to more (CH) maleness in the flowers instead of the qualitative shift from male to female flower production seen in monoecious species.

A phase shift from one floral form to another is analogous to the shift in leaf form observed in the more dramatic cases of leaf heteroblasty where hormones apparently act to cause what are often complex changes in form (Frydman and Wareing, 1974; Grey, 1957; Scurfield and Moore, 1958). Detailed structural study of these induced forms is necessary to determine whether a real phase shift has occurred or whether a mimic has been formed as appears to be the case in the GA_3-induced CH floral forms. Despite our present lack of knowledge of the sites of synthesis and basic mode of action of hormones, a coupling of the study of comparative development with an experimental approach can be useful in the study of floral development.

D. Growth Models

Comparative morphologists are traditionally interested in the history of an organism or an organ and use developmental data to suggest possible phylogenetic relationships. This approach is particularly useful on a microevolutionary scale where the "ancestral" and "derived" forms are available and their differences small enough to make ontogenetic study useful. An example is Kaplan's (1970b) work on the derivation of the ensiform leaf in _Acorus_ from the more common dorsiventral leaf type in the Araceae.

Gould (1977), using data and concepts from the literature on animal development, has provided us with a framework within which to devise models that explain the common features of ontogenetic change in plants. Botanists have often used the concepts of paedomorphosis, neoteny, and arrest to explain the derivation of form in whole organisms, organs, and even tissues (Carlquist, 1961; Goebel, 1904, 1905; Takhtajan, 1972). These concepts have been used on the macroevolutionary scale to explain the origin of the angiospermous flower or leaf (Takhtajan, 1972); hypotheses difficult or impossible to test.

We have found that the model of arrest, as put forth by early workers (Goebel, 1904, 1905; Ritzerow, 1908) to explain CL floral derivation, does not adequately describe the processes by which the shapes of floral organs are modified in the CL flowers of _Lamium_, _Collomia_ or _Viola_. Arrested development implies truncation along the path of shapes representing ontogeny of a fully developed or an ancestral form. The processes of progenesis and neoteny (both leading to juvenilized forms as the CL flower was assumed to be) are two ways in which arrest at an early shape can occur (Gould, 1977). Progenesis is precocious sexual maturation at that

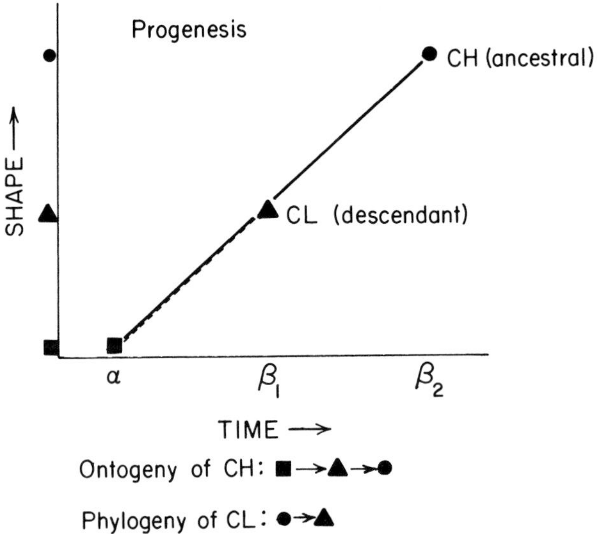

Fig. 8. (a) Model for the ontogeny and phylogeny of the CL floral form (descendant) from the CH floral form (ancestral). (After Alberch et al., 1979). Signs represent developmental stages; α = meiosis in anthers, β = pollination. (b) Test of the above model showing comparative allometric growth plot of corolla length/width ratio vs. calyx length in CL and CH flowers of *Lamium amplexicaule*. (E. M. Lord. 1982b. Bot. Gaz. 143:63–72.)

early shape with no change in developmental rates; neoteny implies slower growth with sexual maturation occurring at the normal time, but in an early shape. Alberch et al. (1979)

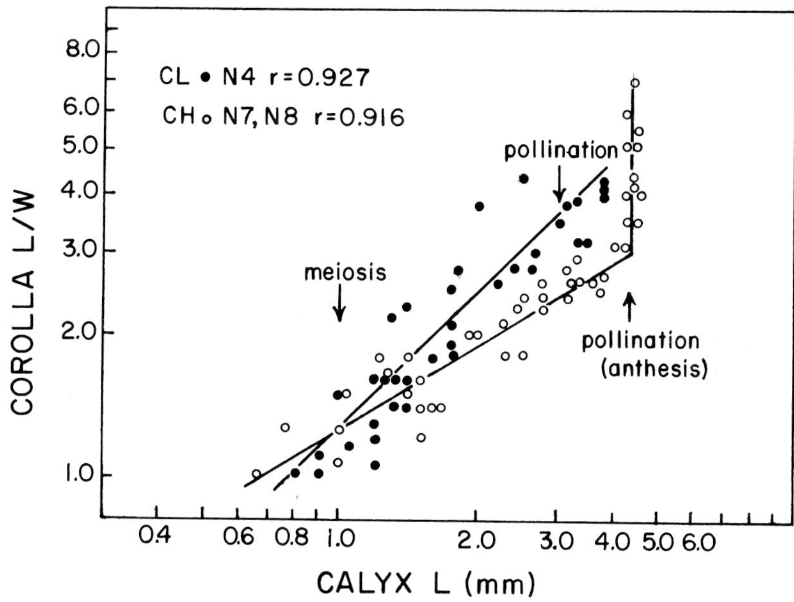

Figure 8 (b).

have presented testable models for predicting how shape changes occur in ontogeny and phylogeny. These models were first used for plants by Guerrant (1982) to test the hypothesis that Delphinium nudicaule, a hummingbird-pollinated larkspur, was evolutionarily derived from a bee-pollinated ancestor.

Since the CL is obviously derived from the CH flower and both forms are produced on an individual, cleistogamy provides a good comparative system for study of developmental processes responsible for floral shape changes that occur throughout the angiosperms. We have proposed a modification of the pure progenesis model, as put forth by Alberch et al. (1979), to explain the ontogeny and phylogeny of the CL floral form in Lamium amplexicaule (Lord, 1982b) (Fig. 8a, b).

Documenting development from inception is necessary in order to adequately test these models. In the flower any developmental markers can be used, but since floral inception, anther meiosis, and pollination (or anthesis) are discrete events in time, they serve as useful ones. The CL and CH flowers of *Lamium* are indistinguishable from inception until meiosis in the anthers. From this point, changes in corolla form of both the CL and CH flowers were measured over time by utilizing length/width ratio as a shape criterion. The CL flower is sexually precocious, showing earlier anther dehiscence and stigma receptivity, and rate of change in the corolla shape is actually accelerated over that of the CH flower (Fig. 8b). When the bud is 2 mm long, true divergence in corolla form occurs in the CL flower due to dissociation between the growth of the apex and base of the corolla. At this point, the CL corolla leaves the path of shapes prescribed by the ancestral CH form with L/W ratio no longer serving as an adequate shape measure. Obscured by the later arrest of cell expansion in the CL corolla, this shape divergence in addition to its acceleration means that the CL flower cannot be viewed as an example of pure progenetic evolution. Rather, the CL flower is more like a progenetic dwarf form of the CH; a scaled-down "adult" with respect to certain organs. The lack of anthesis gives it a "juvenile" form, but the defining feature is the precocious sexual maturation and scaled-down size of component parts. If treated separately, the anthers could be described as progenetic in *Lamium*, but the flower, as a whole, is a mixture of "adult" (CH), "juvenile," and intermediate characters. Dissociation of calyx/ovary and corolla/androecium, as well as the parts of the corolla itself, allows for a new morphology to occur in the CL flower.

The CL flowers of Collomia grandiflora and Viola odorata show greater divergence in form than those of Lamium. Though the CL corolla of Collomia appears to be a neotenous form of the CH corolla, the anthers are quite divergent due to abortion of the two front locules. Considering the same developmental markers, meiosis and pollination, Collomia's CL flower is the reverse of Lamium's with respect to the time of growth acceleration. Flowers are indistinguishable during inception of all organs, but differentiation of the CL anther is precocious, when divergence in shape is already apparent, with meiosis occurring 2 days earlier than in the CH anther. The time from meiosis to pollen mitosis and then to pollination in both flowers is the same. In this case, the crucial developmental events occur prior to anther meiosis, demonstrating that study of form from inception is necessary to detect processes responsible for later post-genital phases of growth. The models are a good starting point but they must be examined critically. In addition, we need better ways to quantify shape changes and present them graphically in three dimensions.

In Viola odorata, floral divergence occurs at inception, and this is reflected in the highly modified, precociously developing, mature CL flower where all organs have unique shapes compared to those of the CH flower (Mayers, 1983).

There are common denominators in the evolution of the CL flowers, such as reduction in, at least, the androecium and corolla and precocious differentiation and function of sexual organs. These modifications are of a simple regulatory nature, and perhaps mechanisms responsible for at least these changes can now be sought.

These same common features, reduction in androecium and corolla, with simultaneous differentiation and function of the sexual organs, characterize the evolution of autogamous flowers throughout the angiosperms. No doubt, developmental processes similar to those responsible for derivation of a CL from a CH flower characterized the derivation of inbreeding species from outcrossing ancestors.

In autogamous species, the small flowers are open and, though predominantly selfing, may effect outcrossing occasionally. This dual role has been separated into the two floral forms in cleistogamous species making autogamy obligatory in the CL flower and xenogamy possible in the CH. There are cases of species with derived autogamous relatives of known genetic relatedness which can be studied using similar comparative developmental techniques.

Questions about how plant organ shapes and sizes change in ontogeny and phylogeny are old (Wardlaw, 1965). In order to adequately devise new models for plant organ development, precise ontogenetic data is needed. In particular, a more three-dimensional view of the developing floral organs is necessary, along with knowledge of the timing of differentiation events and the processes, in terms of cell division and elongation patterns responsible for internal form change in the flower.

E. Function in the Breeding System

The dimorphic floral forms in cleistogamous species show a direct relation to function in the breeding system. The closed, CL flowers set selfed seed and the open, CH flowers provide a means for outcrossing when visited by a vector (Lord, 1981). Of the three species studied, Lamium is the most autogamous, Collomia showing a low percentage of selfed seed set in the CH flowers, and Viola the most xenogamous, with the CH flowers absolutely requiring a vector to effect pollination. Little structural or experimental work has been done on the functional aspects of cleistogamy. Pollen germinates inside the anthers of the CL flower, the tubes growing through the stomium to the stigma nearby in the closed bud. In some cases, tubes actually grow through the wall in an unopened anther. The CL pollen may be reduced in size and have a thinner wall. A common comparative study with cleistogamous species is one on mega- and microsporogenesis, gametogenesis, and embryology. In all cases studied, these processes appear similar in both flower types at the light microscope level. At the EM level, divergence at the microspore stage resulting in greater starch deposits in the CL pollen has been reported in three genera (Pargney and Dexheimer, 1976; Pargney, 1977, 1978). Anderson's (1980) study of CL pollination and fertilization in four genera of Malpighiaceae demonstrated a unique process of pollination whereby in unopened anthers pollen tubes penetrate the staminal filaments and grow down to the receptacle and up to the carpels.

The work of Lee et al. (1978) suggests that the CL pollen may have different germination requirements than CH pollen. The pollen-stigma interaction in cleistogamous species needs further study (Shivanna, 1982).

In _Lamium_, the CH stigma is not receptive until anthesis, when secretions are produced in which the pollen germinates. There are no detectable differences in pollen structure in the two flowers, but the CL pollen germinates in the anther and the tubes emerge at the slightly open stomium, growing to the wet stigma nearby (Fig. 9a–e).

In _Collomia_, pollen germinates only when in contact with the stigma papillae which are placed close to anthers in the tightly closed CL bud (Fig. 10a–d). Tubes penetrate the cuticle of the papillae in this dry stigma type and grow

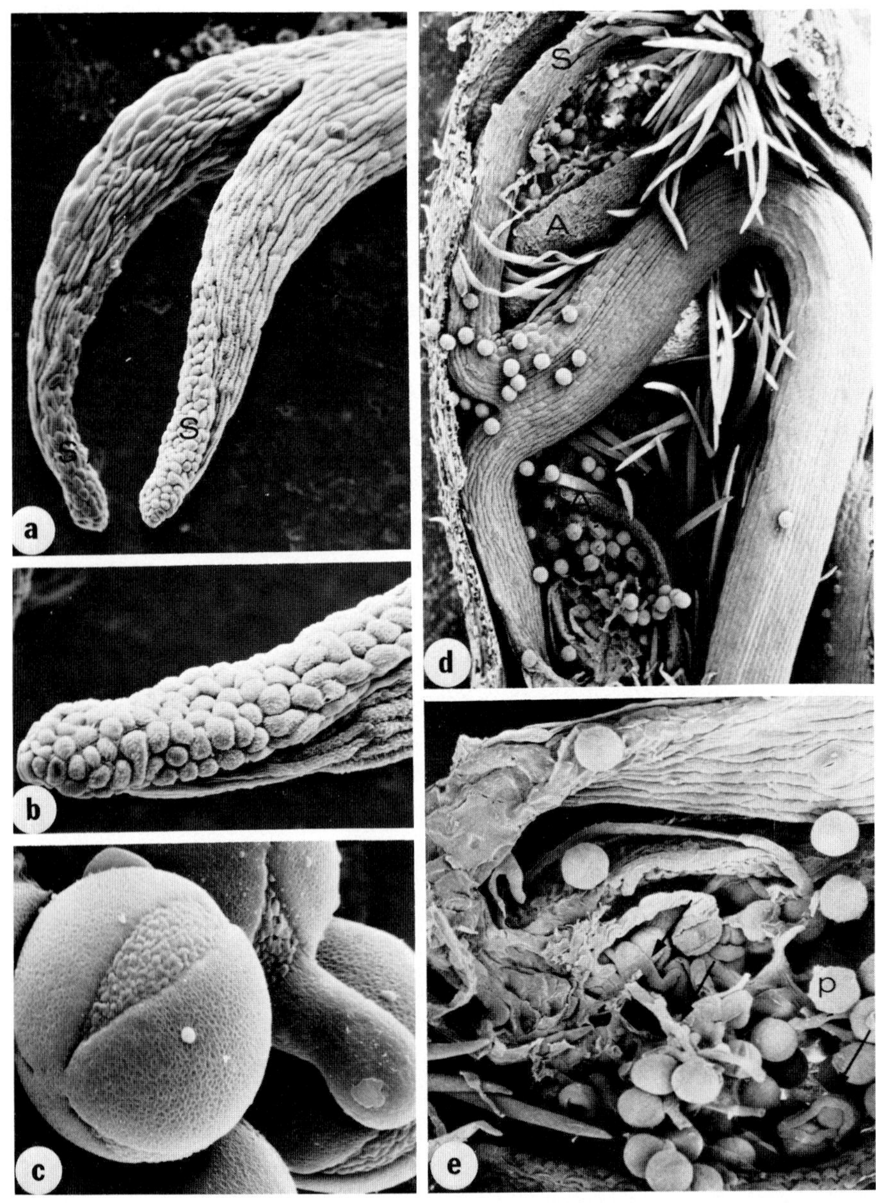
S
S
a
b
c
S
A
A
d
p
e

down through the short, solid style to the ovary. The CL pollen grains are smaller than the CH (Minter and Lord, in press) and show deeper ridges in the exine.

Viola shows the most peculiar pollination process of the three species. Pollen germinates in the unopened anthers of the CL flower and penetrates the wall in the apical portion of the anther sac to emerge and grow down to the recurved stigma (Mayers, 1983). Pollen may be dehydrated to the extent of dry seeds at the time of dispersal (Heslop-Harrison, 1979). The CH pollen must be deposited on the stigma for hydration and germination to occur. Precocious germination of the CL pollen in the anther of Viola indicates that the dormant phase, which characterizes pollen generally, may be absent.

In contrast to the much studied self-incompatibility system in heterostylous species (Shivanna, 1982; Heslop-Harrison, 1978), cleistogamy represents what might be called an ultra self-compatible system whereby the CL flower has evolved mechanisms to ensure precocious self-pollination in the bud. Since the pollination process is often quite different in the two floral forms, it is possible that subtle structural and physiological differences occur between CL and CH pollen and stigmas.

Fig. 9. Pollination in the CL flower of Lamium amplexicaule. (a) Stigmatic lobes of CL flower. X 150. (b) Close-up of (a) showing receptive papillae. X 300. (c) CL pollen. X 1,125. (d) Bud pollination in CL flower. Note proximity of stigma lobes to open anther sacs. X 75. (e) Close-up of (d) showing pollen tubes penetrating wet stigma surface. X 225.

(E. M. Lord. 1979. Bot. Gaz. 140:39-50.)

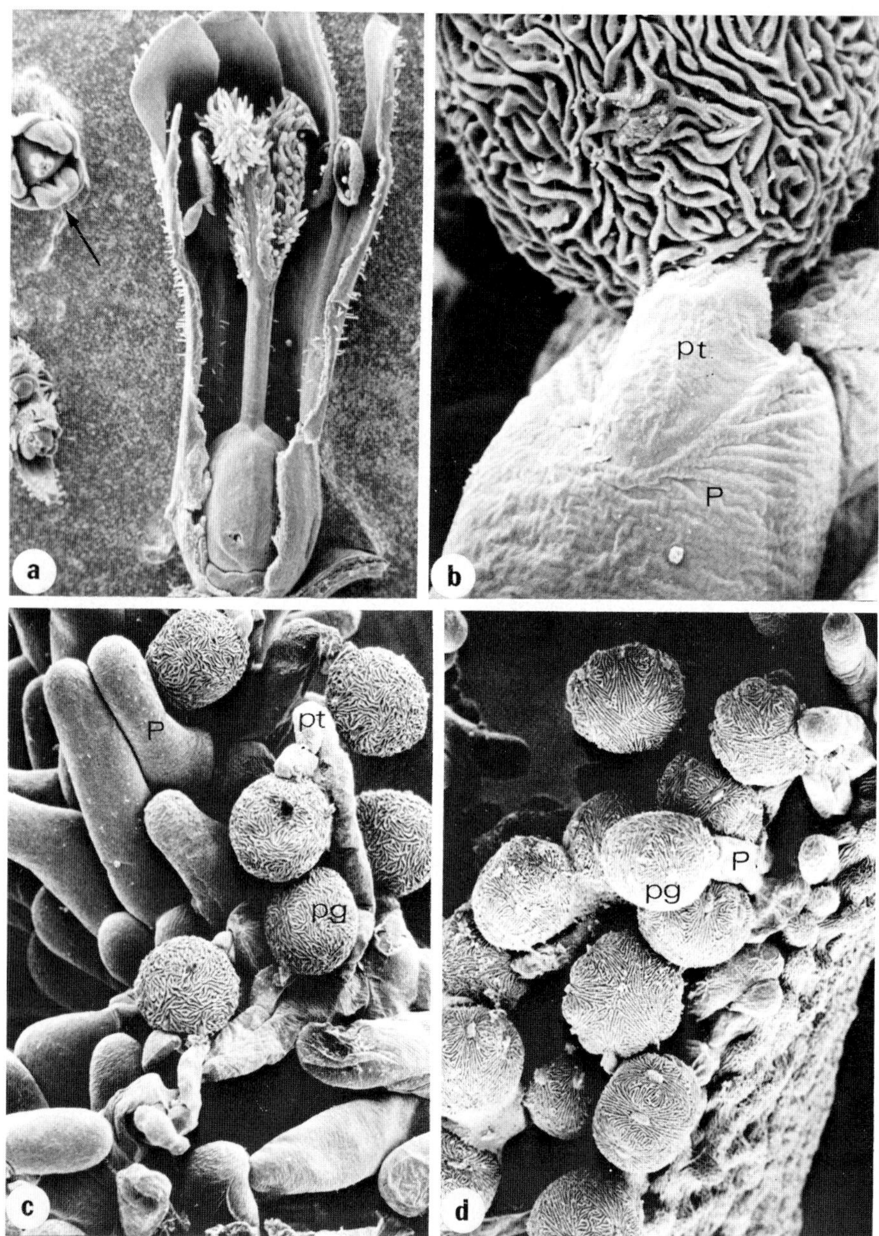
a
b
pt
P
c
P
pt
pg
d
P
pg

Only recently has floral structure been studied in relation to function in the breeding system. Developmental study is essential for an understanding of the dynamics of floral function. In species with self-compatible, bisexual flowers, slight changes in relative positioning and timing of maturation of the sexual parts may result in modifications of the breeding system (Frankel and Galun, 1977; Moore and Lewis, 1965; Rick et al., 1978).

Species are continually being added to the list of angiosperms which have this particular breeding system (cleistogamy) where form and function are clearly related. Self-compatibility of most angiosperms warrants investigation of a system like cleistogamy, where selfing is obligate in the CL flower. It is known that bud pollination in the Cruciferae is a means of overcoming the incompatibility that exists there (Shivanna et al., 1978). If bud pollination, essentially cleistogamy, could be induced in or selected for in self-incompatible crop species, barriers to selfing might be overcome. In self-compatible crops, obligate inbreeding is often sought to reduce the genetic variation that results from outcrossing and to ensure seed set. Cleistogamy has been selected for in a number of agronomic species (Frankel and Galun, 1977). In particular, the grasses and legumes show a preponderance of cleistogamy in these agronomically important families, though essentially

Fig 10. Pollination in Collomia grandiflora. (a) Bud pollination in the CL flower. X 22. (b) Pollen tube penetration of papilla in CL flower. X 1,460. (c) Pollen grains germinating on stigma surface in CL flower. X 292. (d) Pollen grains germinating on stigma surface in CH flower. X 292.

nothing is known about the developmental features that characterize it.

In summary, the basic developmental information derived from study of a comparative system like cleistogamy can be used to address morphogenetic questions about controls on form, phylogenetic questions about the derivation of new forms, as well as the practical questions about how these dimorphic floral types function in the breeding system of an individual.

ACKNOWLEDGMENTS

The research on Collomia and Viola was done in collaboration with Teresa Minter and Anna Mayers, respectively. Kathleen Eckard deserves particular thanks for her technical expertise throughout the project and for her help with editing. The work was supported in part by NSF grant PCM 8021645 and by a state-wide Critical Applied Research Grant.

REFERENCES

Alberch, P., Gould, S.J., Oster, G.F., and Wake D. B. (1979). Size and shape in ontogeny and phylogeny. Paleobiology 5, 296-317.

Anderson, W.R. (1980). Cryptic self-fertilization in the Malpighiaceae. Science 207, 892-893.

Arroyo, M.T.K. de (1973). Chiasma frequency evidence on the evolution of autogamy in Limnanthes floccosa (Limnanthaceae). Evolution 27, 679-688.

Atsmon, D. and Galun, E. (1960). A morphogenetic study of staminate, pistillate and hermaphroditic flowers in Cucumis sativas (L.). Phytomorphology 10, 110-115.

Avery, G. S. (1933). Structure and development of the tobacco leaf. Amer. J. Bot. 20, 565-592.

Bawa, K.S. (1980). Evolution of dioecy in flowering plants. Ann. Rev. Ecol. Syst. 11, 15-39.

Bergdolt, E. (1932). Morphologische und physiologische. Untersuchungen über Viola zugleich zur Losüng des Problems der Kleistogamie. Botan. Abhandl. 20, 42-87.

Boke, N. H. (1948). Development of the perianth in Vinca rosea L. Amer. J. Bot. 35, 413-423.

_____. (1949). Development of the stamens and carpels in Vinca rosea L. Amer. J. Bot. 36, 535-547.

Carlquist, S. (1961). "Comparative Plant Anatomy," Holt, Rinehart and Winston, New York.

Charnov, E. L. (1982). "The theory of sex allocation," Princeton University Press, New Jersey.

Chouard, P. (1948). Diversité des types de comparte-ment au photo- et au thermo-periodisme dans le genre Viola (Violettes et Pensees). Compt. Rend. Hebd. Séances Acad. Sci. 226, 1831-1833.

Correns, C. (1930). Genetische Untersuchungen an Lamium amplexicaule L. IV. Biol. Centrabl. 50, 7-19.

Daniel, E. and Sattler, R. (1978). Development of perianth tubes of Solanum dulcamara: implications for comparative morphology. Phytomorphology 28, 151-171.

Darwin, C. (1897). "The different forms of flowers on plants of the same species," Appleton, New York.

DeMason, D. A., Stolte, K. W., and Tisserat, B. (1982). Floral development in Phoenix dactylifera. Can. J. Bot. 60, 1437-1446.

Erickson, R.O. (1948). Cytological and growth correlations in the flower. Amer. J. Bot. 35, 729-739.

_____. (1976). Modeling of plant growth. Ann. Rev. Plant Physiol. 27, 407-434.

Evans, L.T. (1956). Chasmogamous flowering in Viola palustris L. Nature 178, 1301.

Frankel, R. and Galun, E. (1977). "Pollination mechanisms, reproduction and plant breeding," Springer-Verlag, Berlin.

Frydman, V.M. and Wareing, P.F. (1974). Phase change in Hedera helix L. III. The effects of gibberellins, abscisic acid, and growth retardants on juvenile and adult ivy. Journal of Exp. Bot. 25, 420-499.

Fryxell, P.A. (1957). Mode of reproduction of higher plants. Bot. Rev. 23, 135-155.

Galun, E. (1961). Gibberellic acid as a tool for the estimation of the time interval between physiological and morphological bisexuality of cucumber floral buds. Phyton. 16, 57-62.

Galun, E. and Atsmon, D. (1960). The leaf-floral bud relationship of genetic sexual types in the cucumber plant. Bull. Res. Council of Israel. 9D, 43-50.

Ganders, F.R. (1979). The biology of heterostyly. New Zeal. J. Bot. 17, 607-635.

Goebel, K. (1900). "Organography of Plants," Clarendon Press, Oxford.

_____. (1904). Die kleistogamen Blüten und die anpassungstheorien. Biol. Zentralbl. 24, 673-787.

_____. (1905). Chasmogamie und kleistogamie Blüten bei Viola. Allg. Bot. Z. Syst. 95, 234-239.

Gould, S.J. (1977). "Ontogeny and Phylogeny," Harvard University Press, Cambridge.

Grey, R.A. (1957). Alteration of leaf size and shape and other changes cuased by gibberellins in plants. Amer. J. Bot. 44, 674-681.

Greyson, R. I. and Raman, K. (1975). Differential sensitivity of "double" and "single" flowers of Nigella damascena L. (Ranunculaceae) to emasculation and to GA_3. Amer. J. Bot. 62, 531-536.

Greyson, R.I. and Tepfer, S.S. (1966). An analysis of stamen filament growth of Nigella hispanica. Amer. J. Bot. 53, 485-490.

Guerrant, E. O. (1982). Neotenic evolution of Delphinium nudicaule (Ranunculaceae): a hummingbird-pollinated larkspur. Evolution 36, 699-712.

Heslop-Harrison, J. (1959). Growth substances and flower morphogenesis. Bot. J. Linn. Soc. 56, 269-281.

_____. (1978). Recognition and response in the pollen-stigma interaction. In "Cell-Cell Recognition." (Ed. A.S.G. Curtis), pp. 121-138. Cambridge University Press, England.

_____. (1979). Pollen walls as adaptive systems. Ann. Missouri Bot. Gard. 66, 813-829.

_____. (1982). The reproductive versality of flowering plants: an overview. In "Strategies of Plant Reproduction." (Ed. W. J. Meudt), pp. 3-18. Allenheld, Osmon Publ., London.

Hicks, G.S. (1975). Carpelloids on tobacco stamen primordia in vitro. Can. J. Bot. 53, 77-81.

_____. (1980). Patterns of organ development in plant tissue culture and the problem of organ determination. Bot. Rev. 46, 1-24.

Kaplan, D.R. (1968). Structure and development of the perianth in Downingia bacigalupii. Amer. J. Bot. 55, 406-420.

_____. (1970). Comparative foliar histogenesis in Acorus calamus and its bearing on the phyllode theory of monocotyledonous leaves. Amer. J. Bot. 57, 331-361.

_____. (1973). Comparative developmental analysis of the heteroblastic leaf series of axillary shoots of Acorus calamus L. (Araceae). La Cellule. 69, 253-390.

Larson, P. R. and Isebrands, J. G. (1971). The plastochron index as applied to developmental studies of cottonwood. Can. J. Forest Res. 1, 1-11.

Lazarte, J. E. and Palser, B. F. (1979). Morphology, vascular anatomy and embryology of pistillate and staminate flowers of Asparagus officinalis. Amer. J. Bot. 66, 753-764.

Lee, C. W., Erickson, H. T., and Janick, J. (1978). Chasmogamous and cleistogamous pollination in Salpiglossis sinuata. Physiol. Pl. 43, 225-230.

Letham, D. S., Goodwin, P. B., and Higgins, T. J. V. (1978). "Phytohormones and related compounds: a comprehensive treatise" Vol II, Elsevier/North Holland Press, Oxford.

Lloyd, D. G. (1965). Evolution of self-compatibility and racial differentiation in Leavenworthia (Cruciferae). Contr. Grey Herb. 195, 3-134.

_____. (1975). Breeding systems in Cotula. IV. Reversion from dioecy to monoecy. New Phytol. 74, 124-245.

Lord, E. (1979). The development of cleistogamous and chasmogamous flowers in Lamium amplexicaule (Labiatae): an example of heteroblastic inflorescence development. Bot. Gaz. 140, 49-50.

_____. (1980a). Physiological controls on the production of cleistogamous and chasmogamous flowers in Lamium amplexicaule L. Ann. Bot. 44, 757-766.

_____. (1980b). Intra-inflorescence variability in pollen/ovule ratios in the cleistogamous species Lamium amplexicaule L. (Labiatae). Amer. J. Bot. 67, 529-533.

_____. (1980c). An anatomical basis for the divergent floral forms in the cleistogamous species, Lamium amplexicaule L. (Labiatae). Amer. J. Bot. 67, 1430-1441.

_____. (1981). Cleistogamy: a tool for the study of floral morphogenesis, function and evolution. Bot. Rev. 47, 421-449.

_____. (1982a). Effect of daylength on open flower production in the cleistogamous species Lamium amplexicaule L. Ann. Bot. 49, 261-263.

_____. (1982b). Floral morphogenesis in Lamium amplexicaule L. (Labiatae) with a model for the evolution of the cleistogamous flower. Bot. Gaz. 143, 63-72.

Lord, E. and Mayers, A. M. (1982). Effects of gibberellic acid on floral development in vivo and in vitro in the cleistogamous species, Lamium amplexicaule L. Ann. Bot. 50, 301-307.

Maksymowych, R. (1973). "Analysis of leaf development," Cambridge University Press, England.

Mayers, A.M. (1983). Comparative flower development in cleistogamous species Viola odorata. Ph.D. dissertation, University of California, Riverside.

McHughen, A. (1980). The regulation of tobacco floral organ initiation. Bot. Gaz. 141, 389-395.

Minter, T. C. and Lord, E. M. (1983). Effects of water stress, abscisic acid, and gibberellic acid on flower production and differentiation in the cleistogamous species *Collomia grandiflora* Dougl. x Lindl. (Polemoniaceae). *Amer. J. Bot.* 70, 618-624.

Minter, T. C. and Lord, E. M. A comparison of cleistogamous and chasmogamous floral development in *Collomia grandiflora* Dougl. x Lindl. (Polemoniaceae). *Amer. J. Bot.* (in press).

Moore, D.M. and Lewis, H. (1965). The evolution of self pollination in *Clarkia xanthiana*. *Evolution* 19, 104-114.

Nitsch, J.P. (1965). Physiology of flower and fruit development. In "Encyclopedia of Plant Physiology." (Ed. W. Ruhland), pp. 1537-1647. Springer-Verlag, Berlin.

Nitsch, J. P., Kurtz, E. B., Liverman, J. L., and Went, F. W. (1952). The development of sex expression in cucurbit flowers. *Amer. J. Bot.* 39, 32-43.

Ornduff, R. (1969). Reproductive biology in relation to systematics. *Taxon* 18, 121-144.

Pargney, J. C. (1977). Etude comparée de l'évolution des grains de pollen dans les fleurs cléistogames et les fleurs chasmogames de *Viola* sp. (Violacées). *Cytologia* 42, 233-240.

_____. (1978). Etude ultrastructurale de la gamétogenèse mâle dans une espèce à floraison cléistogame: *Oxalis corniculata*, suivie de quelques considérations générales sur la cléistogamie. *Can. J. Bot.* 56, 1262-1269.

Pargney, J. C. and Dexheimer, J. (1976). Etude comparée de la gamétogenèse mâles dans les fleurs cléistogames et dans les fleurs chasmogames der Streptocarpus nobilis (Gesnériaceae). Rev. Gen. Bot. 83, 201-229.

Quarrie, S. A. and Jones, H. G. (1977). Effects of abscisic acid and water stress on development and morphology of wheat. J. Exp. Bot. 28, 192-203.

Richards, J. H. and Barrett, S. C. H. (1983). Developmental basis of tristyly in Eichornia paniculata Spreng. (Pontederiaceae). Abstr. Bot. Soc. Amer. Series 163.

Rick, C. M., Holle, M., and Thorp, R. W. (1978). Rates of cross pollination in Lycopersicon piminellifolium. Impact of genetic variation in floral characters. Pl. Syst. Evol. 129, 31-44.

Ritzerow, H. (1908). Uber Bau and Befruchtung Kleistogamer Blüten. Flora 98, 163-212.

Rollins, R.C. (1963). The evolution and systematics of Leavenworthia (Cruciferae). Contrib. Grey Herb. 192, 1-98.

Sattler, R. (1973). "Organogenesis of flowers: a photographic text - atlas." University of Toronto Press, Toronto.

Scurfield, G. and Moore, C. W. E. (1958). Effects of gibberellic acid on species of Eucalyptus. Nature 181, 1276-1277.

Shivanna, K.R. (1982). Pollen-pistil interaction and control of fertilization. In "Experimental embryology of vascular plants." (Ed. B. M. Johri), Springer-Verlag, New York.

Shivanna, K. R., Heslop-Harrison, Y., and Heslop-Harrison, J. (1978). The pollen-stigma interaction: bud pollination in the Cruciferae. Acta Bot. Neerl. 27, 107-119.

Stebbins, G. L. (1957). Regularities of transformation in the flower. Evolution 11, 106-108.

Takhtajan, A. (1972). Patterns of ontogenetic alterations in the evolution of higher plants. Phytomorphology 22, 164-171.

Tepfer, S.S. (1953). Floral anatomy and ontogeny in Aquilegia formosa var truncata and Ranunculus repens. University of California Publ. Bot. 25, 513-648.

Thomson, J. D. and Barrett, S. C. H. (1981). Selection for outcrossing, sexual selection and the evolution of dioecy in plants. Amer. Natur. 118, 443-449.

Uhl, N. W. (1966). Developmental studies in Ptychosperma (Palmae). II. The staminate and pistillate flowers. Amer. J. Bot. 63, 97-109.

Uphof, J. C. T. (1938). Cleistogamic flowers. Bot. Rev. 4, 21-49.

Vuilleumier, B. S. (1967). The origin and evolutionary development of heterostylly in the angiosperms. Evolution 21, 210-226.

Wardlaw, C. W. (1965). "Organization and evolution in plants," Longmans, Green, and Co. Ltd., London.

Wilken, D.H. (1982). The balance between chasmogamy and cleistogamy in Collomia grandiflora (Polemoniaceae). Amer. J. Bot. 69, 1326-1333.

Williams, R. F. (1975). "The shoot apex and leaf growth," Cambridge University Press, England.

Yampolsky, E. and Yampolsky, H. (1922). Distribution of sex forms in phanerogamic flora. Bibl. Genet. 3, 1-62.

STRUCTURAL CORRELATIONS AMONG WOOD, LEAVES AND PLANT HABIT

Phillip M. Rury

Botanical Museum
Harvard University
Cambridge, Massachusetts

William C. Dickison

Department of Biology
University of North Carolina
Chapel Hill, North Carolina

Within the phylogenetically unrelated genera *Hibbertia* and *Erythroxylum* multiple correlations among woody anatomy, plant stature, and details of leaf morphology and anatomy are clearly environmentally related. Xeric radiants typically have undergone plant and leaf size reductions, along with the development of foliar xeromorphy, deciduousness, and xeromorphic xylem specializations. Mesic radiants have evolved an increased plant stature and leaf size as well as retaining a primitively mesomorphic xylem structure. Clinal variation patterns are much more extreme and apparent in *Hibbertia* whereas *Erythroxylum* species possess more subtle degrees of habitat related variability. In species of both genera, modifications in leaf morphology and anatomy and the existence of deciduousness, appear to buffer the xylem and permit rather primitive and mesomorphic wood anatomy to exist within seasonally or permanently dry locations. The extent of the buffering influence of leaves upon the xylem

ISBN 0-12-746620-7

tissues is presumably determined by the complex interrelationships between various climatic variables and such features as leaf size, duration, type and degree of scleromorphy, as well as non-foliar characteristics such as architectual forms, degree of stem succulence, and the nature of the root system. In this regard the foliage constitutes perhaps the most significant component of the plant hydrovascular system. Biological and evolutionary interpretations of wood anatomy within an ecological context, therefore, are ill-advised in the absence of correlative leaf structural data.

I. INTRODUCTION

In recent years a renewed awareness has developed among plant morphologists of the need to correlate anatomical diversity with biologically interrelated factors such as species ecological profiles and floristic preferences, plant architecture, various leaf morphological features, and associated differences in hydrovascular physiology. Recent studies by Zimmermann and coworkers of hydraulic architecture in dicotyledonous trees and shrubs have contributed to clarification of the ecologically adaptive significance of whole-vessel morphology, internal distribution patterns and conductive physiology (Zimmermann, 1978, 1982; Zimmermann and Jeje, 1981; Zimmermann and Potter, 1982). An appreciation of the complex directing influences on xylem structure also has resulted in the further documentation of several, ecologically correlated trends in fiber and vessel element morphology (Carlquist, 1975, 1977, 1980a; Baas, 1973, 1982; van der Graaff and Baas, 1974; Dickison et al., 1978; van den Oever et al., 1981; Bissing, 1982).

Major conclusions which have resulted from syntheses of both classical and contemporary wood anatomical studies include the hypothesis that comparatively long and wide, evolutionarily primitive, vessel elements are of adaptive value (or not disadvantageous) only in mesic habitats where soil moisture is not limiting and foliar transpiration is fairly slow and uniform owing to a constantly high relative humidity. Conversely, available ecological and wood anatomical data imply that shorter and narrower, but more numerous vessel elements per unit area, with simple or few-barred, scalariform perforation plates have evolved under drier conditions of both low soil moisture and atmospheric humidity. Narrowness of vessels in classically "xeromorphic" dicotyledonous taxa is generally correlated with an increased number of vessels per unit area of xylem transection. Available evidence from comparative wood anatomy indicates that these and other anatomical trends can occur among diverse woody dicot taxa along ecological gradients (Carlquist, 1975, 1977, 1980a; Dickison et al., 1978; Baas, 1982; Bissing, 1982).

In the cosmopolitan genus *Ilex* (Aquifoliaceae), Baas (1973) found that pronounced variation in wood structure is correlated with altitudinal and latitudinal distribution of the species. Temperate to subtropical species possess: (1) conspicuous growth rings; (2) numerous, narrow and relatively short vessel elements; (3) few bars per scalariform perforation plate and (4) prominent, helical thickenings on both vessel and fiber walls. Tropical lowland woods, conversely, revealed: (1) an absence of growth rings; (2) fewer, longer and wider vessel elements; (3) often with a higher number of bars per scalariform perforation plate and (4) absent or ill-defined, helical wall thickenings. As would be expected, *Ilex* species of tropical montane habitats are wood anatomically similar to their higher latitude, temperate counterparts.

Additional data by van der Graaff and Baas (1974) indicated that, with increasing latitude, woods reveal an increase in pore frequency as well as shorter rays and shorter, narrower fibers and vessel elements. Increasing altitude revealed similar but less conspicuous wood anatomical trends. The number of bars per scalariform perforation plate, however, was found to be not correlated with latitude and altitude. Similar overall trends also were reported for *Symplocos*, in which fiber and vessel wall thickness also decreased with increasing latitude (van den Oever et al., 1981).

Genera either not conforming totally, or in some respect contradictory, to these wood anatomical trends have been described by a number of workers (Novruzova, 1968; van den Outer and Van Veenendaal, 1976; Dickison, 1977, 1979; van den Oever et al., 1981). Furthermore, Carlquist (1975, 1980a) and Baas (1976, 1982) have noted that assumptions of strictly irreversible phylogenetic trends in the xylem are sometimes erroneous. However, exceptions to these patterns of xylem evolutionary specialization may well be the result of (1) leaf structural features which serve to buffer xylem tissues from prevailing ecological conditions (2) plant life forms and architectural features, including such extremes as lignotubers, rosette trees and stem succulents and (3) diverse, mitigating physiological adaptations.

Carlquist (1977) and Carlquist and DeBuhr (1977) have advocated calculating a"vulnerability index" of wood anatomy (VULN), by dividing mean vessel diameter by the mean number of vessels per mm^2. The lowest indices of vulnerability appear in xerophytic taxa with narrow but very numerous vessels per unit area. Theoretically, this xylem formulation ("high vessel redundancy") is much safer than one which consists of wider and fewer vessels per unit area since it will restrict air embolisms to a smaller and more localized portion of the transpiration stream in the event of disruptively high negative pressures, freezing, or other types of injury to the xylem vessels.

Carlquist also introduced the concept for an index of wood anatomical mesomorphy (MESO), in which the vulnerability index (VULN) is multiplied by the mean length of the vessel elements. Plants considered as mesophytic on ecological and macromorphological grounds typically exhibit mesomorphy indices greater than 200, whereas xerophytic taxa rarely possess indices of xylem anatomical mesomorphy in excess of 75 (Carlquist, 1977).

Baas (1976, 1982) has criticized these functional interpretations of xylem anatomy, which view anatomical variation as the adaptive result of selective, environmental pressures acting upon plant populations. Rather, he envisions anatomical diversification as a process that is not necessarily advantageous or disadvantageous to the plant. Baas (1982) notes, however, that ecological trends are of definite predictive value, despite the current disagreement concerning the functional and adaptive significance of individual xylem anatomical features.

Vessel element length is a particularly controversial feature with respect to its hypothetical ecological and functional significance. Zimmermann (1978), for example, has discounted Carlquist's (1975, 1977) interpretation of vessel element length as functionally significant in ecological terms (i.e., shorter vessel elements are less prone to collapse), stating that "the length of the vessel element is functionally meaningless, as far as we know." Nevertheless, Zimmermann (1978, 1982) concurs with Carlquist and other wood anatomists that vessel diameter and frequency (including vulnerability index) are both functionally significant and ecologically adaptive. Zimmermann has expressed the belief that entire vessel length is a more meaningful consideration than vessel element length for integration with the vulnerability index to calculate relative degrees of wood anatomical mesomorphy (Zimmermann, pers. comm.). Zimmermann and coworkers have shown that summerwood vessels are not only narrower, but also much shorter, in comparison to springwood vessels even in the same growth ring. They suggest that xylem with narrow and short vessels is physiologically safer by restricting injury-induced cavitations to a smaller area of the transpiration stream in close proximity to the point of the initial damage (Zimmermann, 1982).

General correlations among wood anatomical variation and ecologically related differences in plant habit, leaf size, duration and anatomy have been presented by several authors (Carlquist, 1978, 1980; Dickison et al., 1978; Gibson, 1973; Michener, 1981, Rury, 1981). These studies of various woody dicot families and genera have revealed diverse, habitat-related, wood and leaf structural differences which may be physiologically significant from the perspective of plant-water relations. Published observations suggest that eco-physiological interpretations of the plant vascular system are possible only when plants are analyzed as dynamic, structural units.

The leaf is often considered the most anatomically variable organ of the plant (Carlquist, 1961) and leaf type frequently has been used as an indicator of plant habitat. The anatomical and ecological literature is replete with examples of foliar structure interpreted in terms of adaptations within sun and shade, as well as physiologically wet or dry microhabitats (e.g., Maximov, 1929; Shields, 1950; Ferri, 1959; Tibbits, 1979). It is now clear, however, that many so-called xeromorphic features in leaves result from soil mineral deficiencies rather than from a significant lack of water (Firbas, 1931; Müller-Stoll, 1947-1948; Beadle, 1966, 1968; Specht, 1972; Steubing and Alberdi, 1973). Beadle (1968) added phosphorus to potted plants of several sclerophyllous South Australian species, such as *Hibbertia sericea* (Dilleniaceae) and *Hakea teretifolia* (Proteaceae), and noted a resultant increase in both foliar size and mesomorphy. Clearly, such leaves should be referred to as scleromorphic, especially since many anatomical traits classically associated with foliar xeromorphy cannot be readily or reliably interpreted, either ecophysiologically or evolutionarily, when considered individually.

Despite renewed interest in various aspects of ecological and functional xylem anatomy, there has been a notable lack of studies which correlate wood and leaf anatomy from an ecological perspective. Noteworthy exceptions are the recent contributions of Michener (1981) and Rury (1981, 1982), but even these comprehensive anatomical studies of individual genera were not integrated with correlative analyses of xylem conduction and foliar transpiration.

Historically, many anatomists have tended to consider leaves and stems as isolated, functionally unrelated plant organs, although Howard (1974) has addressed the importance of interpreting the shoot of dicotyledons as a structural continuum. In order to promote an additional understanding of the various aspects of comparative wood anatomy, both Carlquist (1980a) and Baas (1982) have listed a number of priorities for future research, including the integration of wood anatomical data with information from other parts of the plant, especially roots and leaves. Focal points for future studies include the multiple, interrelated facets of ecology, physiology and plant structure, such as (1) plant ecological profiles (2) morphogenetic interrelationships among roots, stems, leaves (3) the relationship of xylem anatomy to plant architecture and leaf structure (4) altitudinal and latitudinal variation (5) adaptations to fluctuations in both temperature and water availability and (6) the interrelationships of foliar transpiration and plant hydraulic architecture.

Accordingly, our major objectives in this contribution are to integrate xylem and leaf structural data with available information from plant ecology and floristics, and to consider the woody plant body as a structural and functional continuum, in which roots and shoots have evolved presumably in relation to one another and to prevailing environmental conditions. We hope to illustrate the complex nature of these ecological, plant structural, and presumed physiological interrelationships with several diverse examples taken from species and genera of two pantropical, dicotyledonous families - the relatively primitive Dilleniaceae and the more advanced Erythroxylaceae.

For both families indices of wood anatomical vulnerability (VULN) and mesomorphy (MESO) were calculated following the suggestions of Carlquist (1977). Data relating to the Erythroxylaceae were derived from some 350 wood and herbarium specimens, representing 140 species and all four genera of Erythroxylaceae. The list of specimens studied are provided elsewhere (Rury, 1982). Whenever possible, voucher specimens were determined by Dr. Timothy Plowman (Field Museum of Natural History, Chicago), who also generously provided ecological characterizations for the neotropical species of Erythroxylum. Tables IV-VII represent summary data only for neotropical species of known ecological profile, which were represented by both wood and leaf specimens (often from the same plant). Data were derived only from positively identified specimens. Leaf size classes follow Raunkiaer (1934), as defined in TABLE I, and the infrageneric classification of Erythroxylum presented by Schulz (1907, 1931) is used for convenience.

II. OBSERVATIONS

A. Dilleniaceae

The Dilleniaceae are a family of eleven genera distributed pantropically and ranging in habit from trees, shrubs, subshrubs and rosette trees to woody climbers. The family is an excellent example of adaptive radiation and structural diversification at both the generic and specific levels of comparison, from intermediate habitats towards both more mesic and more xeric conditions (Stebbins, 1974). Of particular interest is the remarkable morphological and anatomical diversity encountered in the largest and most primitive genus Hibbertia Andrews. Aspects of this diversity have been the subject of several recent contributions by Dickison (1969, 1970), Stebbins and Hoogland (1976), Rury & Dickison (1977) and Dickison et al. (1978).

Several mesic Australian species of Hibbertia, such as H. cuneiformis and H. serrata, are shrubs up to 3 meters tall, and possess a combination of primitive xylem, trilacunar nodes and mesomorphic toothed leaves. It is noteworthy in this regard that most other genera of Dilleniaceae which possess rather large, toothed leaves also are mesophytes (Rury & Dickison, 1977). The mesomorphic xylem of these plants contains comparatively few, angular and thin-walled vessels with scalariform perforations containing 30 or more bars, among the longest tracheary elements within the genus, and thin-walled imperforate elements. Their leaves lack obvious drought resistant features and have comparatively unspecialized venation patterns (Fig. 1D).

Leaves of H. cuneiformis reveal a heteroblastic, acropetal sequence along single branches. Small, entire, lanceolate leaves occur basally on the branch whereas the more distal leaves are larger and possess vascularized, marginal teeth of the dillenioid type (sensu Hickey and Wolfe, 1975). The venation patterns of all leaf types are identical except that in the latter form teeth are vascularized by tertiary veins which diverge exmedially from the secondary, brochidodromous vein arches which are common to all leaves (Rury & Dickison, 1977). This acropetal transition in foliar form may reflect seasonal changes experienced by the plant or a change from juvenile to adult leaf morphology. Perhaps the leaf morphological plasticity is drought adaptive, in that shoots may develop one leaf form or another in response to prevailing microenvironmental conditions. The rather primitive and mesomorphic xylem anatomy of H. cuneiformis thus would not be disadvantageous during dry periods if the variable foliar forms compensate for seasonal fluctuations in rainfall, temperature and atmospheric humidity.

Major trends of specialization among the Australian species of Hibbertia have been toward reduction in overall plant stature, resulting in small, semi-prostrate or prostrate, thin-stemmed shrubs. Decreased leaf sizes have also accompanied the migration into drier microhabitats and both plant and leaf size reductions are correlated with

TABLE I. Wood Anatomy and Leaf Size Classes of *Hibbertia* in Relation to Habitat.[1]

Habitat[2]	SPP	LS	FL	VEL	PDM	VEL/PDM	PDN	VULN	MESO	BARS	FWT
Mesic	6	3.2	1317	992	65	15.3	68	.96	948	25	t-m
Mesic to Semi-Xeric	2	2.3	1097	778	30	25.9	235	.13	99	16	t-m
Semi-Xeric	17	2.2	882	627	31	20.2	283	.11	69	16	t-m
Semi-Xeric to Xeric	1	2-3	809	534	25	21.4	394	.06	34	10	m-th
Xeric	3	1.3	758	544	16	34.0	452	.04	19	3	m-th

[1]Data recalculated from species means originally published in Dickison et al. (1978).

[2]Habitat categories according to definitions outlined in Dickison et al. (1978).

Symbols: SPP, number of species studied; LS, mean (maximum) leaf size classes, sensu Raunkiaer (1934), LS-1 = 0-25 mm^2, LS 2 = 25-225 mm^2, LS 3 = 225-2025 mm^2, LS 4 = 2025-18222 mm^2; FL, mean fiber length, μm; VEL, mean vessel element length, μm; PDM, mean tangential pore diameter, μm; VEL/PDM, mean vessel element length: width ratio; PDN, mean pore density per mm^2; VULN, index of vulnerability, PDM/PDN; MESO, mean index of mesomorphy, VULN x VEL; BARS, mean number of bars per scalariform vessel perforation; FWT, fiber wall thickness, t = thin, m = medium, th = thick; RDN, mean ray density in tangential section, mm^2; HABIT, mean height in meters; LSMX, mean maximum leaf size; LTK, mean leaf thickness, μm; STFQ, mean stomatal frequency as percentage of leaf surface occupied by stomata.

xeromorphic xylem anatomical specializations. The wood anatomy of *Hibbertia* is compared with leaf size, plant habit, and habitat in TABLES I-III. The woody anatomy of many xeric hibbertias is characterized by growth rings, shorter and thicker-walled tracheary elements (fibers and vessel elements), and narrower, more numerous vessels with a reduced number of bars (under 10) per scalariform perforation plate (Fig. 1B, C, H).

Closely correlated with the changes in plant habit and wood structure that have resulted from extremes in the habitat, leaves of the Australian hibbertias exhibit marked trends in form, texture, thickness, and venation.

FIGURE 1A-J. Wood and leaf anatomy of *Hibbertia*. A, *H. lucens* (SJRw 28402), transverse section of wood showing thin-walled vessels and fibers. B, *H. exutiacies* (Kraehenbuhl 2153), transverse section of wood showing irregularly-shaped growth rings and numerous pores. C, *H. uncinata* (Stebbins & Keighery A-3), transverse section of wood showing growth rings and numerous, narrow pores. D, *H. cuneiformis* (Cult. K s.n.), transverse section of leaf. E-G, *H. virotii* (Dickison 253). E, Longisection of wood showing a scalariform perforation plate with numerous bars. F, Transverse section of leaf. G, Drawing of cleared leaf showing venation pattern. H-J, *H. pungens* (Simpson s.n.). H, Longisection of wood illustrating scalariform perforation plate with very few, thick bars. I, Transverse section of leaf showing ericoid habit. J, Drawing of cleared leaf illustrating venation pattern.

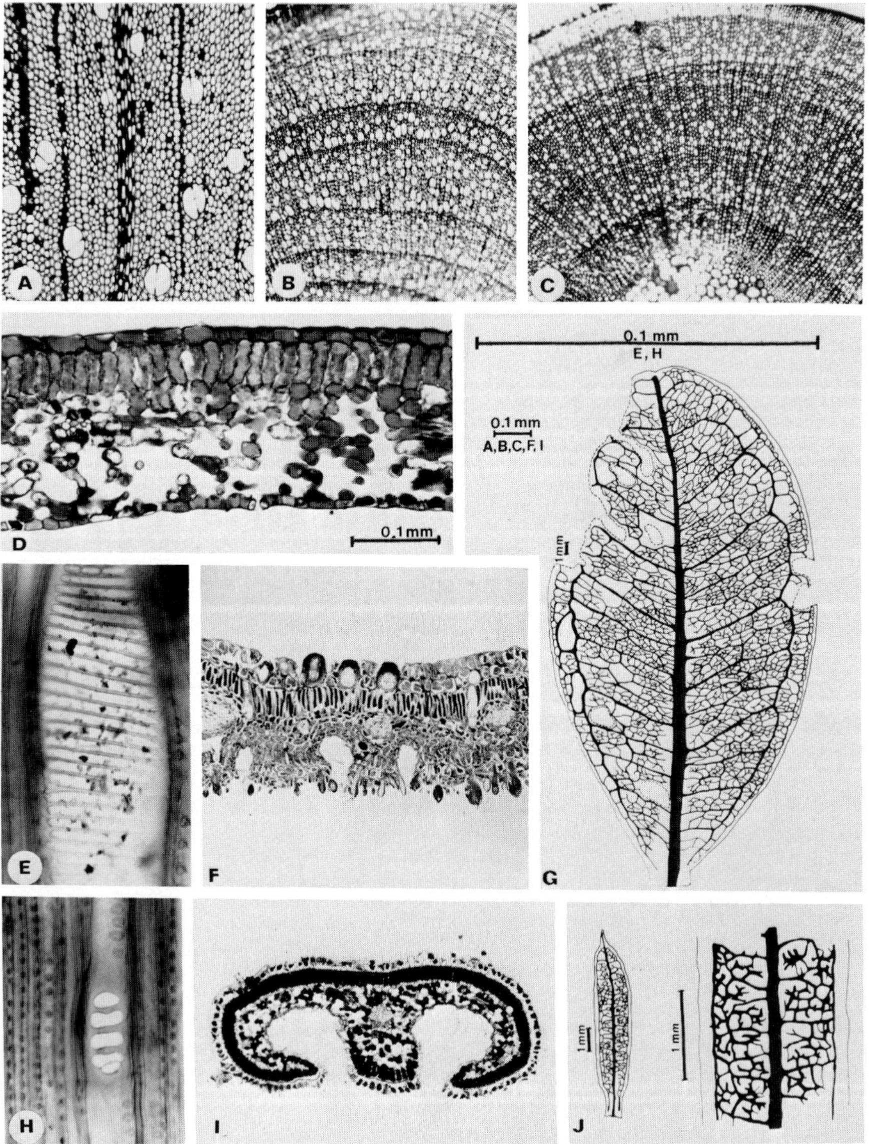
A
B
C
D
0.1 mm
0.1 mm
E, H
0.1 mm
A,B,C,F, I
1 mm
E
F
G
H
I
J
1 mm
1 mm

TABLE II. Wood Anatomy of *Hibbertia* in Relation to Leaf Size.[1]

Leaf Size[2]	SPP	FL	VEL	PDM	VEL/PDM	PDN	VULN	MESO	BARS	FWT
L.S. 1-2	13	753	553	19	29.1	416	.05	25	10	m
L.S. 2-3	3	1228	810	44	18.4	156	.28	228	18	m
L.S. 3-4	13	1150	842	46	18.3	155	.30	250	22	t-m

[1]For key to symbols see Table I.

[2]The category L.S. 2-3 is that of intermediate sized leaves. Species with exclusively L.S. 2 or L.S. 3 leaves are excluded.

TABLE III. Wood Anatomy of *Hibbertia* in Relation to Plant Habit, Habitat and Leaf Size.[1]

Plant Habit and Habitat	SPP	LS	FL	VEL	PDM	VEL/PDM	PDN	VULN	MESO	BARS	FWT
Australian Shrubs, under 1 meter tall	13	1.9	791	585	20	29.3	402	.05	29	12	m-th
Mesic	1	3.0	1434	1228	32	38.4	130	.25	302	59	t-m
Semi-Xeric	9	1.9	730	527	20	26.4	417	.05	25	11	m
Xeric	3	1.3	758	544	16	34.0	452	.04	19	3	m-th
Australian Shrubs, 1-3 meters tall	9	2.5	1023	761	34	22.4	257	.13	91	14	t-m
Mesic	2	3.0	1139	974	-	-	-	-	-	22	-
Mesic to Semi-Xeric[2]	2	2.3	1097	778	30	25.9	235	.13	99	16	t-m
Semi-Xeric	4	2.4	982	703	40	17.6	226	.18	124	11	m
Semi-Xeric to Xeric	1	2-3	809	534	25	21.4	394	.06	34	10	m-th
Australian Mesic Liana	1	3.0	943	731	38	16.2	108	.35	257	9	t
New Caledonian Rosette Shrubs and Trees, 1-10 meters tall	6	3.0	1371	906	61	14.9	44	1.4	1256	29	t-m
Mesic	2	3.5	1623	1023	94	10.9	17	5.7	5828	20	t-m
Semi-Xeric	4	2.7	1203	827	45	18.4	58	0.78	642	33	t-m

[1]For key to symbols see Table I.

[2]Includes *H. coriacea* from Madagascar.

Progressive reduction in leaf size among all xeric and numerous semi-xeric species has resulted in small leaves of diverse, although often acicular, form. Leaf size reduction has been accompanied by the frequent acquisition of such drought-resistant, ericoid features as thick cuticle, woolly vestiture, revolute margins with thick-walled epidermal cells, hypodermis, and stomata confined to abaxial grooves extending the length of the leaf. Leaf venation patterns of xeric plants commonly exhibit massive primary and intramarginal veins, massive tracheary accumulations at the basal termini of the intramarginal veins, and areolation that is lacking or poorly developed. Leaves of *Hibbertia pungens* are an example of this condition (Figs. 1, I, J). Vessel elements in this plant are narrower (21 μm), shorter (572 μm) and possess fewer bars (3.7) per scalariform perforation plate, than those of mesophytic hibbertias with larger, more mesomorphic leaves (Fig. 1H). The fibers of *H. pungens* also are rather short (705 μm) and, as in other needle-leaved hibbertias, have much thicker walls than in their mesomorphic relatives. It is clear that the vegetative morphological and wood anatomical characteristics of this and other xeric hibbertias have evolved, in general, as a single adaptive unit in response to environmental selection.

Whereas the correlation between aspects of vegetative structural specialization and adaptation to drought is strong among the Australian species of *Hibbertia*, evolutionary patterns among the island species of Madagascar, Fiji, New Guinea and New Caledonia are somewhat different. Most of these plants are large woody shrubs or small rosette trees and have large, sessile or petiolate leaves. Despite the fact that many of the New Caledonian, Fijian and Malagasy species occupy dry, exposed sites, none of them possess acicular leaves of the type described above. Rather, there is a divergent trend toward increasingly coriaceous leaves in the more exposed, xeric taxa. This foliage type is distinguished by a woolly vestiture with sclerified trichomes, uniform venation pattern with intensive vein and veinlet sclerification, tendency to become isolateral, a hypodermis and multiple palisade layers, absence of intercellular space in the mesophyll, formation of sclereids and development of abaxial stomatal crypts (Figs. 1F, G).

Although the leaf anatomy of the xerophytic island species is often very scleromorphic, vessel elements and other xylem features have remained relatively primitive, or unspecialized. Comparisons of selected species will illustrate this point. *Hibbertia lucens* is a Fijian and New Caledonian shrub or rosette tree typically found in dry forests on exposed slopes, or in open and fairly dry localities (Smith, 1981). Xylem of *H. lucens* is relatively primitive, with long vessel elements (900+ μm) and scalariform perforations with more than 20 bars (Fig. 1A). Apparently, this plant has become adapted to its environment by virtue of its scleromorphic foliage. Leaves of *H. lucens* possess a woolly abaxial vestiture and stomata that are restricted to abaxial crypts.

Hibbertia trachyphylla is a common element of semi-xeric to xeric localities of the scrubby New Caledonian "maquis" vegetation. Although its leaves are distinctly scleromorphic, once again the wood is unexpectedly primitive and mesomorphic with wide (40 μm) and sparse (under 100 per mm^2), angular, thin-walled, and rather long tracheary elements with scalariform perforations containing over 20 bars.

An extreme example of these contrasting wood and leaf structural profiles is provided by _H. virotii_, a small, semi-prostrate shrub usually less than 50 centimeters tall, which usually occurs in dry, exposed sites. A specimen collected on Mt. Des Sources, New Caledonia possesses leaves which are broad and extremely coriaceous, with a thick cuticle, thick-walled adaxial and abaxial hypodermal cells, very compact mesophyll, highly lignified trichomes, upper epidermal sclereids, stomatal crypts, lignified vein sheaths, and abundant raphide crystals and silica (Fig. 1F, G). Despite the small stature of _H. virotii_ and its preference for dry, exposed habitats it has retained a rather unspecialized and mesomorphic xylem anatomy (Fig. 1E).

A notable exception to these trends are the very large leaves (L.S. 4) of the mesophytic rosette tree _H. baudouinii_. Leaves of this species possess a venation pattern similar to other island species but are membranous in texture. Interestingly, wood of _H. baudouinii_ is both more mesomorphic and specialized than New Caledonian hibbertias from drier habitats, and also possesses fewer (17) bars per scalariform vessel perforation. Semi-xeric New Caledonian species are apparently less specialized, revealing narrower vessel elements with higher length:width ratios, and more numerous bars per scalariform perforation plate, as compared with mesophytic species such as _H. baudouinii_ and _H. ngoyensis_. This suggests that the mesophytic rosette trees were probably derived from semi-xeric ancestors with smaller leaves. Presumably, their migration into wetter habitats promoted increases in both plant and leaf sizes as well as in vessel diameter, and a correlative decrease in bar number per scalariform perforation plate. A comparable phyletic shift to wider vessels with fewer-barred scalariform perforation plates is also evident in the mesic liana *H. scandens*.

B. *Erythroxylaceae*

The Erythroxylaceae are a relatively specialized, woody, dicotyledonous family of four genera and an estimated 200-250 species of shrubs and trees, the vast majority belonging to the genus *Erythroxylum* P. Browne. The family is pantropical in distribution but exhibits maximum ecological and morphological diversity in the neotropics. The generic diversity of the family is greatest in the African-Madagascan tropics, where three additional genera occur. *Nectaropetalum* Engler includes at least nine species, with maximum diversity in Madagascar but extending from tropical Africa southward to Capetown. *Pinacopodium* Exell & Mendonca includes two species that are restricted to tropical Africa whereas the monotypic *Aneulophus africanus* Bentham is known only from equatorial West Africa.

Wood of the Erythroxylaceae is of an advanced type characterized by both solitary vessels and long radial pore multiples. Vessel perforations are exclusively simple in mature wood and intervascular pitting is typically alternate and distinctly bordered. Imperforate tracheary elements conform to the libriform fiber type. Axial parenchyma is quite variable in abundance and distributional patterns, with both apotracheal and paratracheal parenchyma present, sometimes forming paratracheal bands (e.g., *E. hypericifolium*). Crystals of calcium oxalate or silica are typically abundant and usually occur as an apparent reflection of substrate and soil characteristics, viz. calcareous as opposed to siliceous origin (Rury, 1981, 1982).

1. Wood Anatomy in Relation to Plant Habit. Although little information is available regarding the diversity of architectural forms within the Erythroxylaceae, both published and unpublished information concerning plant habit (i.e., maximum height), leaf duration, and ecological preferences were gathered for nearly 50 selected neotropical species. This information was combined with wood anatomical data to produce TABLES IV-VII, in which wood anatomy is analyzed according to several habitat, habit, and leaf structural categories.

Striking correlations of wood anatomy with plant habit emerge from this analysis. Species of diminutive stature (0-2 m) consistently contain (1) the shortest and most thick-walled tracheary elements (2) the lowest vessel element length:width ratios (3) the narrowest and most numerous vessels (4) high fiber:vessel element length ratios and (5) the lowest indices of wood anatomical vulnerability and mesomorphy (Fig. 3A-D).

Conversely, truly arborescent erythroxylums (15-30 m) contain (1) among the longest and thinnest-walled tracheary elements within the family (2) the widest, least numerous vessels and (3) the highest wood anatomical indices of vulnerability and mesomorphy (Fig. 2F, I). Similar correlations are very obvious in many Old World species as well. Rainforest trees (to 37 m) of the paleotropics, such as *E. cuneatum*, *E. ecarinatum* and *E. mannii*, possess woods with atypically long, wide and sparse vessel elements, features which contribute to the highest vulnerability and mesomorphy indices observed within the family (Fig. 2F). As

TABLE IV. Wood Anatomy of Neotropical Species of *Erythroxylum* in Relation to Plant Habit.[1].

Habit	SPP	FL	VEL	PDM	PDN	VEL/PDM	FL/VEL	VULN	MESO
0-2 meters	5	1129	471	43	116	11.0	2.4	.37	175
2-5 meters	17	1145	491	44	112	11.2	2.3	.39	193
5-10 meters	17	1295	534	47	95	11.4	2.4	.49	264
10-15 meters	5	1249	595	45	104	13.2	2.1	.43	257
15-30 meters	4	1269	641	53	39	12.1	2.0	.90	576
Neotropical mean	48	1217	546	46	97	11.9	2.2	.47	259

[1]For key to symbols see Table I.

TABLE V. Wood Anatomy of Neotropical Species of *Erythroxylum* in Relation to Leaf Size[1]

Leaf Size	SPP	FL	VEL	PDM	PDN	RDN	VEL/PDM	FL/VEL	VULN	MESO
L.S. 2	2	966	385	43	121	51	8.5	2.5	.36	137
L.S. 3	14	1158	486	42	119	30	11.6	2.4	.35	172
L.S. 4	28	1204	511	46	87	24	11.1	2.4	.53	270
L.S. 5	3	1405	674	49	79	18	13.8	2.1	.62	418
Neotropical mean:										
L.S. 3.6	47	1183	514	45	102	31	11.4	2.3	.44	227

[1]For key to symbols see Table I.

TABLE VI. Wood and Leaf Anatomy of Neotropical Species of *Erythroxylum* in Relation to Leaf Duration.[1]

Section and Leaf Duration	FL	VEL	PDM	PDN	VEL/PDM	FL/VEL	VULN	MESO	LTK	STFQ
Macrocalyx										
All evergreen	1349	646	51	76	12.7	2.1	.67	433	237	17%
Rhabdophyllum										
Evergreen	1232	635	51	73	12.5	1.9	.70	444	213	17%
Deciduous	1119	434	53	70	8.2	2.6	.76	329	200	25%
Leptogramme										
All evergreen	1343	512	48	63	10.7	2.6	.76	390	197	24%
Heterogyne										
Evergreen	981	396	40	170	9.9	2.5	.24	93	221	24%
Deciduous	1103	457	44	97	10.4	2.4	.45	207	200	19%
Archerythroxylum										
Evergreen	1245	522	46	130	11.3	2.4	.35	185	182	19%
Deciduous	1162	485	43	119	11.3	2.4	.29	175	186	18%
Microphyllum										
All evergreen	1167	463	39	136	11.9	2.5	.29	133	208	14%
Neotropical means										
Evergreen	1219	529	46	108	11.5	2.3	.43	225	210	19%
Deciduous	1128	459	47	95	9.8	2.5	.49	227	195	21%

[1]For key to symbols see Table I.

might be expected, Old World erythroxylums which rarely exceed the stature of shrubs or very small trees (e.g., *E. australe*, *E. ellipticum*), reveal a wood anatomical profile most similar to that of their smaller (0-2 m), neotropical counterparts (Fig. 3A-D).

Erythroxylum macrocnemium occurs as a small (1-2 m), evergreen, typically unbranched rosette tree in the humid, shady understory of the Peruvian primary forests ("alto bosque"), and bears the largest leaves (L.S. 5) known within

FIGURE 2A-K. Wood and leaf anatomy of *Erythroxylum*. A, *E. macrocnemium* (MADw 28835), transverse section of wood showing thin-walled and angular vessel elements. B, *E. macrocnemium* (Plowman & Kennedy 5800), transverse section of leaf midvein illustrating medullated vascular cylinder and large size of midrib in relation to thin lamina. C-E, *E. cuatrecasasii* (Cuatrecasas 15736). C, Transverse section of wood. D, Cleared leaf. E, Transverse section of leaf showing thin, membranous lamina. F, *E. ecarinatum* (Uw 18165), transverse section of wood showing wide, thin-walled vessels with rounded outlines. G-H, *E. ecarinatum* (Clemens 3933). G, Cleared leaf. H, Transverse section of same leaf as illustrated in G. I, *E. macrophyllum* (Uw 1828), transverse section of wood. J, *E. macrophyllum* (Plowman 5797), cleared leaf. K, *E. macrophyllum* (Guedes 1.300), transverse section of leaf photographed with partially polarized light. Note the fibrous vein sheathing and ramified, epidermally-bracing fibrosclereids.

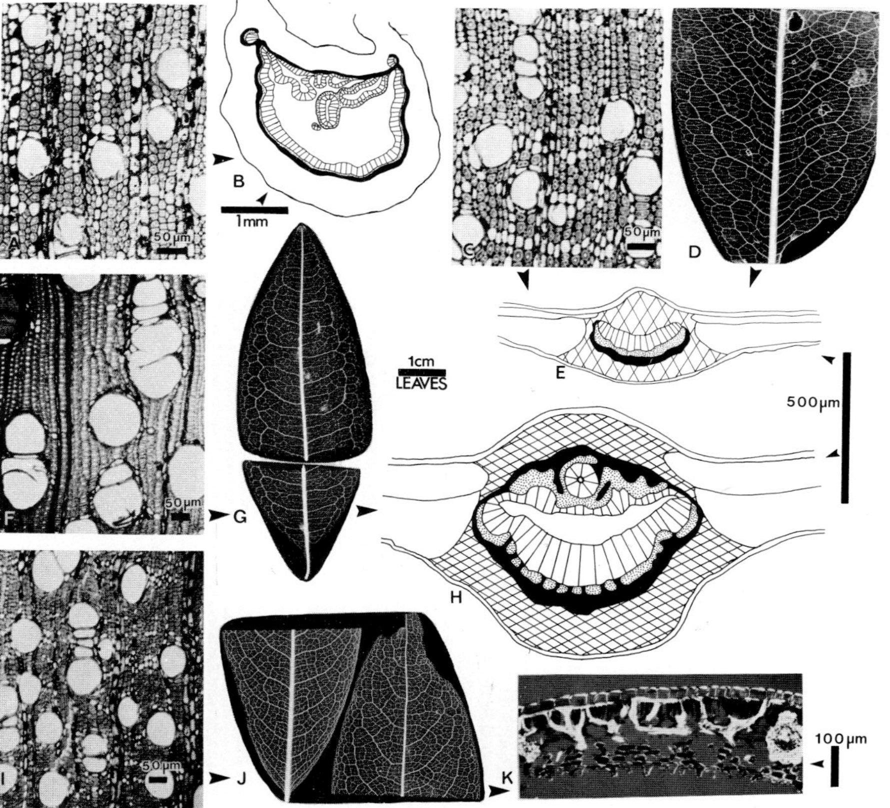
A
50 µm
B
1mm
C
50 µm
D
E
500µm
F
50 µm
G
1cm
LEAVES
H
I
50 µm
J
K
100 µm

TABLE VII. Wood anatomy of Neotropical Species of *Erythroxylum* in Relation to Plant Habit, Leaf Structure and Habitat.[1]

Category	SPP	HABIT	LSMX	FL	VEL	PDM	VEL/PDM	PDN	VULN	MESO
Evergreen with										
foliar sclereids	12	10.4	4.1	1203	605	49	12.3	83	.59	357
Mesic	8	11.9	4.3	1300	663	49	13.6	79	.62	411
Semi-Xeric	3	8.7	3.7	1196	573	47	12.2	90	.52	299
Xeric[2]	1	3.0	4.0	1114	580	52	11.2	79	.66	383
Evergreen without										
foliar sclereids	17	5.7	3.4	1121	460	41	11.2	119	.34	158
Mesic	10	6.1	3.8	1285	567	36	15.8	109	.33	187
Semi-Xeric	6	5.9	2.9	1086	445	43	10.3	126	.34	152
Xeric[3]	1	0.6	2.0	991	367	43	8.6	121	.36	130
Deciduous without										
foliar sclereids										
Semi-Xeric	18	5.5	3.6	1168	489	47	10.4	103	.46	223
Overall	47	6.8	3.7	1163	517	44	11.8	105	.42	217
Mesic	18	8.6	3.9	1293	615	43	14.3	94	.46	281
Semi-Xeric	27	5.9	3.5	1150	502	46	10.9	106	.43	218
Xeric	2	1.8	3.0	1053	474	48	9.9	100	.48	228

[1] For key to symbols see Table I.

[2] E. *testaceum*, a possible phreatophyte of the Brazilian cerrado.

[3] E. *microphyllum*, a tiny-leaved, small shrub from Argentina and the Brazilian cerrado.

the family (Fig. 2B). As described by Carlquist (1975, 1977, 1980a), rosette shrubs and trees normally possess both primitive and mesomorphic xylem anatomy, with rather wide, angular and thin-walled vessels, as well as long, often thin-walled and unspecialized imperforate elements. Wood anatomy of E. macrocnemium clearly fits this characterization Fig. 2A). Carlquist also noted that rosette trees appear to transpire slowly and steadily within their continually wet and humid habitats and thus are not hindered by the possession of the more primitive xylem lacking xeromorphic specializations.

Preliminary studies were made by the senior author of wild and cultivated erythroxylums in Peru during November and December 1981 and October and November 1982, in an around the primary forests of Tingo Maria and Tocache Nuevo (Dept. San Martin). One community included plants of E. macrocnemium, E. macrophyllum (Figs. 2I-K) and E. mucronatum (Fig. 3I), all of which were monitored for transpiration rates and subsequently collected for anatomical study. Despite a basically similar, sclereid-rich leaf anatomy, the large leaves of the rosette trees (E. macrocnemium) transpired more slowly and consistently than did the smaller, sun-form foliage of E. mucronatum growing nearby at the forest margins. Significantly, wood of E. mucronatum contains more specialized, rounded and thick-walled vessels than the wood of E. macrocnemium.

2. Wood Anatomy in Relation to Leaf Size. Wood anatomical variation is strongly interrelated with leaf size differences among neotropical Erythroxylum species (TABLE V). Very distinct xylem anatomical trends occur along the gradient from very small to very large leaves. Furthermore, these

correlations parallel the changes in xylem anatomy in relation to increasing plant stature. As in *Erythroxylum* species of diminutive stature, small-leaved erythroxylums (e.g., *E. microphyllum*, *E. minutifolium*) always contain more numerous rays per unit area and the shortest, thickest-walled, most narrow and numerous vessel elements within the family (Fig. 3A,B,D,G). Enlarged leaf sizes occur in conjunction with increased fiber and vessel element dimensions, vessel element length:width ratios (VEL:PDM) and indices of wood anatomical vulnerability (VULN) and mesomorphy (MESO). On the other hand, both tracheary element wall thickness and length

FIGURE 3A–L. Wood and leaf anatomy of *Erythroxylum*. A, *E. ellipticum* (Aw 26620), transverse section of wood showing relatively narrow, thick-walled vessels and tendency to form radial pore multiples. B, *E. australe* (Kw s.n.), transverse section of wood. C, *E. australe* (Lazarides 6891), cleared leaves. D, *E. brevipes* (Sintenis 6751), wood transection and cleared leaves. Note very xeromorphic wood structure and slender venation of leaves. E, *E. rufum* (Plowman 7750), transverse section of wood. F, *E. rufum* (Wagner 1806), cleared leaf. G, *E. rotundifolium* (Aw 29508), transverse section of wood showing rather narrow, thin-walled vessels of rounded outline and faint growth increment. H, *E. rotundifolium* (Molina 1524), cleared leaf. I, *E. mucronatum* (Plowman 5685), cleared leaf illustrating very regular venation of stout gauge. J, *E. campestre* (Gentry et al. 3537), cleared leaf. K, *E. deciduum* (Machado 60699), cleared leaf showing comparatively homogeneous reticulum of slender, low and high order veins. L, *E. haughtii* (Plowman & Vaughan 5254), cleared leaf.

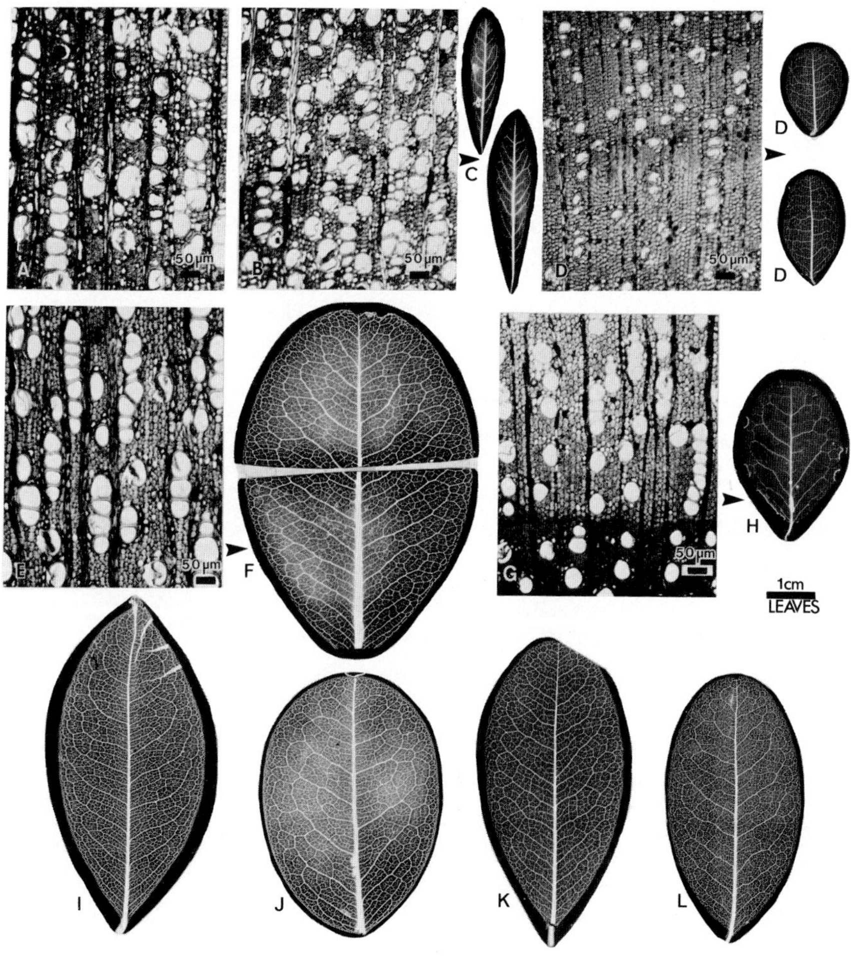
A
50 µm
B
50 µm
C
D
50 µm
D
D
E
50 µm
F
G
50 µm
H
1cm
LEAVES
I
J
K
L

differentials between fibers and vessel elements (FL:VEL) decrease as leaf sizes increase. The lowest FL:VEL ratios appear among the most mesophytic species with very large leaves, such as E. amplum, E. cuatrecasasii (Fig. 2C-E) and E. macrocnemium (Fig. 2A-B). Although leaf size data for Old World taxa is sparse, when available information is compared with wood structure similar xylem anatomical trends in relation to leaf size are evident. Most paleotropical species possess moderately sized leaves (leaf size classes 3-4, sensu Raunkiaer, 1934) and relatively long, wide and sparse vessel elements (Fig. 2F-H). Small-leaved species, such as E. australe and E. ellipticum, produce xylem containing numerous, thick-walled vessel elements of short length and narrow diameter (Fig. 3A-D).

3. Wood Anatomy in Relation to Leaf Duration. Neotropical species also reveal strong correlations between wood anatomy and leaf duration, although the extent and nature of these variation patterns may differ among various sections of Erythroxylum (TABLE VI). Qualitative differences between woods of deciduous and evergreen erythroxylums include wood crystal type, presence of growth rings, as well as vessel outline, wall thickness and distribution patterns. Crystal type reveals perhaps the strongest correlation with leaf duration. Silica grains are usually restricted to xylem rays of evergreen species with sclereid-rich leaves, as in Sections Macrocalyx and Rhabdophyllum. In contrast, deciduous species lack silica in the ray cells, possessing instead calcium oxalate prisms in the axial wood parenchyma.

Growth rings are often related with leaf duration, occurring most frequently in species of seasonally-dry habitats. Of those species investigated, erythroxylums with very distinct growth rings are mostly deciduous, and often reveal a high incidence of radial vessel multiples, as in *E. rufum* (Fig. 3E). Despite this general correlation, some deciduous species may lack growth rings entirely (e.g., *E. brevipes* and *E. rufum*; Fig. 3D, E), and weakly defined rings may occur in a few evergreen species, also of seasonal habitats, such as *E. amplifolium*, *E. cuneifolium*, *E. kapplerianum* and *E. rotundifolium* (Fig. 3G).

Quantitative wood anatomical characters are even more strongly correlated with leaf duration than are qualitative features (TABLE VI). Evergreen species of Sections *Archerythroxylum* and *Rhabdophyllum*, for example, contain longer tracheary elements than do their deciduous counterparts. Deciduous erythroxylums generally possess the shortest fibers and vessel elements and most specialized wood anatomy, often with a decidedly mesomorphic structure. This specialized condition is evidenced by the low VEL:PDM and high FL:VEL ratios in taxa belonging to Section *Rhabdophyllum* and for New World species overall. Evergreen and deciduous species of Section *Archerythroxylum* are nearly identical, however, in their statistical wood anatomy. Most *Archerythroxylum* species occupy mesic habitats and bear mesomorphic, often shade-form leaves that differ only in duration and minor venation characters (Fig. 2C–E). Greater xylem anatomical differences occur, however, between deciduous and evergreen species of Section *Rhabdophyllum*. This is presumably a reflection of the remarkable ecological, architectural, and leaf structural diversity encountered among *Rhabdophyllum* species.

The above analyses of xylem anatomy in relation to presumed degrees of foliar duration are admittedly simplistic. Since evergreen erythroxylums are both ecologically, architecturally, and leaf-structurally more diverse than their deciduous counterparts, more informative correlations between wood and leaf structure might best be obtained through section by section comparative analyses according to both specific ecological preferences and leaf structural types. Careful attention to these ecological and plant structural factors will be valuable in future comparisons of hydraulic architecture, foliar anatomy and transpiration among these plants.

4. Wood Anatomy in Relation to Leaf Structural Type. Evergreen species of Erythroxylaceae are considerably more diverse than deciduous erythroxylums with respect to ecological preferences, architecture, and wood and leaf structure. Evergreen leaves contain diverse and often taxonomically useful combinations of characters such as size, form, texture (thickness), epidermal features, vein gauge and pattern, midvein vasculation, vein sheathing and girder formation, and occurrence of sclereids (Rury, 1981). The most structurally diverse of the evergreen *Erythroxylum* leaves are those which lack ramified fibrosclereids (Fig. 2D,E,G,H; 3H), whereas leaves with sclereids are, for the most part, anatomically uniform and restricted to species of Sections *Macrocalyx* (Fig. 2B,J,K) and *Rhabdophyllum* (Figs. 3I,J).

Deciduous species of *Erythroxylum* are more uniformly mesomorphic in both wood and leaf structure than are their evergreen counterparts. Although several rare exceptions do occur, the following vegetative character correlations usually are evident within each major neotropical section of *Erythroxylum*.

Wood crystal type is strongly correlated with leaf structural type since nearly all erythroxylums containing silica in wood rays possess evergreen leaves with stout, camptodromous venation and highly-ramified, vein-associated fibrosclereids (Fig. 2J,K; 3I,J). All species with wood-ray silica and sclereid-rich, evergreen leaves also possess abundantly striated stipules (e.g., Sections *Macrocalyx* and *Rhabdophyllum*). In contrast, evergreen and deciduous species which contain prisms of calcium oxalate in their wood parenchyma typically lack foliar fibrosclereids and striated stipules.

The only exceptions to this generalization are *E. squamatum*, with sclereid-rich, evergreen leaves but calcium oxalate prisms in the axial parenchyma, and *E. amplifolium*, which contains silica in its xylem rays but possesses sclereid-free, evergreen leaves with a form, texture and venation pattern identical to taxa with foliar fibrosclereids. In these respects both species represent morphological intermediates.

The intermediate structural conditions interconnect species of *Erythroxylum* with different leaf structural types into a morphological continuum. The deciduous habit, for example, seems to have arisen more than once within the genus and appears to represent an evolutionary endpoint. Species of diverse sections of *Erythroxylum* reveal clinal variation in leaf size, form, venation and anatomy along a gradient from evergreen mesophytes with coriaceous leaves to deciduous xerophytes with membranous leaf texture. Along this gradient leaf blades become progressively thinner, with more slender veins and an increasingly regular, homogeneous and orthogonal reticulum of low and high order venation. Examples of this leaf morphocline are provided by (1) sclereid-rich, evergreen leaves of E. macrophyllum (Figs. 2J,K), E. mucronatum (Fig. 3I) and E. campestre (Fig. 3J) (2) sclereid-free, evergreen leaves such as E. amplifolium and E. ecarinatum (Fig. 2G) and (3) sclereid-free, membranous, deciduous leaves with a progressively finer and more homogeneous reticulum of slender, low and high order veins, such as E. rufum (Fig. 3F), E. deciduum (Fig. 3K) and E. haughtii (Fig. 3L).

Among neotropical Erythroxylum species quantitative aspects of wood anatomy also are correlated with leaf structural type (TABLE VII). Evergreen, sclerophyllous taxa clearly are the most wood anatomically mesomorphic as evidenced by longer tracheary elements, thinner tracheary element walls, wider and less numerous vessels, higher VEL:PDM ratios, and much higher indices of wood anatomical vulnerability (VULN) and mesomorphy (MESO) (Fig. 2A,I). Wood anatomical mesomorphy is also greater for species with sclereid-rich, evergreen leaves within each of the habitat subcategories. Sclerophyllous erythroxylums are more wood anatomically uniform than species with evergreen sclereid-free leaves despite the equally broad ecological ranges occupied by species with each leaf structural type. Even the length of the tracheary elements and VEL:PDM ratios decrease only slightly along a decreasing moisture gradient, in contrast to a more linear decrease in the same xylem features among non-sclerophyllous, evergreen species along seemingly similar ecological gradients.

Evergreen erythroxylums which lack foliar sclereids represent the most wood anatomically specialized group, both overall and within each habitat category (TABLES IV & VII). The non-sclerophyllous Erythroxylum species consistently possess the shortest tracheary elements, narrowest and most numerous vessels and the lowest indices of xylem anatomical vulnerability and mesomorphy. Trends of xeromorphic xylem specialization along a decreasing moisture gradient are also more pronounced among non-sclerophyllous evergreen species than among sclerophyllous taxa. Additionally, the totality of wood anatomical features among erythroxylums with evergreen, sclereid-free leaves constitutes a more xeromorphic profile than is present in deciduous or evergreen and sclerophyllous plants. Deciduous erythroxylums tend to be intermediate in quantitative aspects of wood anatomy between evergreen species with foliar sclereids and those which lack sclereids.

More quantitative disparity in xylem anatomy among species with different leaf structural types is more significant than that observed between ecologically distinct species with identical foliar anatomy. Differences in pore diameter and density, as well as overall vulnerability and mesomorphy indices, are especially significant. Interestingly, these data indicate that wood anatomy, irrespective of the taxonomic relationships of the species involved, is more often strongly correlated with leaf structure than with the specific ecological preference of the plant.

III. DISCUSSION AND CONCLUSIONS

The genera described provide excellent examples of structural correlations between plant habit and vegetative structure. The multiple correlations among plant stature and both qualitative and quantitative features of wood and leaf anatomy are clearly environmentally related in both *Hibbertia* and *Erythroxylum*. These correlated features, termed "adaptive character syndromes" by Dickison et al. (1978), appear as multiple facets of the plant hydrovascular system or continuum, which presumably have evolved as structural and functional units in response to selective environmental pressures, and, therefore, reflect the ecological preferences of individual species.

Despite the crude statistical nature of the data presented, the patterns and basic evolutionary trends in plant habit, vegetative structure, and xylem anatomy among taxa of Dilleniaceae and Erythroxylaceae are evident. Future improvements of both sampling techniques and statistical methodology presumably will further refine, rather than invalidate, the multiple character correlations and redundant ecophyletic trends observed in these two dicotyledonous families.

Both the Dilleniaceae and Erythroxylaceae include species in which distinct anatomical trends have occurred in response to seasonal drought, resulting in the formation of xylem growth rings, decrease in vessel element length and diameter, increase in pore frequency per unit area, shortening of imperforate tracheary elements along with an increase in wall thickness, and in the case of Hibbertia, reduction in the number of bars on the scalariform perforation plate. The trends, correlations, and extreme adaptations are often easily observed among Western Australian species of Hibbertia, whereas in a genus such as Erythroxylum the anatomical and morphological trends and correlations are more subtle and become evident only following careful, comprehensive study of the entire genus.

It appears, moreover, that within the genus Hibbertia at least two major strategies have evolved in dry habitat plants. The first is a situation of overall plant reduction in which the xylem and foliage have evolved together as a unit and show correlated adaptations to environmental extremes. The result is a plant of small size bearing linear or needle-like leaves that are supplied by distinctly xeromorphic xylem tissues. Leaves have developed many xeromorphic features including thick cuticles, inrolled margins, stomatal grooves, and a regression in level of specialization of venation. A second pattern, exemplified by many island inhabiting taxa, is one in which only the leaves have become strongly modified whereas the xylem has remained at a low level of specialization. In these plants the leaves develop a broad lamina that is structurally very sclerophyllous and otherwise xeromorphic.

A comparable situation exists among species of *Erythroxylum*, in which the wood anatomy of species with evergreen, sclereid-rich leaves is uniformly mesomorphic, even when the plants grow in rather xeric microhabitats. Possible explanations for this phenomenon are that the leaves of *Hibbertia* and *Erythroxylum* have coadapted with the xylem tissues in a mosaic fashion as one component of the plant hydrovascular system, or that the foliage buffers xylem tissues from prevailing environmental conditions.

Several cerrado species of *Erythroxylum* from Brazil have been described as phreatophytes which have deep tap roots that reach the water table. This would permit the sclerophyllous leaves of such species as *E. suberosum* to transpire freely and continuously in the dry, hot and sunny aerial microenvironment, even during the dry season (Rawitscher, 1948; Ferri and Labouriau, 1952; Ferri, 1953; Coutinho and Ferri, 1956; Eiten, 1972). Therefore, roots of such plants apparently occupy a comparatively mesic micro-environment whereas the scleromorphic foliage is subjected to more extreme surroundings. This perhaps explains the slightly wider vessels and unexpectedly mesomorphic xylem anatomy in many cerrado erythroxylums with sclereid-rich leaves, such as *E. compestre*, *E. citrifolium* and *E. testaceum*. Eiten (loc. cit.), Ferri (loc. cit.), and Goodland (1971) have suggested that the high degree of sclerophylly among cerrado taxa has resulted from mineral deficiencies and aluminotoxicity of the siliceous soil. Nevertheless, erythroxylums with evergreen, sclereid-rich foliage of this type rarely possess xeromorphic wood anatomy, suggesting that even in the absence of a deep root system, this foliar type may serve to buffer the xylem tissues.

The concept of foliar buffering also seems applicable to many deciduous erythroxylums in which mesomorphic foliage occurs together with mesomorphic xylem anatomy. The deciduous habit may provide an alternative to the development of xeromorphic wood specializations.

Carlquist (1975, 1977, 1980a) has drawn attention to this phenomenon within particular taxa, and introduced the idea of foliar buffering of the secondary xylem as often responsible for the retention of very mesomorphic wood structure within seemingly incongruous, xeric habitats. Carlquist (1975) invoked this concept in his discussion of the broad-leaved, vesselless families Winteraceae and Trochodendraceae as well as Balanopaceae (1980b) and Buxaceae (1982). It has also been applied by Michener (1981) in his interpretations of wood and leaf structure in the genus *Keckiella* (Scrophulariaceae).

This appears to be, nevertheless, an overlooked aspect of comparative anatomical studies and one which may explain the presence of mesomorphic xylem in plants where it would not be expected. The extent of the buffering influence of leaves upon the xylem tissues is presumably determined by the complex interrelationships between various climatic variables and such features such as leaf size, duration, type and degree of scleromorphy, as well as non-foliar characteristics such as architectural forms, degrees of stem succulence, and the nature of root systems. Although the principal environmental factors producing foliar scleromorphy may differ, such as decreased water or nutrient availability, the resulting influences on xylem tissues would appear to be the same.

In order to more fully understand the adaptive significance and probable evolutionary origins of the diverse combinations of architectural, leaf structural and xylem anatomical characteristics exhibited among woody plants careful attention must be paid to the interrelationships of foliar, shoot, and xylem morphogenesis. Comprehensive studies of character correlations within woody plants serve only to illustrate the nature of plant structural variation patterns. Correlative studies of foliar and xylem morphogenesis in relation to environmental factors are needed to determine the functionally adaptive significance of these variation patterns.

Foliar transpiration is the primary determinant of water demands made by the leaves upon the xylem of shoots and roots. The influence of the foliage upon xylem structure and function has been stressed by Larson (1964). He noted that environmentally induced modifications in foliar growth and development are accompanied by commensurate changes in the tracheary cells. Larson concluded that the environment exerts its influence directly on the crown and only indirectly on the formation of wood.

Perhaps the most widely-recognized example of the influence of foliage on xylem structure occurs in temperate, ring-porous, dicotyledonous trees in which wide, earlywood vessels are formed in response to springtime bud break and the resultant basipetal transport of auxin from rapidly elongating shoots and expanding leaves. Also, decreasing levels of auxin output from shoots are thought to result in the differentiation of latewood tissues (Larson, 1964; Zimmermann and Brown, 1971; Kramer and Kozlowski, 1972).

Carlquist (1980a) interpreted the different types of growth rings in dicotyledonous woods from an ecologically adaptive perspective. In many plants of seasonal habitats leaves developed under conditions of water stress are more xeromorphic than those formed during wetter conditions. It would be useful to examine xylem structure and growth ring occurrence in relation to changing ecological conditions and foliar morphology in such plants.

In view of the many relationships between the xylem and foliage, considerations of the possible ecophysiological significance and adaptive value of diverse xylem structures among woody taxa are unwise in the absence of leaf structural comparisons of the species under study.

A collaborative effort among plant anatomists, ecologists, and physiologists is needed in order to more completely understand the plant hydrovascular system.

REFERENCES

Baas, P. (1973). The wood anatomy of *Ilex* (Aquifoliaceae) and its ecological and phylogenetic significance. *Blumea* 21: 193-258.

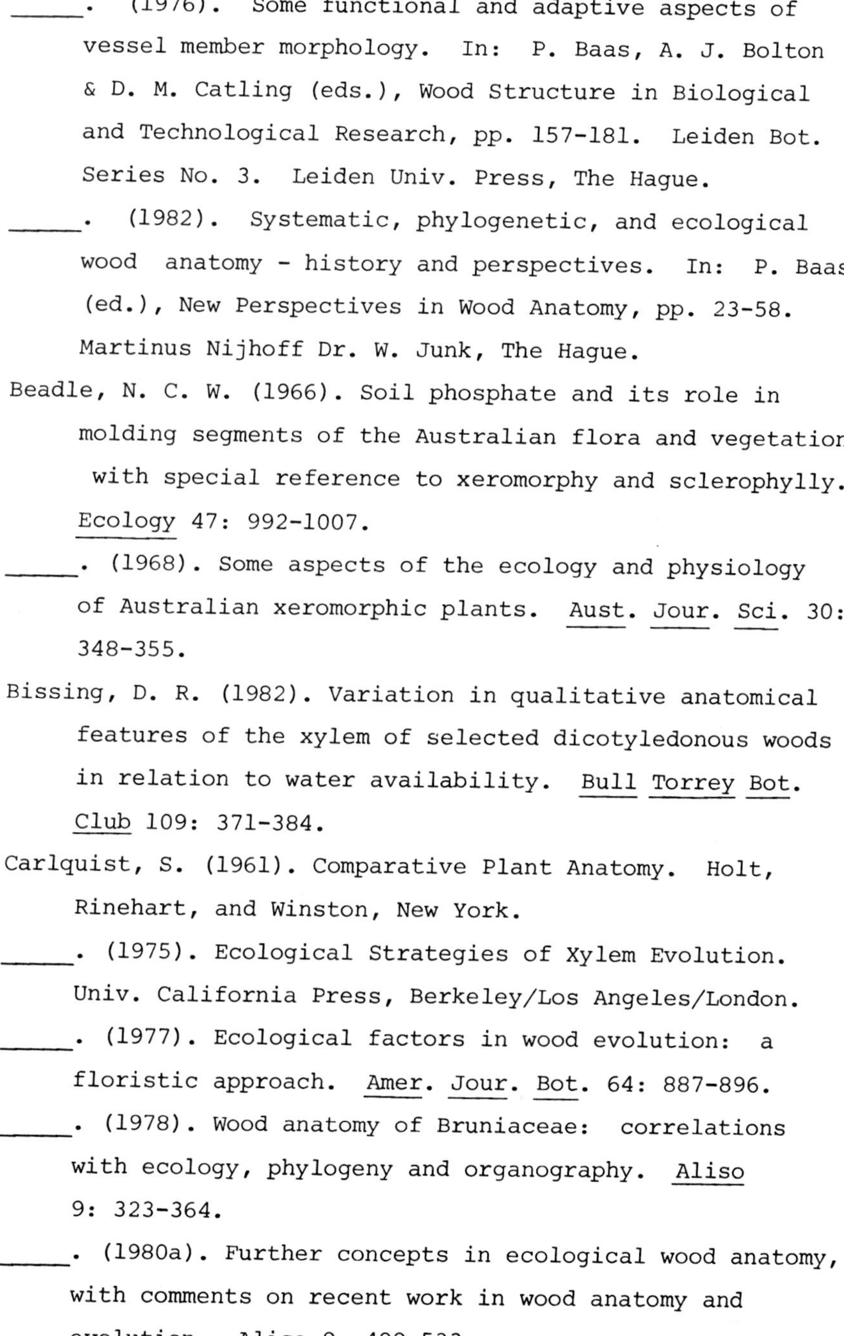

______. (1976). Some functional and adaptive aspects of vessel member morphology. In: P. Baas, A. J. Bolton & D. M. Catling (eds.), Wood Structure in Biological and Technological Research, pp. 157-181. Leiden Bot. Series No. 3. Leiden Univ. Press, The Hague.

______. (1982). Systematic, phylogenetic, and ecological wood anatomy - history and perspectives. In: P. Baas (ed.), New Perspectives in Wood Anatomy, pp. 23-58. Martinus Nijhoff Dr. W. Junk, The Hague.

Beadle, N. C. W. (1966). Soil phosphate and its role in molding segments of the Australian flora and vegetation with special reference to xeromorphy and sclerophylly. *Ecology* 47: 992-1007.

______. (1968). Some aspects of the ecology and physiology of Australian xeromorphic plants. *Aust. Jour. Sci.* 30: 348-355.

Bissing, D. R. (1982). Variation in qualitative anatomical features of the xylem of selected dicotyledonous woods in relation to water availability. *Bull Torrey Bot. Club* 109: 371-384.

Carlquist, S. (1961). Comparative Plant Anatomy. Holt, Rinehart, and Winston, New York.

______. (1975). Ecological Strategies of Xylem Evolution. Univ. California Press, Berkeley/Los Angeles/London.

______. (1977). Ecological factors in wood evolution: a floristic approach. *Amer. Jour. Bot.* 64: 887-896.

______. (1978). Wood anatomy of Bruniaceae: correlations with ecology, phylogeny and organography. *Aliso* 9: 323-364.

______. (1980a). Further concepts in ecological wood anatomy, with comments on recent work in wood anatomy and evolution. *Aliso* 9: 499-533.

_____. (1980b). Anatomy and systematics of Balanopaceae. Allertonia 2: 191-246.

_____. (1982). Wood anatomy of Buxaceae: correlations with ecology and phylogeny. Flora 172: 463-491.

Carlquist, S. and L. Debuhr. (1977). Wood anatomy of Peneaceae (Myrtales): comparative, phylogenetic, and ecological implications. Bot. Jour. Linnean Soc. 75: 211-227.

Coutinho, L.M., and M.G. Ferri. (1956). Transpiracao de plantas permanentes do carrado na Estacao das Chuvas. Rev. Brasil. Biol. 16: 501-518.

Dickison, W.C. (1969). Comparative morphological studies in Dilleniaceae, IV. Anatomy of the node and vascularization of the leaf. Jour. Arnold Arb. 50: 384-410.

_____. (1970). Comparative morphological studies in Dilleniaceae, V. Leaf anatomy. Jour. Arnold Arb. 51: 89-113.

_____. (1977). Wood anatomy of Weinmannia (Cunoniaceae). Bull. Torrey Bot. Club 104: 12-23.

_____. (1979). A note on the wood anatomy of Dillenia (Dilleniaceae). I. A. W. A. Bull. 1979/ 2 & 3: 57-60.

Dickison, W.C., P.M. Rury, and G.L. Stebbins. (1978). Xylem anatomy of Hibbertia (Dilleniaceae) in relation to ecology and evolution. Jour. Arnold Arb. 59: 32-49.

Eiten, G. (1972). The cerrado vegetation of Brazil. Bot. Rev. 38: 201-341.

Ferri, M.G. (1953). Water balance of plants from the "caatinga." II. Further information on transpiration and stomatal behavior. Rev. Brasil. Biol. 13: 237-244.

_____. (1959). Problems of water relations of some Brazilian vegetation types, with special consideration of the concepts of xeromorphy and xerophytism. In: Plant Water Relations in Arid and Semi-Arid Conditions, pp. 191-195. UNESCO Symposium, Madrid.

Ferri, M. G., and L. G. Labouriau. (1952). Water balance of plants from the "caatinga". I. Transpiration of some of the most frequent species of the "caatinga" of Paulo Afonso (Bahia) in the rainy season. Rev. Brasil. Biol. 12: 301-312.

Firbas, F. (1931). Untersuchungen über den Wasserhaushalt der Hochmoorpflanzen. Jb. Wiss. Bot. 74: 459-496.

Gibson, A. C. (1973). Wood anatomy of the Cactoideae (Cactaceae). Biotropica 5: 29-65.

Goodland, R. (1971). Oligotrofismo e alumino no cerrado. In: M. G. Ferri (ed.), III Simposio sobre o Cerrado, pp. 44-60. Univ. de Sao Paulo.

Graaff, N. A. van der, and P. Baas. (1974). Wood anatomical variation in relation to latitude and altitude. Blumea 22: 101-121.

Hickey, L. J., and J. A. Wolfe. (1975). The bases of angiosperm phylogeny: vegetative morphology. Ann. Missouri Bot. Gard. 62: 538-589.

Howard, R. A. (1974). The stem-node-leaf continuum of the Dicotyledoneae. Jour. Arnold Arb. 55: 125-181.

Kramer, P. J., and T. T. Kozlowski. (1979). Physiology of Woody Plants. Academic Press, New York.

Larson, P. R. (1964). Some indirect effects of environment on wood formation. In: M. H. Zimmermann (ed.), The Formation of Wood in Forest Trees, pp. 345-365. Academic Press, New York.

Maximov, N. A. (1929). The Plant in Relation to Water. (Engl. trans. by R.H. Yapp). G. Allen & Unwin Ltd., London.

Michener, D. C. (1981). Wood and leaf anatomy of Keckiella (Scrophulariaceae): ecological considerations. Aliso 10: 39-57.

Müller-Stoll, W.R. (1947-48). Der Einfluss der Ernährung auf die Xeromorphie der Hochmoorpflanzen. Planta 35: 225-251.

Novruzova, Z. A. (1968). The water-conducting system of trees and shrubs in relation to ecology. Baku. Izd. An. Azerb. SSR (in Russian).

Oever, L. van den, P. Baas, and M. Zandee. (1981). Comparative wood anatomy of Symplocos and latitude and altitude of provenance. I. A. W. A. Bull. n.s. 2: 3-24.

Outer, R. W. den, and W. L. H. van Veenendaal. (1976). Variation in wood anatomy of species with a distribution covering both rain forest and savanna areas of the Ivory Coast, West-Africa. In: P. Baas, A.J. Bolton, and D. M. Catling (eds.), Wood Structure in Biological and Technological Research, pp. 182-195. Leiden Bot. Series No. 3. Leiden Univ. Press, The Hague.

Raunkiaer, C. (1934). The Life Forms of Plants and Statistical Plant Geography. The Clarendon Press, Oxford.

Rawitscher, F. (1948). The water economy of the vegetation of the "campos cerrados" in southern Brazil. Jour. Ecol. 36: 237-268.

Rury, P. M. (1981). Systematic anatomy of Erythroxylum P. Browne: practical and evolutionary implications for the cultivated cocas. Jour. Ethnopharmacology 3: 229-264.

_____. (1982). Systematic Anatomy of the Erythroxylaceae. Ph.D. Diss., The Univ. of North Carolina, Chapel Hill.

Rury, P. M., and W. C. Dickison. (1977). Leaf venation patterns of the genus Hibbertia (Dilleniaceae). Jour. Arnold Arb. 58: 209-256.

Schulz, O. E. von. (1907). Erythroxylaceae. In: A. Engler (ed.), Das Pflanzenreich 4, 164 pp. W. Engelmann, Leipzig.

_____. (1931). Erythroxylaceae. In: A. Engler and K. Prantl (eds.), Die Natürlichen Pflanzenfamilien, 2nd ed., 19a: 130-143. W. Engelmann, Leipzig.

Shields, L.M. (1950). Leaf xeromorphy as related to physiological and structural influences. Bot. Rev. 16: 399-447.

Smith, A.C. (1981). Flora Vitiensis Nova. Vol. 2. Lawai, Kauai, Hawaii.

Specht, R.L. (1972). Vegetation of South Australia. 2nd ed. A. B. James, Adelaide.

Stebbins, G. L. (1974). Flowering Plants: Evolution Above the Species Level. Belknap Press, Cambridge, Mass.

Stebbins, G. L., and R. D. Hoogland. (1976). Species diversity, ecology and evolution in a primitive angiosperm genus: Hibbertia (Dilleniaceae). Plant Syst. Evol. 125: 139-154.

Steubing, L., and M. Alberdi. (1973). The influence of phosphorus efficiency on the sclerophylly. Oecol. Plant. 8: 211-218.

Tibbits, T. W. (1979). Humidity and plants. BioScience 29: 358-365.

Zimmermann, M.H. (1978). Hydraulic architecture of some diffuse-porous trees. Canadian Jour. Bot. 56: 2286-2295.

_____. (1982). Functional xylem anatomy of angiosperm trees. In: P. Baas (ed.), New Perspectives in Wood Anatomy, pp. 59-70. Martinus Nijhoff/Dr. W. Junk, The Hague.

Zimmermann, M. H., and C. L. Brown. (1971). Trees: Structure and Function. Springer Verlag, New York.

Zimmermann, M. H., and A. A. Jeje. (1981). Vessel-length distributions in stems of some American woody plants. Canadian Jour. Bot. 59: 1882-1892.

Zimmermann, M.H., and D. Potter. (1982). Vessel-length distribution in branches, stem and roots of *Acer rubrum* L. I. A. W. A. Bull. n.s. 3: 103-110.

TREE ARCHITECTURE: RELATIONSHIPS BETWEEN STRUCTURE AND FUNCTION

Jack B. Fisher

Fairchild Tropical Garden
Miami, Florida

The mechanisms of tree architecture are based upon deterministic plans or models of development which may later be modified by reiteration and other opportunistic changes in structure. Crown form is a result of many variables as shown in two species of *Alstonia* (Apocynaceae). The effect of scale upon observed structural and functional relationships is stressed. Published optimization and functional studies of mechanical structure, branching pattern, axis orientation, bifurcation ratio, spatial distribution of leaves, and life-form are reviewed. Analogues of branches and compound leaves are discussed as are the strengths and weaknesses of computer modeling of tree architecture. The problems of assigning evolutionary significance to observable structure and form are noted.

I. INTRODUCTION

Since the beginning of the scientific study of plants, botanists have been interested in *how* the structure which is characteristic for a species arises. In addition, they have asked the question, so familiar

ISBN 0-12-746620-7

to any instructor of plant anatomy or morphology, why does a species have its particular recognizable form; what is the purpose of a distinctive structure or shape? Using examples from woody plants under the general concept of tree architecture, I hope to clarify the more general problem of assigning adaptive significance (i.e. functional relationship) to features of anatomy (structure) and morphology (form).

The general concept of tree architecture, expanded fully by Hallé, Oldeman and Tomlinson (1978), is the holistic and dynamic description of tree form from the seedling to the aged, senescent individual. Because this analysis is based on the diversity found in tropical woody plants, which appears to represent the full range of present day tree form, it has crystallized our knowledge of organization and development in all trees. For too long, botany has been largely temperate-oriented. The clarification of the mechanisms of tree growth by these authors has led to a better understanding of how tree form develops and how such large, complex and long-lived organisms are built. It seems appropriate at this time to turn to the question of what ecological and/or evolutionary forces, if any, direct or modify the mechanisms of tree architecture. Before tackling the "whys" of tree structure and form, I will first review the "hows", the mechanisms and variables that determine architecture. Then, aspects of structure and function relationships will be reviewed in detail. Finally, the question of the evolutionary significance of tree architecture will be examined.

II. MECHANISMS OF TREE ARCHITECTURE

Hallé et al. (1978) presented a scheme for classifying the observed diversity of tree form (using the term broadly to include most aerial plants) into 23 architectural models. Such models are dynamic descriptions of the basic, inherited blueprint for a plant's growth throughout its life cycle. This is seen most easily in a plant growing in a protected habitat where environmental stress is at a minimum, as in a nursery. Although essentially typological, each model is based on the developmental mechanisms that result in defined patterns of organization. These processes mainly involve the primary meristems and include the following: 1) one or more apical meristems producing monaxial or polyaxial trees; 2) reproductive and vegetative behavior which produces monopodial or sympodial branches and trunks; 3) differentiation of lateral branches and an erect leader to form a distinct trunk; 4) gravimorphic behavior of apices yielding distinctly orthotropic or plagiotropic branches or axes which change in their behavior over time; and 5) periodicity of meristem activity (continuous or rhythmic growth and branching). Thus, older concepts of apical dominance, apical control, and correlative inhibition have been incorporated into a chronological description of growth.

Since plants display open growth with the prolonged production of a theoretically indeterminate number of organs, they can be considered metameric or modular organisms (White 1979). Leaves, internodes and buds are always produced, even if distinct modules or branch units are absent. While the organs of the shoot (leaves, internodes, and flowers or cones) are repeated and organized according to the plan of the model of the species, the plan itself can be repeated or reiterated within an individual. Architectural reiteration adds another dimension to the complexity and variability of tree form (Hallé et al. 1978, de Castro e Santos 1980). Reiteration is the rule for most forest trees growing under natural conditions in which they are stressed and damaged by biological and physical agents. Hallé et al. (1978) called this opportunistic tree architecture. It can easily confuse the observer and has too often in the past overwhelmed morphologists who sought regular branching patterns in mature trees. In the system of architectural models, we now have a means of identifying order within the apparent irregularity of mature tree crowns.

More recently, Hallé and Ng (1981) described examples of fundamental architectural transformations in the crowns of certain dipterocarp trees which occurs during their growth from juvenile but tall understory trees into large forest emergents. This predictable change was termed a metamorphosis, an architectural change that seems to parallel (but is not equivalent to) a juvenile to adult phase change. The subject is still clearly in a state of flux as new studies modify concepts and generalities.

Although a species displays deterministic architecture in its model, it can have a range of crown forms as a result of opportunistic responses. Architectural plasticity was clearly shown in several species which displayed the same model (Fisher and Hibbs 1982). The degree of reiteration and subsequent loss of the tiered, pagoda-like arrangement of lateral branches varied between two species of *Terminalia*. The leaves in both are distributed throughout the volume of the crown since new leaves are produced by both central and peripheral branch units in the original and reiterated tiers. On the other hand, new leaves in *Manilkara* are initiated only at the distal ends of the original and reiterated tiers; thus the crown forms a hemispherical shell of leaves. Unidirectional illumination occurring along riverbanks or in forest canopy gaps greatly distorts crown form by promoting growth on one side. Thus, a single model can serve as the basis for crowns of widely differing geometries of branches and leaves. The way in which branches and leaf surfaces are described and interpreted is to a large degree determined by the relative scale of observation, as will be noted later in part IV.

III. VARIABLES AFFECTING CROWN FORM

The unique aspect or "Gestalt" of any particular tree -- the size, geometry and texture that are characteristic for a species and allows identification from afar -- is a product of a relatively consistant combination of structural features. Foremost are the woody frame and leaf surface.

The woody frame or skeleton is determined by the following: 1) degree of apical dominance (correlative inhibition) of the branches and apical control by the leader (this affects the frequency of branching); 2) axis symmetry based on the organization of primary growth at or near the shoot apex (phyllotaxy and often a reorientation of leaves); 3) relative vigor and dominance of lateral branches which establishes the pattern of branch ordering (this varies with the age or life span of a branch, short vs. long shoot [common in temperate trees] and proleptic vs. sylleptic branch development [the latter common in tropical trees]); 4) branch geometry (the inclination and azimuthal orientation of branch segments and their lengths); and 5) reorientation of axes (the changing of branch position after secondary growth has started may be a part of the normal development in some trees). In such species reorientation is an important aspect of trunk or lateral branch formation and is correlated with reaction wood formation but not always with the presence of gelatinous fibers (Fisher and Stevenson 1981). Axis reorientation is also a response to a repositioning of the crown and may not always involve the induction of reaction wood (Fisher and Mueller, 1983).

The leaf surface of the crown is determined by the following: 1) phyllotaxy of the shoot, in which leader and branches may differ; 2) leaf orientation (often the leaf blade is horizontal in a dorsiventral branch because of a bending of the petiole or rachis); 3) clustering of new leaves on short shoots; 4) variations in internode length (leaf rosettes or pseudowhorls may be present if there are congested nodes); and 5) longitudinal distribution of leaves along a branch (the relative birth and death rates of leaves determines the distribution of leaves behind the branch apex in evergreen trees).

Both the woody frame and the leaf surface of the crown are affected by reiteration and other forms of structural plasticity. For example, reiterations form subunits of the crown that may have physical characteristics of minicrowns (Fisher and Hibbs 1982), e.g. autonomous movement in the wind or leaf flushing. Suppressed saplings under the forest canopy have a reduced crown with little branching and sometimes juvenile leaf form. Leaf shape and orientation may vary greatly between sun and shade conditions (Boojh and Ramakrishnan 1982).

Currently, I am contrasting the growth and geometry of two species of _Alstonia_ (Apocynaceae) which have different crown forms when open-grown and display different models. My goal is to define exactly which architectural parameters directly influence crown geometry. The study is simplified because these tropical forest trees are evergreen and distinctly modular in construction. Both species have whorled phyllotaxis. The trunk and branches are built up by modules. The apical meristem of each module produces several nodes and then either loses its meristematic activity and matures into parenchyma tissue or produces a terminal inflorescence. Four lateral buds at the last node of the module rapidly develop during this change of the apical meristem (Fig. 1A,B) and grow out as equivalent new modules. My colleague, Dr. Richard J. Mueller, is now examining the development of these buds. Although each of the new modules has the potential to repeat this pattern, there are differences in their later growth. When a leader module of *A. scholaris* terminates its growth, all the new modules grow horizontally to form a tier of four lateral branch complexes. The leader axis is continued by the proleptic outgrowth of a resting bud on the old leader directly below the branch tier (Fig. 1E). This type of sympodial trunk is a feature of Prévost's model. In *A. macrophylla* the four new modules produced by the leader grow out unequally. The most vigorous one quickly bends up and forms the continuation of the trunk. The remaining modules grow horizontally and form a tier of three branch complexes (Fig. 1D). This reorientation of a potential lateral branch to form the sympodial trunk is a feature of Koriba's model. In all other qualitative aspects, the mechanism of crown formation is similar in both species. The lateral branch modules are radially symmetrical in early organization, but become dorsiventral by a higher rate of growth and frequency of branching in the one or two modules that are oriented (apparently randomly) on the lower side of the branch complex (Fig. 1C,F). Anisophylly is present with larger mature leaves on the lower side of a node. The number of nodes per module is relatively constant in *A. scholaris*, with two foliage and one scale leaf node per lateral module and more in the leader module. A greater and more variable number of nodes per module occurs in *A. macrophylla*.

Morphometric and growth data are now being collected from

f
s
b
A
f
s
b
a
b
s
B
2
3
I
C
D
E
3
2
1
F

field grown specimens of these trees. Their well-defined modular branch construction and similar growth patterns (sympodial trunk and repetitive production of four initially symmetrical branches in the lateral complex) will make these ideal trees to analyze the quantitative changes that produce different crown form. Old trees of A. scholaris tend to have broad, hemispherical crowns (Fig. 2A,B), while those of A. macrophylla have tall cylindrical crowns with more obvious branch tiers (Fig. 2C,D). Preliminary study indicates the following mechanisms are responsible for differences in mature crown form: 1) relatively shorter module length in A. scholaris which produces a more planar tier of foliage; 2) greater lateral branch dominance and hence a flatter top in A. scholaris (Fig. 2B); 3) few orders of branching in lateral complexes and a slower growth rate in A. macrophylla (Fig. 2D); 4) shorter retention of later complexes on the trunk in A. scholaris resulting in few tiers per crown; and 5) symmetrical tier of four branch complexes in A. scholaris; asymmetrical tier of three in A.

Fig. 1. Alstonia. (A,B). Apex of a module of A. scholaris after its apical meristem (a) matures and four lateral buds (b) have grown out from the axils of scale leaves (s) to form four new modules. Colleters occur at the bases of scale and foliage (f) leaves. Both at same magnification (bar = 0.5 mm). (C). Lateral branch of A. macrophylla, modules are numbered in sequence (1-3). (D). Trunk of A. macrophylla, hand points to the vertical module with remaining three modules forming the tier. (E). Trunk of A. scholaris, hand points to the new leader axis arising below the tier of four branches. (F). Lateral branch modules of A. scholaris, modules are numbered in sequence (1-3). A and B courtesy of R. J. Mueller.

A
B
C
D

macrophylla with no apparent relationship in the position of the reoriented branch module (successive tier asymmetries). This last feature is mainly noticeable in small crowns. It is still unclear if the position of leaf surface is affected by phyllotaxy, leaf size and leaf orientation. Reiteration does not seem to be a significant feature affecting crown form.

In conclusion, the differences in crown shape seem to be mainly influenced by vigor and longevity of lateral branches, leaf arrangement, and the relative difference in growth rates of leader and lateral modules. The difference in crown shape is not a direct consequence of the architectural model.

IV. FUNCTIONAL TREE ARCHITECTURE

A. *The Difficulties*

Generally, function is implied by the close association between a particular structure and a known physiological or ecological process. Specialized structures that diverge in form or organization from the typical ones are most clearly linked to a modified function for those same structures. This is obvious in modifications of organs, e.g. typical absorbing

Fig. 2. Alstonia. (A,B). Mature crown of A. scholaris viewed from the side and from below at the base of the trunk. Note four large lower branches derived from the original four modules of one tier. (C,D). Mature crowns of A. macrophylla viewed from the side and from below at the base of the trunk. Note many small lateral branches without a well-defined lower tier as in B.

roots, attaching adventitious roots, enlarged storage roots, supporting stilt roots, and protecting spine roots.

The tightness of the correlation between a structure and a function is based to a large degree on the relative scale of view (Balik and Hutchinson 1982). A reasonable function is constrained by the level of observation and vice versa. As an example, structure/function relationships are listed in Table I according to increasing size in which the main function is energy capture for photosynthesis.

The smallest elements of tree architecture, organs and organ systems, have been historically the main subject of comparative anatomy and morphology. By the beginning of the 20th century, G. Haberlandt and K. Goebel had established the close linkage between structure and function at the level of cells and organs. At a larger scale the modifications of the entire plant body in relation to extreme habitats (succulents, epiphytes, aquatics), were emphasized in the expanding field of ecology. However, it has only been in the last few decades that the spatial distribution of branches and leaves at the scale of the individual have been analyzed qualitatively and from a physiological/ecological view point (Horn 1971, Ashton 1978). The canopy was described mostly from the view point of succession and community structure as evidenced by studies of forest profiles and stratification of leaf surfaces. Long time scale changes also take on importance. At an even larger scale the canopy takes on the properties of a film or surface layer that can be studied by remote sensing. This is an active area of study, and recently characteristics of structure have been used to identify tree species in tropical rain forests (Myers 1982). The pixel (individual picture element) may replace the leaf of the earthbound botanist as the elemental

TABLE I. *The influence of scale on structure/function relationships when photosynthesis is emphasized.*

Scale	Structure	Function
Atom	Electron orbitals	Photon capture
Molecule	Pigment	Electron transfer
Polymer	Enzyme configuration	Specificity of chemical reaction
Membrane	Thylakoid	Framework for pigments and enzyme systems
Organelle	Chloroplast	Compartmentalizing photosynthesis, unit of replication
Cell	Chlorenchyma cell	Maintaining large plastid population
Organ	Leaf	Large surface with vascular tissue and control of gas exchange by stomata
Organ system	Shoot or twig	Spatial arrangement of leaves, leaf production
Organism	Tree crown	Integration of leaf surface, prolonging life of surface (= life span of individual), establishment of new surfaces via dispersal units
Population	Forest canopy	Interacting surfaces that vary with time (phenology), succession, large scale energy capture affecting climate (environmental feedback)

unit, and the functional questions raised are quite different (e.g., meteorological, land surface exposure, vegetation and land use patterns).

In linking structure and function, biologists commonly assign the biological utility of the relationship under the general and often vague term -- adaptation. In the broadest sense adaptation means being fit for a particular role. But adaptation has been defined and interpreted in two fundamentally different ways according to Gould and Vrba (1982): historical genesis and current utility. Biologists have viewed natural selection as so dominant among evolutionary mechanisms that historical process and current product are usually assumed to be identical. The term exaptation was introduced for a feature that now enhances fitness but was not built by natural selection for its current role (Gould and Vrba 1982). Thus, adaptations (*sensu stricto*) are only those features built by the process of natural selection for their current role. Using this more precise terminology, many structures long interpreted as evolutionarily derived adaptations might be exaptations. The functional relationship is the same but its biological interpretation differs. I will return to this distinction when the evolutionary significance of architecture is discussed in Part V.

Lange et al. (1981) suggested that the physiologist tends to be preoccupied with states and processes within relatively small scales of space and time (e.g. Table 1). Adaptations which together determine fitness are understood in terms of the way "processes are fitted together for optimal performance of the organism in a particular habitat." On the other hand, ecologists (and I would include evolutionists) view adaptation as "a product of any property of the organism which confers a reproductive advantage relative to its competitors." The expanding field of physiological ecology is bridging these two views. At their extremes, the former is strictly mechanistic and includes both adaptations and exaptations, while the latter view is evolutionary and too often assumes that all useful features enhancing fitness are true adaptations. Recent advances in technology (especially portable field instrumentation) now offer a means to quantitatively test hypotheses and speculations regarding adaptive significance, the conferring of superior fitness in one genotype by some trait as compared to another (Lange et al. 1981).

Closely associated with interpretations of adaptive significance are more recent studies of optimization. Rosen (1967) stated that biological organisms tend to assume (display) characteristics that are optimal (most economical) for their circumstances. Thus, he felt that biological structures which are optimal with respect to natural selection (an assumption that would include exaptations) also minimize some cost function derived from the engineering characteristics of the situation. The problem is, as he stressed, that the appropriate cost function is usually not obvious, making optimization studies an "art". Maynard Smith (1978) reviewed this problem and noted that three assumptions underlie any optimization model (to date most dealt with animal behavior): 1) the kinds of phenotypes or strategies possible; 2) what is being maximized, usually one component of an individual's fitness; and 3) the mode of inheritance and the population structure. This last assumption is often tacit. He argues that in such an approach the first assumption is critical. It establishes the constraints, the structural or phenotypic set, upon which selection can operate.

B. Theoretically Optimal Values and Real Trees

I will now turn to some recent studies of plant architecture which deal with the economy of structure as related to function. The diameter and length of tree boles and branches have been analyzed in light of the mechanical properties of columns and beams. In these studies, structure is treated independently from physiological function, e.g. water transport, spatial positioning of photosynthetic surface. Early works were essentially theoretical engineering efforts to model the optimal shape for mechanical support. In a later and commonly overlooked paper, Yamakoshi et al. (1976) found a close correlation between real and theoretically optimal mechanical features of branch structure (size, angle and weight) which result in minimum energy loss defined simply as construction (wood volume) plus support (elastic strain) energy. A more detailed study by McMahon and Kronauer (1976) used a tapered cantilever beam in a mechanical model and compared this with morphometric data and natural oscillation frequencies of limbs from several tree species. They found that an elastically similar model fit the real data closely and speculated that the possible biological mechanisms for this basic principle of the mechanical design of trees might be: a constancy of branching pattern (stationarity), a feedback or sensory system which allows a tree to alter its proportions during growth, and constraints of necessary proportions imposed upon the trunk being carried over to the branches.

Their approach was carried a step further by King and Loucks (1978) who examined the hypothesis that mechanical support in trees must approach the optimal use of wood, i.e. stems and branches will have optimal form with respect to the amount of support tissue. Thus, support costs will be minimized assuming costs are constant. They are explicit in their assumption that natural selection acts upon plant form to maximize survivorship, given certain constraints of the species. Using data from closely grown Populus, they found that trunks have 1.6 times as much wood as the theoretical minimum (based upon trunk diameter) required to prevent buckling. The cost of wood in excess of the theoretical is presumed to offer the tree a margin of safety (expanded upon by King 1981). It may be related to increased reiteration in old or open-grown trees (noted by Hallé et al. 1978 pp. 296-298). Branch diameters were also close to the theoretical minimum required to maintain their position. King and Loucks (1978) concluded that their original hypothesis that observed tree form is an optimal design cannot be rejected by their results. This conclusion is similar to the earlier work of Cobble (1971) on trunk volume. The problem, as I see it, is that the hypothesis can never be disproved because any lack of an exact fit with a theoretical value can be explained too easily by added margins of safety, unmeasured environmental effects, etc.

Several studies have shown that certain features of geometry and structure are close to optimal values. The overriding difficulty in determining the theoretically optimal form is its complexity and variability and the number of factors that have an effect on form. In most plant studies the component of fitness that is maximized is the capacity for photosynthesis, which is dependent upon such structural factors as light interception by leaf surface, shading, heat load, etc. Paltridge (1973) showed in a preliminary way how constraints of crown morphology (diameter, height, simple solid shapes) would affect net photosynthesis as influenced by leaf water potential and light intensity. His models were simple but indicated reasonable trends in crown shape (cylinders, hemispheres) which maximized photosynthesis.

More detailed quantitative studies of shape have been made

of cacti which can be considered as geometrically simple trees in which branch geometry and leaf surface are identical. For the barrel cactus (Ferocactus), Lewis and Nobel (1977) used a computer model to study the real and theoretical interactions between thermal energy exchange (heat loss and solar heating) and water loss (with resulting cooling). Extensive field measurements were incorporated into the model so that calculated and observed values were nearly the same. Using this highly predictive model and being able to simulate the effects of removing structures, they showed that the following structural features were linked to function: 1) spines moderate diurnal temperature changes of the stem, especially near the apical meristems; and 2) ribs (or stem fluting) increased convective cooling. In other cacti (Nobel 1978) the diurnal temperature fluctuation of stems was reduced by spines, pubescence and increased stem diameter. Nobel (1980, 1982) extended his studies to the effects of stem morphology and orientation upon the interception of photosynthetically active radiation (PAR) and the physiology of these CAM plants. Increased surface area does not increase carbon gain because less PAR is intercepted by the necessarily deeper grooves. Ribs also tend to increase daytime temperature (in apparent contradiction to the results of Lewis and Nobel 1977). Some cacti show an increase in height that is related to competitive shading by neighboring vegetation. Cladodes (flattened branches) also tend strongly to face in the direction that maximizes PAR interception at the time the cladodes develop (Nobel 1982). This last feature appears to be an example of adaptive plasticity in the spatial arrangement of branches.

Woodhouse et al. (1980) used a computer simulation to study the relationship between PAR interception and leaf orientation in Agave. The methods used in studying this geometrically complex rosette of leaves might be applied successfully to complex tree crowns. They showed that the observed low frequency of plants on north-facing slopes, as compared to south-facing, could be explained by PAR limitations, especially during winter when maximum PAR interception occurs on south-facing slopes. In this example a structure/function relationship accounted for topographical plant distribution. In another simulation study of leaf distribution in trees, Oikawa (1977) and Oikawa and Saeki (1977) demonstrated the major effect phyllotaxy, leaf orientation and leaf spacing have on theoretical light interception in a group of trees.

More complex geometrical patterns in trees have been correlated to components of fitness. *Terminalia* is a tropical tree with distinct tiers of horizontal branches. Lateral branch complexes are constructed from regular sympodial branch units arranged in a single plane. The regular and predictable branching pattern makes both the measurement and theoretical simulation of branch positions relatively simple as compared to more spatially complex patterns of branching in most trees. In addition, leaves are clustered at the tips of branch units and can be reasonably approximated by horizontal discs. A computer simulation of branch and leaf distribution was made which could calculate total leaf area and unshaded effective leaf area (EA) of the horizontal branch tiers (Fig. 3A). Then, components of branch geometry were varied and values which resulted in maximum EA were determined while the basic pattern (the phenotypic constraint) was held constant. The theoretical optimal asymmetric branching angles (the angles of forking within the plane of the branch tier) were essentially identical to the average values observed in nature (Honda and Fisher 1978). At first glance (Fig. 3B), it seems that real branch angles maximize EA within the constraints of a planar branching pattern of this species. However, a significant difference in adaptive terms between real and theoretical values is somewhat arbitrary. The range of possible real values that are relatively "near" the maximum EA can be quite large depending on the acceptable range in theoretical optimal values. A reasonable range is subjective because we don't know how a given change in leaf area will affect the fitness of a tree. We can only conclude that the optimum value for branch angles is within the range of real values. Maximum EA (most efficient PAR interception) must be balanced with increased heat load, water loss, photosynthetic efficiency, mechanical stability, etc.

In *Terminalia* there are most frequently five branch complexes per tier ($\bar{x}$ = 4.7). When branch complexes are added to a tier, the simulation indicates that there is a major increase in shaded area and a decline in additional EA in a tier of four complexes compared to one of six complexes. Thus, a tier of five complexes can be considered optimal packing in balancing increased surface against branch overlap (Fisher and Honda 1979). There is a gradual shortening of branch units from the trunk to the crown periphery in *Terminalia*, as in many trees. The theoretical ratios of successive branch lengths that produce an equitable distribution of leaf clusters (minimum variance in EA per cluster) are similar to those in nature (Honda and Fisher 1979). These results supported the view that two basic parameters of branch geometry (branch angle and branch length ratio) are presently balanced, such that a maximum EA is produced with the most uniform distribution of leaf clusters, thus distributing biophysical costs throughout the system. Honda and Fisher (1979) assumed that these features have been evolutionarily selected for the most efficient presentation of leaf surface within the constraints of a deterministic pattern of branching (i.e. a true adaptation). However, an alternate hypothesis that these are the results of mechanical constraints which are not phylogeneti-

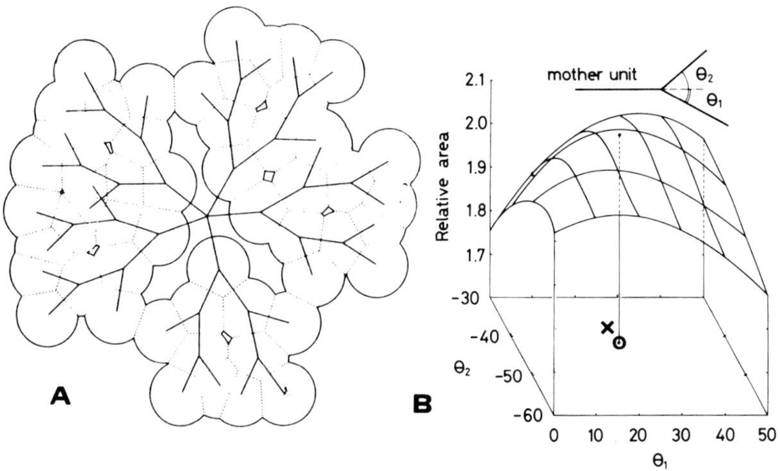

Fig. 3. Effective leaf area in computer simulations of Terminalia catappa. (A). Branch tier as viewed from above using real values for branching angles (from Honda and Fisher 1979). (B). Graph of relative effective leaf area plotted against branching angles, θ_1 and θ_2. o, theoretical optimal angles; x, real average angles (from Honda and Fisher 1978).

cally derived, i.e. the exaptations of Gould and Vrba (1982), cannot be rejected. Lewontin's criticism (cited in Maynard Smith 1978) that hypotheses of adaptation are untestable thus rears its accusing face. (For a recent review of this problem see Brady 1982.) Nevertheless, these studies do show that morphological variables affect form (and therefore the structural basis for function) and to what degree they do so, even if they cannot demonstrate the ultimate mechanism that accounts for a present day feature. Optimality studies certainly are useful in comparing species and in understanding how changes in an environmental parameter can change the optimal point. Horn (1979) presented a lively defense of this approach.

Similar theoretical strengths and speculative weaknesses are found in recent papers dealing with optimal shapes and sizes of leaves (Givnish 1978, Parkhurst and Loucks 1972), but they will not be covered here.

C. Crown Shapes and Life-Forms

Another approach in establishing the function of an architectural feature is to identify those features which affect branching pattern when they are varied. Honda (1971) simulated a wide diversity of theoretical but realistic crown shapes by using only two variables, branching angle and relative branch lengths. Seemingly insignificantly small changes in angle, length, or asymmetry of forking produced major changes in crown shape after many orders of branching (Fig. 4). Complex branching patterns in *Cornus* and *Terminalia* were simulated by limiting distal bifurcation with a hypothetical zone of branch interaction possibly based on light competition or mechanical stress (Honda et al. 1981). Unfortunately, there was no objective way to measure pattern (spatial) similarity, and thus no way to evaluate the biological similarities of this or other models. As will be seen later, the mathematical comparisons of trees using bifurcation ratios are of limited biological value.

Changes in branching angles of rhizomes have significant effects on pattern formation in these long-lived clonal individuals. Rhizomes are of architectural interest because they are essentially horizontal, two-dimensional trees. Several basic patterns of branching are present in rhizomes: octago-

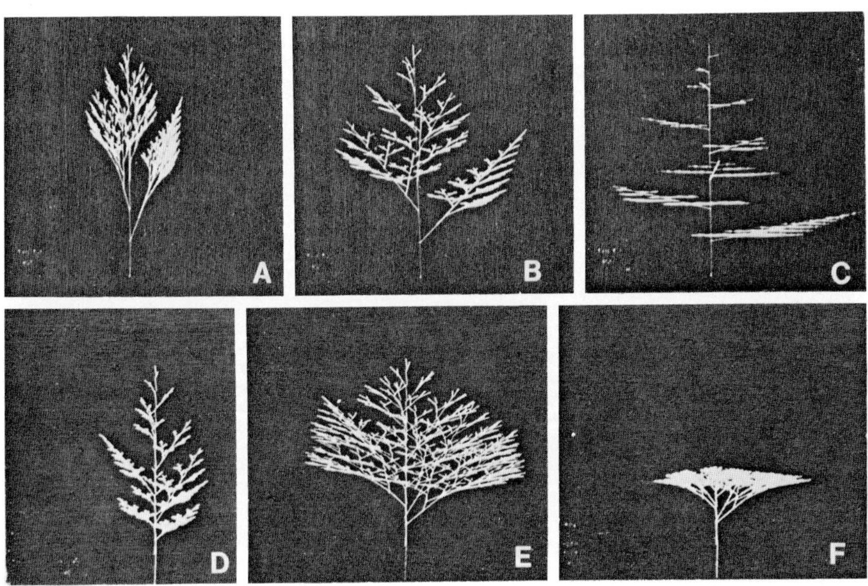

Fig. 4. Computer simulation of theoretical trees using a deterministic model. Changes in crown shapes due to changes in one branching angle or in relative branch length. (A-C). All parameters held constant except one of the two branching angles (θ_2) = -20°, -45°, and -80°, respectively. (D-F). All parameters held constant except for changes in relative branch lengths, R_1 and R_2 , which = 0.9 and 0.6, 0.9 and 0.9, 0.7 and 0.9, respectively (rearranged from Honda 1971).

nal grids (divergence angle ±45°), hexagonal grids (Y-shaped branching), linear systems, and mixed systems combining grids and linear systems (Bell 1979, Bell and Tomlinson 1980, Bell et al. 1979). What is significant is that the manner of growth and clone expansion via rhizomes (clonal strategy) is directly related to the intrinsic architectural pattern. This is particularly clear from simulations in which branching is carried through many generations to show the long term effects of architectural changes on the individual, i.e. the shape of ramets that may no longer be interconnected in the aged rhizome system. This assumes that architecture remains constant (stationary). However, the adaptive significance of a particular pattern per se, e.g. monopodial vs. sympodial or hexagonal vs. linear in the Zingiberales (Bell and Tomlinson 1980), is not clear since two or more patterns can occur in adjacently growing species.

Turning away from attempts at simulating and optimizing architecture, let us examine efforts to correlate overall plant form with its environment, starting with general qualitative studies and ending with quantitative ones. The concept of plant life-forms emphasizes the relative positions of renewal (perennating) buds or the general biological "type" (xerophyte, epiphyte, etc.) and was successfully used by C. Raunkiaer and others (Böcher 1977) to establish strong correlations between physiognomy and climate, i.e. examples of convergence. This approach has been used in comparative systematic studies where trends in life-forms or growth patterns aid in understanding taxonomic affinities (reviewed by Böcher 1977). This is an especially active field of research in the USSR where temperate herbs and shrubs are emphasized appropriately (Serebryakova 1976, 1981). Although plasuible phylogenetic trends in structural complexity or reduction are deduced in these studies (i.e., supposed convergent evolution), few clear adaptive advantages of one form over another are demonstrated. An exception is the correlation of protected perennating buds (geophytes and rosettes) with stressful climates (severe cold and/or drought) and frequently burned habitats.

A more integrated approach to a description of tree form

was presented by Brunig (1976). His goal was to relate the diversity of tree crown forms to major physical environmental variables: solar radiation, water supply, wind, and temperature. He emphasized the aerodynamic properties of leaves and the exposure of leaves to light and wind. As a result, he presented a set of tree crown ideotypes (Fig. 5), which he presumes would be ideally fit to particular situations (preferred habitats and successional status). However, few empirical data are given to support this scheme.

D. Architecture and Succession

The study of tree geometry from an ecological (adaptive) point of view was revitalized by Horn (1971). He stressed the great significance of leaf surface distribution with regard to shading within the crown and to the competitive success of a tree. A diffuse or multilayered distribution of leaves is associated with early successional, full sun species; a peripheral or monolayered distribution is associated with late successional, understory species.

Ashton (1978) was also concerned with relationships between crown shape and ecological factors. He presented quantitative data on leaf distribution within the crown and found major differences in species occupying similar habitats (forest gaps) as measured by leaf area density (leaf surface per unit volume). The wide diversity of leaf area densities in pioneer species indicates that more than leaf distribution is involved in successional status and, by extension, in fitness.

Several workers have attempted to quantify architecture and compare species of different successional status. Kempf and Pickett (1981) found a range of branch angles and branch lengths among shrubs growing in an early succession, open habitat vs. other shrubs in old, closed canopy forests. No consistant relationship was found between habitat and geometrical parameters in this limited study of only four species. The authors observed that the understory shrubs had a monolayered leaf distribution, but this was achieved by different architectural parameters.

In a similar but more extensive morphological and phenological study, Boojh and Ramakrishnan (1982) found more consistant differences between early and late successional trees. Early ones have more vertical and indeterminate growth with a dominance of sylleptic branching over proleptic. Late successional trees display more horizontal and determinate growth with proleptic branching. Their data show that both branches and leaves were more horizontal in forest-grown than in open-grown specimens and also were more horizontal within the crown than at the periphery. Similar variation within a crown was also reported by Pickett and Kempf (1980).

Another method to quantify branching pattern is the bifurcation ratio (R_b), the ratio of the number of branches in successive branch orders. This mathematical index describes the degree or amount of branching. The ratio showed a great deal of promise in early studies of trees which indicated (Oohata and Shidei 1971) or assumed (Leopold 1971, McMahon and Kronauer 1976) that the R_b was constant (stationary) within a tree and for a species. Whitney's (1976) results indicated that R_b was related to successional status. However, the later findings that R_b varies within the same crown (Borchert and Slade 1981, Steingraeber 1982) and within the same species growing in different habitats (Boojh and Ramakrishnan 1982, Pickett and Kempf 1980, Steingraeber et al. 1979, Veres and Pickett 1982) limited the usefulness of this index. The lack of close correspondence between a single <u>overall</u> R_b and crown geometry is clearly seen by similar R_b values in typical trees

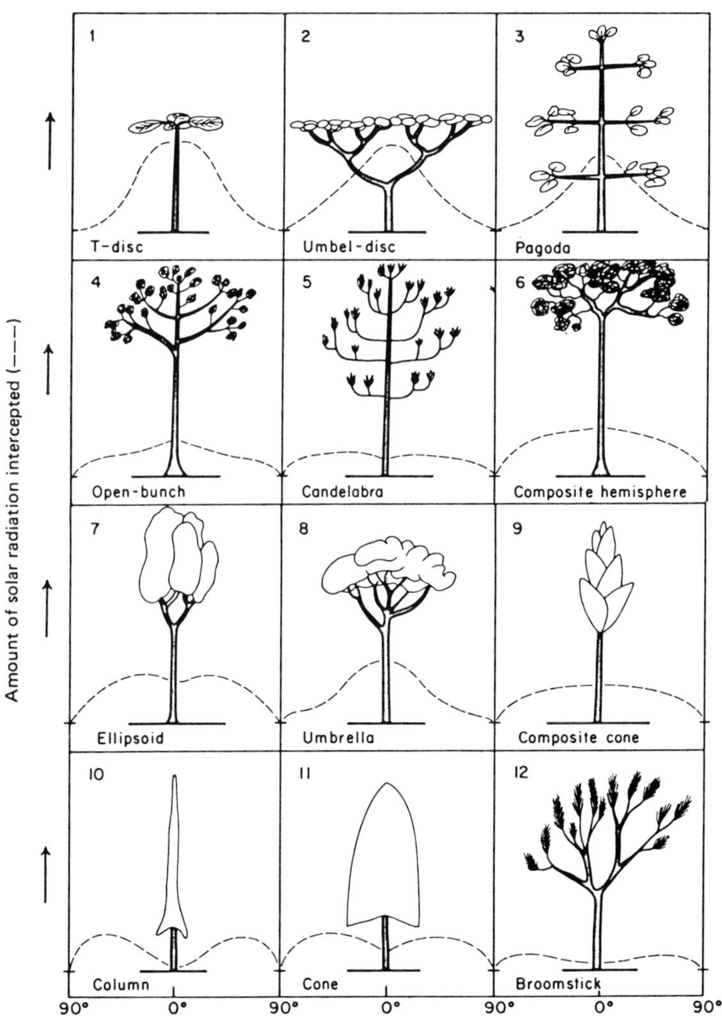

Fig. 5. Crown shape ideotypes of Brunig (1976). Broken lines indicate relative amount of solar radiation intercepted by the crowns at the equator where the sun is at 0° at noon.

and divaricating shrubs (Tomlinson 1978). These results led Borchert and Slade (1981) to reject the appropriateness of using R_b in describing or comparing botanical trees. Steingraeber (1982) and Waller and Steingraeber (in press) point out that the use of stepwise R_b's can be biologically meaningful when the data are interpreted in terms of actual shoot construction (e.g., long and short-shoot dimorphism, pseudodichotomy, etc.). The presence of most leaves on terminal twigs and short shoots (all first order branches) in a late successional tree like *Fagus* yields a high R_b. A high value is also found in much branched early successional species (Whitney 1976) emphasizing the limited usefulness of R_b in ecological work.

E. Branch and Leaf Analogues

As a final example of structure/function correlation, I turn to the many examples of convergence of form in which different organs have a similar function and appearance. A relevant architectural example is the occurrence of compound leaf and branch analogues. The functional and often structural similarities between compound leaves and relatively short-lived branches (twigs) have long been noted and were recently emphasized in ecological terms by Givnish (1978). He hypothesized that compound leaves are metabolically "cheap" throwaway branches. This functional interpretation is based on correlations between the high frequency of compound leaves in seasonally dry climates and rain forests and their decreased frequency in floras as one moves northward from the tropics (substantiated by Stowe and Brown 1981). Givnish (1978) argued *a priori* that such throwaway branches are effective in decreasing water loss, in covering a large area quickly, and in assisting rapid growth in height. Unfortunately, there are few hard facts to support or refute his views for a highly adaptive structural feature. This is an ideal area for collaborative research in which the methodologies of comparative morphology and physiological ecology could reinforce one another.

There are many interesting examples of leaflike branches and branchlike leaves among tropical dicotyledonous trees. The compound leaves of most seed plants are determinate and relatively short-lived (Fig. 6A); most branches are basically indeterminate and long-lived (Fig. 6C). However, divergent forms of both organs occur and seem to be analogous in function. Thus, phyllomorphic branches (short-lived, plagiotropic, and dorsiventral) are found in _Phyllanthus_ (Fig. 6D) and other diverse species (Hallé et al. 1978). Among dicotyledons, species of _Guarea_ (Fig. 6B) and _Chisocheton_ of the Meliaceae have indeterminate, branchlike leaves. New pinnae are produced periodically from a leaf tip bud. In addition, the rachis has a complete vascular cambium forming a woody axis with all the structural features of a twig. Most workers have accepted Sonntag's (1887) conclusion that new pinnae are preformed in _Guarea_ and that the divergent morphology of the leaf is merely a result of delayed pinnae expansion. However, Dr. David A. Steingraeber and I (in preparation) have found that there is an active meristem within the leaf tip bud which continues to initiate new pinnae primordia over a period of two or more years. The leaf is behaving like a branch, although all features of initiation, growth and symmetry are similar to leaves of other Meliaceae. We hope to continue studies of compound leaves, branches and their analogues in an effort to better understand their structural and functional relationships. Comparative developmental studies will answer questions of homology and convergence of form. A metabolic cost/income analysis (to compare relative "cheapness" of structure) and the results of changing light conditions (to compare structural and physiological plasticity of these different organs) will be needed to evaluate Givnish's hypothesis. The questions of adaptive significance of tree and leaf architecture are becoming more inseparable.

F. *Computer Modeling*

Computer simulations of branching patterns and crown geometry have played an important role in architectural studies and will undoubtedly be more widely used in the future for simulations of complex growth patterns. It is now possible to vary or eliminate "experimentally" the branching parameters of a model and to observe their effects on architecture or physiological processes immediately, rather than waiting many years for the crown to develop. Simulation models of trees and other modular organisms are critically reviewed in detail by Waller and Steingraeber (in press). They compared the fundamental characteristics and limitations of the models described to date. Spatial models are of architectural interest, and these are distinguished by different biological assumptions. Deterministic (precise and repeated development) models, while not realistic in their invariability, have been useful in testing the effects of changes of branching parameters on form (Honda 1971, Honda and Fisher 1978, 1979, Honda et al. 1981, 1982, Fisher and Honda 1979). Stochastic (i.e., including probability effects in development) models account for the variability observed in nature by introducing randomness into certain parameters or rules. Probabilities are based on frequencies observed in real organisms, resulting in a population of realistic simulations (Bell et al. 1979, Cochrane and Ford

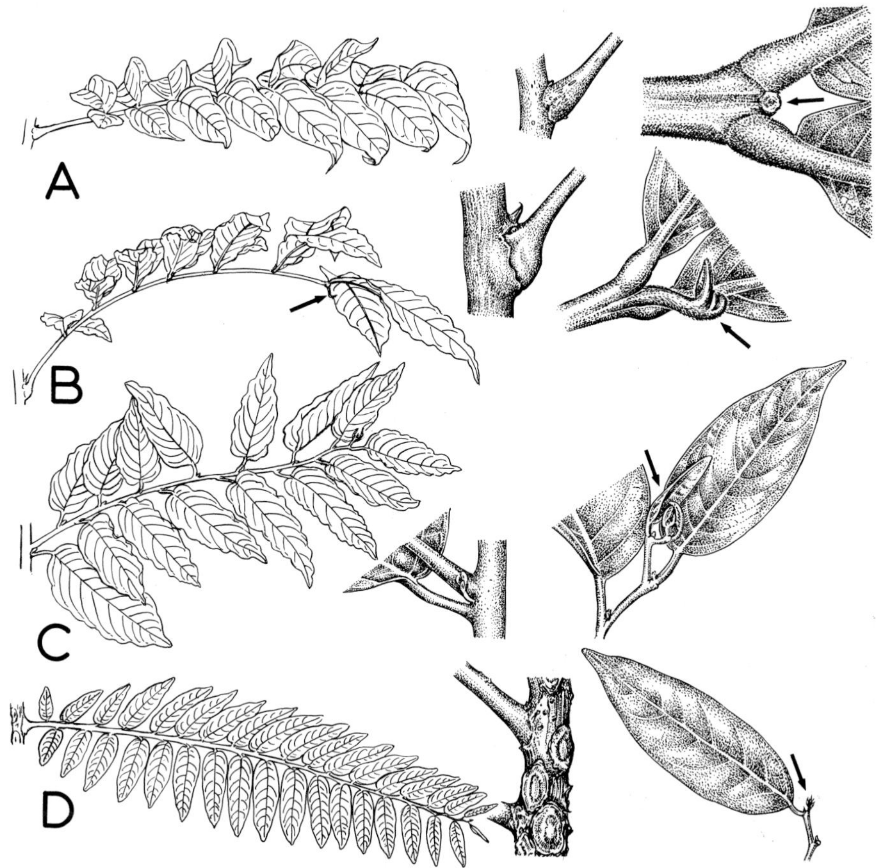

Fig. 6. Typical and divergent forms of compound leaves and branches. Detailed views of the base and tip of each are presented. (A). Typical compound leaf, Chukrasia tabularis, Meliaceae; leaf tip (arrow) abscises. (B). Branchlike leaf, Guarea guidonia, Meliaceae; with a leaf tip bud (arrow). (C). Typical lateral dorsiventral branch, developing from the upper of two axillary buds, Cananga odorata, Annonaceae; indeterminate growing apex (arrow). (D). Phyllomorphic lateral branch, produces branch scars after abscission, Phyllanthus juglandifolius, Euphorbiaceae; apex (arrow) aborts. Drawn by P. Fawcett.

1978, Harper and Bell 1979, Reffye 1981, Reffye and Snoeck 1976, Remphrey et al. in press). There is a danger that such models may lead to a circularity of reasoning if one attempts to validate the model by citing its similarity to the original real data; for this reason Waller and Steingraeber criticized Niklas' (1982) work on branching patterns in fossil land plants.

Another characteristic of a model is whether the rules of growth or branching remain constant during increase in size or branch order (a stationary model) or whether the rules change (a non-stationary model). In real trees non-stationarity is widespread. The previously discussed differences in bifurcation ratio within a tree is evidence of non-stationarity. Frequency of branching decreases with branch order, otherwise branches would overlap unrealistically (Fisher and Honda 1977, Borchert and Slade 1981). Branch interactions and the introduction of flow rates in models also yield realistic, non-stationary simulations (Honda et al. 1981).

Although stochastic, non-stationary tree models produce realistic simulations, there seems to be a fundamental problem in using such models to infer adaptive significance. The probabilities of bifurcation and other architectural events (renewal bud outgrowth, flowering, branch loss, branch growth in length) are randomly distributed in time and/or space (i.e. the growing crown). Although theoretical Poisson (Waller and Steingraeber in press) or other distributions (Niklas 1982) may closely fit the pooled real distribution of an architectural feature of a tree or population, each individual meristem and module (branch unit) is responding to a wide variety of exogenous and endogenous influences, each presumably with its own probability of responses by the plant. As an example let us take branch extinction, which affects the shape of a crown. Apical bud death might display a random spatial distribution in the crown with a particular frequency if caused by herbivory or physical injury (ice, fire or wind damage). Of relevance is Zimmermann's (1978) finding that internal hydraulic constrictions in some trees result in uneven (non-random) effects of water stress within the crown and would result in a patterned die back of branches. Branch extinction is also spatially non-random when due to low light intensity caused by internal or external shading. The probability of branch extinction is also related endogenously to the age and size of the branch. All these extinction responses might be average for a realistic stochastic simulation yet would represent a biologically inappropriate model. Very realistic trees can be uniquely imaged by computer graphics using a non-uniform placement of branches and leaves that is based on the range of probabilities for a particular species (Marshall et al. 1980). Such trees are realistic only in the fact that they vary and are recognizable as a particular tree type (i.e. conifer, broadleaf, etc.). Such an averaged model obscures a direct correspondence between a structural feature (e.g. branch death) and an adaptive function which is a specific programmed response (e.g. elimination of uneconomical shaded branches). Even greater complications arise when the phenomenon of reiteration is introduced into a model of crown geometry. At present, the development of reiterative meristems is unpredictable (de Castro e Santos 1980), but this may only be a consequence of a lack of quantitative studies that would permit us to distinguish causes and the probability of different responses.

G. *Demography of Architectural Elements*

The principles of plant demography have been applied to populations of organs and meristems of trees and other modular plants (Harper and Bell 1979, Noble et al. 1979, White 1979). If this approach can be placed in a spatial context, the dynamic aspect of architecture should be clarified and more easily understood in functional terms. Maillette (1982) presented a detailed quantitative description of the dynamic changes in structure (i.e. bud and branch growth) of *Betula* which result in the cylindrical crown on a free-grown tree. There is a strong basipetal gradient within the crown in which lower lateral branches have decreased branching, higher frequency of short shoots, and shorter long shoots. In this species upper and lower branches form quite different populations as a result of non-stationarity effects, and these impact directly upon crown form. Future demographic studies of tree parts must be carried out with such possible subpopulations kept in mind.

V. IS ARCHITECTURE OF EVOLUTIONARY SIGNIFICANCE?

There is a widespread and often implicit assumption in many papers dealing with architecture that the observed morphology of a species is an adaptation to the environment or life history strategy of the species and, therefore, is a result of natural selection for the most fit structure. Darwin's original concept and the problems in identifying the reasons for fitness are reviewed by Brady (1982). Alternate hypotheses that a structure may be non-adaptive or, if shown to be adaptive, may not be a result of selection are rarely considered. Gould and Lewontin (1979) forcefully argue that "the immediate utility of an organic structure often says nothing at all about the reason for its being." They point out that a good fit between organism and environment is insufficient evidence that natural selection has been active. This is because too often phenotypic changes due to exogenous factors, which physiologists call adaptations, are confused with truly genetic and Darwinian adaptations. The capacity to develop physiological adaptations is genetically based, but the adaptation itself is not. As noted earlier, characters which are now functionally beneficial and promote fitness but which are not a direct consequence of past natural selection can be called exaptations (Gould and Vrba 1982). Can we determine whether the particular architecture of a species is a product of evolutionary forces?

Many aspects of tree architecture are clearly genetically based. Tomlinson (1982) emphasized the different influence of deterministic and opportunistic aspects of plant construction. The developmental plan as based on organographic pattern, architectural model, phenology, relative size and growth rate is genetically determined and constant over a broad range of environments. This fact has permitted plant morphologists to establish general principles and ground plans of organization. Selected horticultural tree forms remain constant no matter where the clone is grown. However, the scale of observation determines the apparent extent of variability of form. At differing levels of observation the variability in response to environmental effects may change. Thus, there seems to be greater variability in the crown shapes within a population of one species than in the observed phyllotaxy. Tomlinson (1982) refers to this variation caused by environmental chance as opportunistic features of construction. In any seedling population there is genetic variation in architectural features that are phenotypically constant as seen in the crown forms of horticultural oddities or in tree crops (e.g. Nelson et al. 1981). Genetic variation in opportunistic capacity is also present as seen in reaction wood production (e.g. Krempl 1975) or potential for plasticity (Fisher and Hibbs 1982). Thus, there is genetic variability upon which natural selection could operate. Most trees are long-lived and fundamentally "K strategists". Thus, vegetative architectural features would be very important aspects of fitness for reproductive success and therefore under natural selective pressure. Is there evidence for such selective pressure on architecture?

The trends in branching patterns of early land plants were compared with simulations of random branching as a first approximation by Niklas (1982). He concluded that while random (non-deterministic) models could describe the observed patterns, certain trends are observed in the fossils. These trends can be interpreted as: 1) directly related to an increase in size; 2) transitions from regular to geometric branching; and 3) canalization of reiterative branching patterns as seen in the decrease in branching variance over time. Even if future studies support Niklas' conclusions, it is still not possible to account for these architectural changes as an enhancement of fitness or as simply structural limitations in the mechanical design of aerial axes.

Among extant plants there is greater architectural diversity in phylogenetically advanced groups; the dicotyledons are the standard with all 23 models (100%) of Hallé et al. (1978) represented. Tomlinson (1978) gives the following architectural indices: dicotyledons 100%, monocotyledons 35%, conifers 21.5%, and cycads 8.5%. However, the correlation between architectural diversity and evolutionary level does not necessarily imply that natural selection has directly influenced this diveristy. The number of models may only reflect the limited ways in which the plant body can be organized regardless of selective pressure. Perhaps increased diversity is a consequence of widespread neotony among angiosperms or the greater plasticity of meristem organization, as noted by Dr. W. Hagemann in this volume.

Architectural variations are often associated with closely related species or populations (ecotypes) (see Böcher 1977). Provenance trials have shown that racial or clonal differences in branch orientation or crown shape are relatively stable (Cannell et al. 1976, Combe and Plessix 1974, Farmer 1976). Some differences in crown form are a direct result of greater leader growth and, thus, seem independent of any directly inherited regulation of form (Cannell 1974). This is not surprising since we have already seen in part III that crown form is greatly modified by changes in smaller scale parameters. Thus, small genetic changes affecting such features as rates of growth or senescence, gravitropic response of a bud, apical dominance, etc. would have a major effect on the crown shape. Genetic regulation is then not acting directly at the higher level of crown shape.

Phenotypic effects are often hard to distinguish from genotypic ones. The dwarfing of trees in alpine and exposed habitats (Lawton 1982) is usually assumed to be an environmental modification. However, genetic differences in geographically distinct populations were demonstrated for crown structure in _Acacia_ (McMillan 1973) and for trunk shape (taper) in _Pinus_ (La Farge 1974). Form differences can be related to distribution, but often it is unclear if these are true ecotypes (genotypic variations) as in _Picea_ (Alexandrov 1971) and other examples cited by Böcher (1977).

Several cases of questionable adaptive architecture and exaptations come to mind. Crown architecture has a direct effect of the funneling of rainwater down the trunk, and rainforest species differ greatly in the amount of throughfall that is intercepted and funneled to the base of the trunk (Herwitz 1982a, b). Although this has a significant effect on redistribution of rainfall during rare extreme events, it would seem to be an insignificant feature as far as fitness and survivorship of the crown shape in question assuming nutrient runoff is not important in heavy rainfall events.

The orientation and curvature of many axes are a result of the mechanical properties of the axis and the load on it (Niklas and O'Rourke 1982). The predictable shape of the curvature is a result of physical factors acting upon a weakly supported orthotropic stem. The rate of growth or the production of support tissues may affect fitness and be selected for, but the shape of flexure cannot be.

The trunk buttresses common in tropical trees have interested biologists, and a variety of hypotheses have been offered to explain their cause and function. Smith (1972) reversed the usual question about the adaptive significance of these structures in the tropics. Instead, he asked why temperate trees tend to lack buttresses. He concluded that thin bark and buttressing are closely associated (later supported by Smith 1979) and that these features are selected against in more seasonal, temperate climates. Although buttresses appear to play a mechanical role in support, they lack the specialized reaction wood structure which does occur in contractile aerial roots of *Ficus* (Fisher 1982).

Phyllotaxy (and hence lateral branch bud distribution) as well as other deterministic features of organization have not been shown to be of adaptive significance. To me this is not unexpected since other structural features, leaf size and internode length, would appear to have a far greater influence on exposure of leaf surface. In particular, the secondary reorientation of leaves by the petiole or the presence of anisophylly override the geometrical effects of phyllotaxis. Planar or dorsiventral leaf surfaces occur in both distichous and decussate species. Again the scale of observation and the interpretation of function are linked. A potentially fruitful approach is with simulations in which the effects that variations in phyllotaxy have upon light penetration and upon other biologically significant features can be measured. An initial attempt using simple changes in arrangement, density and orientation of leaves indicated that light penetration is affected by phyllotaxy when plant populations are considered (Oikawa 1977, Oikawa and Saeki 1977).

Exaptations may be seen in the present spatial distribution of flowers and fruits with respect to pollinators and dispersal agents. In these cases, architectural features which are non-adaptive or were selected for another feature of fitness may have been co-opted for particular reproductive or dispersal functions. Crown shape may well be advantageous now but is a result of elemental processes which are the likely units of selection. Lower orders of construction and organization constrain and canalize further development. Unfortunately, we may never be able to resolve the question of whether or not crown form (*in toto* or in part) is a feature of evolutionary selection, a true adaptation.

VI. CONCLUSION

With ever increasing quantitative detail we are learning how woody plant architecture is formed and what physiological/environmental factors can influence it. The wider appreciation of tropical diversity and availability of computer technology have resulted in an area of plant morphology that is exciting and rapidly expanding. We now have the means to understand better the higher levels of structural organization and complexity as represented by forest trees.

It appears that the architectural models of Hallé et al. (1978) are important for deterministic development, but a model per se need not be of adaptive significance. Crown shape, including branch and leaf geometry, is produced by the iteration of small scale elements and by the process of reiteration. Thus, seemingly small changes in the elemental parameters (branch angle, branch length) may have major effects on overall crown shape. What seems to be of major biological and evolutionary significance is the capacity for architectural change, i.e., plasticity. By this mechanism a tree can respond opportunistically to changes in its environment. Natural selection may act upon architectural structure or upon the capacity for plasticity, but this is difficult to prove since functional utility is not necessarily evidence for true adaptation in the evolutionary sense.

Several directions for future research would seem fruitful: 1) detailed quantitative description and experimental modification of crown form is needed in order to establish exactly what variables impact upon the crown, and the results extended to geometrical modeling; 2) continued correlation of physiology with structure, especially cost analyses of various organs that show convergence in form; 3) expanded demographic studies of plant organs but including a spatial element with the temporal one; and 4) expanded emphasis on the great woody plant diversity in the tropics instead of in temperate or in extreme habitats.

ACKNOWLEDGMENTS

I thank D.M. Waller and D.A. Steingraeber for providing an advance copy of their manuscript, R.J. Mueller for photomicrographs of Alstonia, and D.E. Hibbs, D.W. Lee, J.H. Richards, A.P. Smith and D.A. Steingraeber for their critical and helpful comments on drafts of this manuscript. This study was supported in part by N.S.F. Grant DEB 79-14635.

REFERENCES

Alexandrov, A. (1971). The occurrence of forms of Norway spruce based on branching habit. Silvae Genet. 20, 204-208.

Ashton, P.S. (1978). Crown characteristics of tropical trees. In "Tropical trees as living systems" (Eds. P.B. Tomlinson and M.H. Zimmermann), pp. 591-615. Cambridge University Press, Cambridge.

Balick, L.K. and Hutchison, B.A., Eds. (1982). Summary of a workshop on plant canopy structure. Techn. Rep. EL-82-5. U.S. Army Engineer Waterways Expt. Sta., Vicksburg.

Bell, A.D. (1979). The hexagonal branching pattern of rhizomes of Alpinia speciosa L. (Zingiberaceae). Ann. Bot. 43, 209-223.

Bell, A.D. and Tomlinson, P.B. (1980). Adaptive architecture in rhizomatous plants. Bot. J. Linn. Soc. 80, 125-160.

Bell, A.D., Roberts, D. and Smith, A. (1979). Branching patterns: the simulation of plant architecture. J. Theor. Biol. 81, 351-375.

Böcher, T.W. (1977). Convergence as an evolutionary process. Bot. J. Linn. Soc. 75, 1-19.

Boojh, R. and Ramakrishnan, P.S. (1982). Growth strategy of trees related to successional status. I. Architecture and extension growth. For. Ecol. Manag. 4, 359-374.

Borchert, R. and Slade, N.A. (1981). Bifurcation ratios and the adaptive geometry of trees. Bot. Gaz. 142, 394-401.

Brady, R.H. (1982). Dogma and doubt. Biol. J. Linn. Soc. 17, 79-96.

Brunig, E.F. (1976). Tree forms in relation to environmental conditions: an ecological viewpoint. In "Tree physiology and yield improvement", (Eds. M.G.R. Cannell and F.T. Last), pp. 139-156. Academic Press, London.

Cannell, M.G.R. (1974). Production of branches and foliage by young trees of Pinus contorta and Picea sitchensis: provenance differences and their simulation. J. Appl. Ecol. 11, 1091-1115.

Cannell, M.G.R., Thompson, S. and Lines, R. (1976). An analysis of inherited differences in shoot growth within some north temperate conifers. In "Tree physiology and yield improvement", (Eds. M.G.R. Cannell and F.T. Last), pp. 173-205. Academic Press, London.

de Castro e Santos, A. (1980). Essai de classification des arbres tropicaux selon leur capacité de réitération. Biotropica 12, 187-194.

Cobble, M.H. (1971). The shape of plant stems. Amer. Midl. Natural. 86, 371-378.

Cochrane, L.A. and Ford, E.D. (1978). Growth of a Sitka spruce plantation: analysis and stochastic description of the development of the branching structure. J. Appl. Ecol. 15, 227-244.

Combe, J.-C. and du Plessix, C.-J. (1974). Étude du développement morphologique de la couronne de Hévéa brasiliensis. Ann. Sci. Forest. 31, 207-228.

Farmer, R.E. Jr. (1976). Relationships between genetic differences in yield of deciduous tree species and variation in canopy size, structure and duration. In "Tree physiology and yield improvement", (Eds. M.G.R. Cannell and F.T. Last), pp. 119-137. Academic Press, London.

Fisher, J.B. (1982). A survey of buttresses and aerial roots of tropical trees for presence of reaction wood. Biotropica 14, 56-61.

Fisher, J.B. and Hibbs, D.E. (1982). Plasticity of tree architecture: specific and ecological variations found in Aubréville's model. Amer. J. Bot. 69, 690-702.

Fisher, J.B. and Honda, H. (1977). Computer simulation of branching pattern and geometry in Terminalia (Combretaceae), a tropical tree. Bot. Gaz. 138, 377-384.

Fisher, J.B. and Honda, H. (1979). Branch geometry and effective leaf area: a study of Terminalia-branching pattern. I. Theoretical trees. II. Survey of real trees. Amer. J. Bot. 66, 633-644, 645-655.

Fisher, J.B. and Mueller, R.J. (1983). Reaction anatomy and reorientation in leaning stems of balsa (Ochroma) and papaya (Carica). Can. J. Bot. 61, 880-887

Fisher, J.B. and Stevenson, J.W. (1981). Occurrence of reaction wood in branches of dicotyledons and its role in tree architecture. Bot. Gaz. 142, 82-95.

Givnish, T.J. (1978). On the adaptive significance of compound leaves, with particular reference to tropical trees. In "Tropical trees as living systems", (Eds. P.B. Tomlinson and M.H. Zimmermann), pp. 351-380. Cambridge University Press, Cambridge.

Gould, S.J. and Lewontin, R.C. (1979). The spandrels of San Marco and the Panglossian paradigm: a critique of the adaptationist programme. Proc. Roy. Soc. Lond. B205, 581-598.

Gould, S.J. and Vrba, E.S. (1982). Exaptation - a missing term in the science of form. Paleobiology 8, 4-15.

Hallé, F. and Ng, F.S.P. (1981). Crown construction in mature dipterocarp trees. Malay. For. 44, 222-233.

Hallé, F., Oldeman, R.A.A. and Tomlinson, P.B. (1978). Tropical trees and forests: an architectural analysis. Springer-Verlag, Berlin.

Harper, J.L. and Bell, A.D. (1979). The population dynamics of growth form in organisms with modular construction. Chapter 2. In "Population dynamics", (Eds. R.M. Anderson, B.D. Turner and L.R. Taylor), pp. 29-52. Blackwells, London.

Herwitz, S.R. (1982a). The redistribution of rainfall by tropical rainforest canopy tree species. First Nat. Symp. Forest Hydrol. (Melbourne), pp. 26-29.

__________ (1982b). Tropical rainforest influences on rainwater flux. Ph.D. thesis, Australian National University, Canberra. 263 pp.

Honda, H. (1971). Description of the form of trees by the parameters of the tree-like body: effects of the branching angle and the branch length in the shape of the tree-like body. J. Theor. Biol. 31, 331-338.

Honda, H. and Fisher, J.B. (1978). Tree branch angle: maximizing effective leaf area. Science 199, 888-890.

Honda, H. and Fisher, J.B. (1979). Ratio of tree branch lengths: the equitable distribution of leaf clusters on branches. Proc. Nat. Acad. Sci. (USA) 76, 3875-3879.

Honda, H., Tomlinson, P.B. and Fisher, J.B. (1981). Computer simulation of branch interaction and regulation by unequal flow rates in botanical trees. Amer. J. Bot. 68, 569-585.

Honda, H., Tomlinson, P.B. and Fisher, J.B. (1982). Two geometrical models of branching in botanical trees. Ann. Bot. 49, 1-11.

Horn, H.S. (1971). The adaptive geometry of trees. Princeton University Press, Princeton.

__________ (1979). Adaptation from the perspective of optimality. In "Topics in plant population biology", (Eds. O.T. Solbrig et al.), pp. 48-61. Columbia University Press, New York.

Kempf, J.S. and Pickett, S.T.A. (1981). The role of branch length and angle in branching pattern of forest shrubs along a successional gradient. New Phytol. 88, 111-116.

King, D. (1981). Tree dimensions: maximizing the rate of height growth in dense stands. Oecologia 51, 351-356.

King, D. and Loucks, O.L. (1978). The theory of tree bole and branch form. Rad. Environ. Biophys. 15, 141-165.

Krempl, H. (1975). Unterschiede in Zugholzanteil verschiedener Pappelsorten. Holzforsch. u. Holzverwert. 27, 131-137.

La Farge, T. (1974). Genetic differences in stem form of ponderosa pine grown in Michigan. Silvae Genet. 23, 211-213.

Lange, O.L., Nobel, P.S., Osmond, C.B. and Ziegler, H. (1981). Introduction: perspectives in ecological plant physiology. Encycl. Plant Physiol. n.s. 12A, 1-9.

Lawton, R.O. (1982). Wind stress and elfin stature in a montane rain forest tree: an adaptive explantation. Amer. J. Bot. 69, 1224-1230.

Leopold, L.B. (1971). Trees and streams: the efficiency of branching patterns. J. Theor. Biol. 31, 339-354.

Lewis, D.A. and Nobel, P.S. (1977). Thermal energy exchange model and water loss of a barrel cactus, Ferocactus acanthodes. Plant Physiol. 60, 609-616.

Maillette, L. (1982). Structural dynamics of silver birch. I. The fate of buds. II. A matrix model of the bud population. J. Appl. Ecol. 19, 203-218, 219-238.

Marshall, R., Wilson, R. and Carlson, W. (1980). Procedure models for generating three-dimensional terrain. Comput. Graph. 14, 154-162.

Maynard Smith, J. (1978). Optimization theory in evolution. Ann. Rev. Ecol. Syst. 9, 31-56.

McMahon, T.A. and Kronauer, R.E. (1976). Tree structure: deducing the principle of mechanical design. J. Theor. Biol. 59, 443-466.

McMillan, C. (1973). Prostrateness in *Acacia farnesiana* from the western coast of Mexico under uniform environmental conditions. Madroño 22, 140-144.

Myers, B.J. (1982). Guide to the identification of some tropical rainforest species from large-scale colour aerial photographs. Aust. For. 45, 28-41.

Nelson, N.D., Burk, T. and Isebrands, J.G. (1981). Crown architecture of short-rotation, intensively cultured *Populus*. I. Effects of clone and spacing on first-order branch characteristics. Can. J. For. Res. 11, 73-81.

Niklas, K. (1982). Computer simulations of early land plant branching morphologies: canalization of patterns during evolution? Paleobiology 8, 196-210.

Niklas, K. and O'Rourke, T.D. (1982). Growth patterns of plants that maximize vertical growth and minimize internal stresses. Amer. J. Bot. 69, 1367-1374.

Nobel, P.S. (1977). Water relations and photosynthesis of a barrel cactus, *Ferocactus acanthodes*, in the Colorado Desert. Oecologia 27, 117-133.

__________ (1978). Surface temperatures of cacti - influences of environmental and morphological factors. Ecology 59, 986-996.

__________ (1980). Interception of photosynthetically active radiation by cacti of different morphology. Oecologia 45, 160-166.

__________ (1982). Orientation, PAR interception, and nocturnal acidity increases for terminal cladodes of a widely cultivated cactus, Optunia ficus-indica. Amer. J. Bot. 69, 1462-1469.

Noble, J.C., Bell, A.D. and Harper, J.L. (1979). The population biology of plants with clonal growth. I. The morphology and structural demography of Carex arenaria. J. Ecol. 67, 983-1008.

Oikawa, T. (1977). Light regime in relation to plant population geometry. II. Light penetration in a square-planted population. Bot. Mag. (Tokyo) 90, 11-22.

Oikawa, T. and Saeki, T. (1977). Light regime in relation to plant population geometry. I. A Monte Carlo simulation of light microclimates within a random distribution foliage. Bot. Mag. (Tokyo) 90, 1-10.

Oohata, S. and Shidei, T. (1971). Studies on the branching structures of trees. I. Bifurcation ratio of trees in Horton's law. Jap. J. Ecol. 21, 7-14.

Paltridge, G.W. (1973). On the shape of trees. J. Theor. Biol. 38, 111-137.

Parkhurst, D.F. and Loucks, O.L. (1972). Optimal leaf size in relation to environment. J. Ecol. 60, 505-537.

Pickett, S.T.A. and Kempf, J.S. (1980). Branching patterns in forest shrubs and understory trees in relation to habitat. New Phytol. 86, 219-228.

Reffye, P. de (1981). Modèle mathématique aléatoire et simulation de la croissance et de l'architecture du caféier robusta. II. Étude de la mortalité des méristèmes plagiotropes. Café Cacao Thé 25, 219-230.

Reffye, P. de and Snoeck, J. (1976). Modèle mathématique de base pour l'étude et la simulation de la croissance et de l'architecture du Coffea robusta. Café Cacao Thé 20, 11-32.

Remphrey, W.R., Neal, B.R. and Steeves, T.A. (In press). The morphology and growth of *Arctostaphylos uva-ursi* (L.) Spreng. (bearberry): an architectural model simulating colonizing growth. *Can. J. Bot.*

Rosen, R. (1967). Optimality principles in biology. Butterworths, London.

Sonntag, P. (1887). Ueber Dauer des Scheitelwachsthums und Entwicklungsgeschichte des Blattes. *Jahrb. Wiss. Bot.* 18, 236-262.

Serebryakova, T.I., Ed. (1976). Problems of the ecological morphology of plants. (in Russ. with Engl. Abstr.) *Trans. Moscow Soc. Natural., Biol. Ser., Sect. Bot.* (Trudy Moskovskogo Obshchestva Ispytatelei Prirody, Otdel Biol., Sek. Bot.) 42, 1-304.

__________________, Ed. (1981). Life forms: structure, spectra and evolution. (in Russ. with Engl. abstr.) *ibid.* 56, 1-287.

Smith, A.P. (1979). Buttressing of tropical trees in relation to bark thickness in Dominica, B.W.I. *Biotropica* 11, 159-160.

__________ (1972). Buttressing of tropical trees. *Amer. Nat.* 106, 32-46.

Steingraeber, D.A. (1982). Phenotypic plasticity of branching pattern in sugar maple (*Acer saccharum*). *Amer. J. Bot.* 69, 638-640.

Steingraeber, D.A., Kascht, L.J. and Franck, D.H. (1979). Variation of shoot morphology and bifurcation ratio in sugar maple (*Acer saccharum*) saplings. *Amer. J. Bot.* 66, 441-445.

Stowe, L. and Brown, J. (1981). A geographic perspective on the ecology of compound leaves. *Evolution* 35, 818-821.

Tomlinson, P.B. (1978). Some qualitative and quantitative aspects of New Zealand divaricating shrubs. *N.Z. J. Bot.* 16, 299-309.

_______________ (1982). Chance and design in the construction of plants. In "Axioms and principles of plant construction", (Ed. R. Sattler), pp. 162-183. (Acta Biotheoretica 31a) M. Nijhoff/W. Junk, The Hague.

Veres, J.S. and Pickett, S.T.A. (1982). Branching patterns of Lindera benzoin beneath gaps and closed canopies. New Phytol. 91, 767-772.

Waller, D.M. and Steingraeber, D.A. (In press). Branching and modular growth: theoretical models and empirical patterns. In "The biology of clonal organisms", (Eds. R. Cook, L. Buss and J. Jackson). Yale University Press, New Haven.

White, J. (1979). The plant as a metapopulation. Ann Rev. Ecol. Syst. 10, 109-145.

Whitney, G.G. (1976). The bifurcation ratio as an indicator of adaptive strategy in woody plant species. Bull. Torrey Bot. Club 103, 67-72.

Woodhouse, R.M., Williams, J.G. and Nobel, P.S. (1980). Leaf orientation, radiation interception, and nocturnal acidity increases by the CAM plant Agave deserti (Agavaceae). Amer. J. Bot. 67, 1179-1185.

Yamakoshi, K., et al. (1976). Optimality in mechanical properties of branching structure in trees. (in Jap. with Engl. abstr.) Iyo-denshi To Seitai-kogaku 14, 296-302.

Zimmermann, M.H. (1978). Hydraulic architecture of some diffuse-porous trees. Can. J. Bot. 56, 2286-2295.

INDEX

C

D

R